Soil Biology

Volume 27

Series Editor
Ajit Varma, Amity Institute of Microbial Technology, Amity University Uttar Pradesh, Noida, UP, India

For further volumes:
http://www.springer.com/series/5138

Niall A. Logan • Paul De Vos
Editors

Endospore-forming Soil Bacteria

Editors
Prof. Niall A. Logan
Dept. Biological & Biomedical Sciences
Glasgow Caledonian University
Cowcaddens Road
Glasgow G4 0BA
United Kingdom
nalo@gcu.ac.uk

Prof. Paul De Vos
Lab. Microbiology
University of Gent
K.L. Ledeganckstraat 35
9000 Gent
Belgium
Paul.DeVos@UGent.be

ISSN 1613-3382
ISBN 978-3-642-19576-1 e-ISBN 978-3-642-19577-8
DOI 10.1007/978-3-642-19577-8
Springer Heidelberg Dordrecht London New York

Library of Congress Control Number: 2011931521

Cover design: deblik, Berlin

Printed on acid-free paper

Springer is part of Springer Science+Business Media (www.springer.com)

This book is dedicated to the memory of
Roger C. W. Berkeley (1937–2010),
who made important contributions to the study
of the aerobic endospore-formers

Preface

This book was conceived following the completion of *Microbiology of Extreme Soils* in this *Soil Biology* series. That volume was characterized by the remarkable diversity of the microorganisms it considered, and by the wide range of extreme soil environments that they inhabited. As bacteriologists with a special interest in the taxonomy of *Bacillus* and its relatives, this set us thinking that, while soils are generally considered to be the main habitats of the aerobic endospore-formers, our understanding of their behaviour in soil environments seems to be greatly outweighed by our knowledge of individual species as studied, exploited, or combated by human activities. Most laboratory-based studies of the properties and behaviour of aerobic endospore-forming bacteria have been in-depth studies of single species or strains – especially *Bacillus subtilis*, which is the most extensively studied Gram-positive bacterium and probably the best understood prokaryote after *Escherichia coli* – and it seems that the diversities of these various organisms' habitats and natural existences have often been given scant attention by investigators.

While a small number of aerobic endospore-forming species are opportunistic or obligate pathogens of animals – including humans, other mammals and insects – most species are saprophytes widely distributed in the natural environment. Their habitats are soils of all kinds, ranging from acid through neutral to alkaline, hot to cold and fertile to desert, and the water columns and bottom deposits of fresh and marine waters. It is well known that endospores confer special properties of dormancy, survival and longevity upon their owners in their natural environments, play dominant parts in their life cycles and dispersal and influence our detection and cultivation of the organisms. We no longer believe that these organisms simply exist in soil as dormant spores, or that their vegetative forms are free-living organisms; we recognize that they are active and dynamic members of the soil flora and that they may interact with other soil inhabitants such as fungi, the roots of plants, insects and nematodes.

Much has been written about these bacteria, on their spores, their genetics, their importance in medical and food microbiology and their industrial applications; some individual species such as *B. subtilis*, *B. cereus* and *B. anthracis* merit their own books or reviews. Much has also been written about soil ecology, including the contribution of microbes to soil fertility, interactions of soil microorganisms with plants, the control of pests and diseases, the floras of extreme environments and the impacts of heavy metals – and these and other such topics have been the subjects of several notable titles in the *Soil Biology* series. There is, however, little literature that brings together the different aspects of behaviour and character of aerobic endospore-formers with their contributions to the lives of soils and the lives of the plants and animals that occupy them.

This book is an attempt to fill that gap to some extent, in a volume having four broad and loosely defined themes (1) diversity, (2) growth, spores and survival, (3) interactions and (4) contributions to soil ecosystems. When considering diversity, we need to appreciate the huge expansion there has been in the number of aerobic, endospore-forming taxa in the last 20 years, and our opening chapter outlines these developments in classification and nomenclature, in order to bring the reader up to date, and provides a listing of the new genus names that we have to learn. Chapter 2 is focused on endospore-former diversity and community composition in soil and the mechanisms that contribute to the diversity. Notwithstanding the many novel species cultivated from soils in recent years, it is clear that many further taxa await detection and cultivation, and Chap. 3 gives an account of approaches to studying endospore-former diversity in soil without cultivation and describes some current protocols for such studies. However, Chap. 4 emphasizes that cultivation-based methods continue to be of value, and it reviews such approaches – both with and without spore selection – including immunocapture or selective media in combination with molecular techniques. Turning to growth, spores and survival, Chap. 5 considers the life cycles of *Bacillus* species, especially *B. cereus*, in soil, and these organisms' parts in the horizontal gene pool, while Chap. 6 is devoted to the methods available for undertaking such studies, including the use of soil extracts, artificial soil microcosms and proteomics. Given that endospores can survive dispersal from soils to a wide variety of other habitats, Chap. 7 looks at their contamination of foods, drinks and animal feeds, and the implications for food safety and quality. The next three chapters examine interactions of aerobic endospore-formers with other members of the soil flora. Chapter 8 describes a positive contribution of these organisms to agriculture, describes the protection of plants from fungal diseases and includes protocols for the study of these effects. Further interactions with eukaryotes are considered in both Chap. 9, with a timely review of the remarkable nematode-parasitic organism *Pasteuria penetrans* and the potential for its exploitation in the protection of plants, and Chap. 10, which gives an account of endospore-formers as members of the intestinal microbial communities of soil invertebrates. The subject of Chap. 11, the diversity of *Bacillus thuringiensis* in soil and phylloplane, straddles several of our themes. The remaining five chapters look at the roles of some aerobic endospore-formers in soil ecosystems. While the presence of spores in a soil may not indicate that the organisms are growing there, isolations of organisms showing special adaptations to extreme environments in which they are found implies that they are metabolically active in these niches. However, spores of such organisms are often to be found in non-extreme soils. Chapter 13 considers the surprisingly widespread presence of thermophiles in temperate soils and reviews the potentials of *Geobacillus* species in the bioremediation of oil contamination. Most of our knowledge of halophiles and haloalkaliphiles has emerged from studies of aquatic habitats, but Chap. 16 surveys the presence of these organisms in soils and sediments and also notes their unexpected frequency in non-saline soils. With another example of bioremediation, Chap. 12 describes the potential applications of *Brevibacillus* species as plant growth-promoting rhizobacteria (PGPR) in metal-polluted soils, their associations with arbuscular mycorrhizal fungi and how these phenomena may be studied. Our remaining two chapters deal with two opposed aspects of the nitrogen cycle and methods for their study: Chap. 15 covers the nitrogen-fixing activities of *Paenibacillus* species and their roles as PGPR and contributors to soil fertility, while Chap. 14 examines denitrification by species of *Bacillus* and related genera, a process that deserves much further study.

It is clear from these chapters that many questions in the field remain unanswered, and we hope that this book will stimulate further studies into aerobic endospore-forming bacteria as important members of the soil microbiota. We take this opportunity to record our warmest thanks to the book's contributors, for their chapters, cooperation and patience. We also extend our special gratitude to Dr Jutta Lindenborn at Springer, for her generous support during the book's gestation, and Prof. Ajit Varma, Editor of the *Soil Biology* series, for inviting us to prepare this book.

Glasgow, Scotland Niall A. Logan
Gent, Belgium Paul De Vos

Contents

Contributors

Rosario Azcón Departamento de Microbiología del Suelo y Sistemas Simbióticos, Estación Experimental del Zaidín (CSIC), Prof. Albareda, 1, 18008 Granada, Spain, rosario.azcon@eez.csic.es

Ibrahim M. Banat School of Biomedical Sciences, Faculty of Life and Health Sciences, University of Ulster, Coleraine, BT52 1SA Northern Ireland, UK, IM.Banat@ulster.ac.uk

O. Berge CEA, IBEB, Lab Ecol Microb Rhizosphere & Environ Extrem (LEMiRE), Saint-Paul-lez-Durance, 13108, France; CNRS, UMR 6191, Saint-Paul-lez-Durance 13108, France; Aix-Marseille Université, Saint-Paul-lez-Durance 13108, France; Centre de Recherche d'Avignon, INRA, UR407 Pathologie Végétale, Domaine Saint Maurice BP 94, 84143 Montfavet cedex, France, odile.berge@avignon.inra.fr

Alistair H. Bishop Detection Department, Defence Science and Technology Laboratory, Building 4, Room 12, Salisbury, SP4 0JQ, UK, AHBishop@dstl.gov.uk

Volker S. Brözel Department of Biology and Microbiology, South Dakota State University, SNP252B, Rotunda Lane, Box 2140D, Brookings, SD 57007, USA, volker.brozel@sdstate.edu

Antonio de Vicente Grupo de Microbiología y Patología Vegetal, Departamento de Microbiología, Facultad de Ciencias, Universidad de Málaga, Bulevar Louis Pasteur-Campus Universitario de Teatinos s/n, 29071 Málaga, Spain

Paul De Vos Laboratory for Microbiology, University of Gent, K.L. Ledeganckstraat 35, 9000 Gent, Belgium, Paul.DeVos@UGent.be

Gillian Halket Department of Biological and Biomedical Sciences, Glasgow Caledonian University, Cowcaddens Road, Glasgow G4 0BA, UK

T. Heulin CEA, DSV, IBEB, SBVME, Lab Ecol Microb Rhizosphere & Environ Extrem (LEMiRE), Saint-Paul-lez-Durance 13108, France; CNRS, UMR 6191, Saint-Paul-lez-Durance 13108, France; Aix-Marseille Université, Saint-Paul-lez-Durance 13108, France, thierry.heulin@cea.fr

Marc Heyndrickx Institute for Agricultural and Fisheries Research (ILVO), Technology and Food Science Unit, Brusselsesteenweg 370, 9090 Melle, Belgium, Marc.Heyndrickx@ilvo.vlaanderen.be

Xiaomin Hu Key Laboratory of Agricultural and Environmental Microbiology, Wuhan Institute of Virology, Chinese Academy of Sciences, Wuhan 430071, China

Helmut König Institut für Mikrobiologie und Weinforschung (Institute of Microbiology and Wine Research), Johannes Gutenberg Universität (Johannes Gutenberg University), Becherweg 15, 55128 Mainz, Germany, hkoenig@uni-mainz.de

Niall A. Logan Department of Biological and Biomedical Sciences, Glasgow Caledonian University, Cowcaddens Road, Glasgow G4 0BA, UK, nalo@gcu.ac.uk

Yun Luo Department of Microbiology, Cornell University, Ithaca, NY, USA, yl423@cornell.edu

M. Carmen Márquez Dept Microbiology and Parasitology, Faculty of Pharmacy, University of Sevilla, Calle Professor Garcia Gonzalez 2, 41012 Sevilla, Spain

Jacques Mahillon Laboratory of Food and Environmental Microbiology – Earth and Life Institute, Université catholique de Louvain, Croix du Sud, 2/12, 1348 Louvain-la-Neuve, Belgium, Jacques.mahillon@uclouvain.be

Ines Mandic-Mulec Biotechnical Faculty, Department of Food Science and Technology, University of Ljubljana, Vecna pot 111, Ljubljana, Slovenia

Roger Marchant School of Biomedical Sciences, Faculty of Life and Health Sciences, University of Ulster, Coleraine, BT52 1SA Northern Ireland, UK

P. Mavingui CNRS, UMR 5557, Laboratoire d'Ecologie Microbienne, Université de Lyon, 69000 Lyon, France; Université Lyon 1, 69622 Villeurbanne, France, patrick.mavingui@univ-lyon1.fr

Michio Ohba Graduate School of Bioresource and Bioenvironmental Sciences, Kyushu University, Fukuoka 812-8581, Japan; Hakata-eki-mae 4-22-25-805, Hakata-ku, Fukuoka 812-0011, Japan, ohba@brs.kyushu-u.ac.jp, michio-ohba1901 @m3.gyao.ne.jp

Alejandro Pérez-García Departamento de Microbiología, Universidad de Málaga, Instituto de Fruticultura Subtropical y Mediterránea (IHSM-UMA-CSIC), Bulevar Louis Pasteur-Campus Universitario de Teatinos s/n, 29071 Málaga, Spain, aperez@uma.es

James I. Prosser Institute of Biological and Environmental Sciences, University of Aberdeen, Cruickshank Building, St. Machar Drive, Aberdeen, AB24 3UU, UK, j.prosser@abdn.ac.uk

Diego Romero Grupo de Microbiología y Patología Vegetal, Departamento de Microbiología, Facultad de Ciencias, Universidad de Málaga, Bulevar Louis Pasteur-Campus Universitario de Teatinos s/n, 29071 Málaga, Spain

Juan Manuel Ruiz-Lozano Departamento de Microbiología del Suelo y Sistemas Simbióticos, Estación Experimental del Zaidín (CSIC), Prof. Albareda, 1, 18008 Granada, Spain

Cristina Sánchez-Porro Dept Microbiology and Parasitology, Faculty of Pharmacy, University of Sevilla, Calle Professor Garcia Gonzalez 2, 41012 Sevilla, Spain

Lucy Seldin Laboratório de Genética Microbiana, Instituto de Microbiologia Prof. Paulo de Góes, CCS – Centro de Ciências da Saúde – Bloco I, Avenida Carlos Chagas Filho, 373, Cidade Universitária, Ilha do Fundão, CEP. 21941-902 Rio de Janeiro, RJ, Brazil, lseldin@micro.ufrj.br, lucy@seldin.com.br

Antonio Ventosa Dept Microbiology and Parasitology, Faculty of Pharmacy, University of Sevilla, Calle Professor Garcia Gonzalez 2, 41012 Sevilla, Spain, ventosa@us.es

Ines Verbaendert Laboratory for Microbiology, University of Gent, K.L. Ledeganckstraat 35, 9000 Gent, Belgium, Ines.Verbaendert@UGent.be

Sebastien Vilain Laboratoire de Biotechnologie des Protéines Recombinantes à Visée Santé (EA4135), ESTBB, Université Victor Segalen Bordeaux 2, 33076 Bordeaux cedex, France, Sebastien.vilain@u-bordeaux2

Houda Zeriouh Grupo de Microbiología y Patología Vegetal, Departamento de Microbiología, Facultad de Ciencias, Universidad de Málaga, Bulevar Louis Pasteur-Campus Universitario de Teatinos s/n, 29071 Málaga, Spain

Chapter 1
Developments in the Taxonomy of Aerobic, Endospore-forming Bacteria

Niall A. Logan and Gillian Halket

1.1 Introduction

The genus *Bacillus* was of key importance in the early history of microbiology, for with observations upon *Bacillus subtilis* and its spores Cohn was finally able to discredit the theory of spontaneous generation, and Koch's study of the life history of *B. anthracis* marked the genesis of clinical bacteriology. Aerobic endospore-formers have been and continue to be important in many fields of basic research, and long-term research into the sporulation process in *B. subtilis* has led to its being probably the best understood developmental system.

The term "aerobic endospore-forming bacteria" is used to embrace *Bacillus* species and related genera, and the production of resistant endospores in the presence of oxygen has long been their defining feature. They may be aerobic or facultatively anaerobic, and are also expected to possess Gram-positive cell wall structures (but staining reactions, even in young cultures, may be Gram-variable or frankly Gram-negative). These characters formed part of the definition of the group for some 75 years, following the final report of Winslow et al. (1920) on the classification of bacterial types. Recently, however, some exceptions have emerged, and the genus *Bacillus* now contains strict anaerobes, non-sporulating organisms and cocci.

The erosion of the definition has occurred because 16S rRNA gene sequence analysis has permitted the recognition of genus boundaries whereas previously genera were defined phenotypically, as pragmatic collections of species sharing key (i.e., diagnostic) features. It is important that genera should not be defined by levels of 16S rRNA gene sequence relatedness alone, regardless of levels of shared phenotype. Species, however, often tend to be delineated on the basis of 16S rRNA gene relatedness and DNA–DNA relatedness, and phenotypic descriptions may be brief – often limited to observations of a single isolate – and so of little practical

N.A. Logan (✉) and G. Halket
Department of Biological and Biomedical Sciences, Glasgow Caledonian University, Cowcaddens Road, Glasgow G4 0BA, UK
e-mail: nalo@gcu.ac.uk

N.A. Logan and P. De Vos (eds.), *Endospore-forming Soil Bacteria*, Soil Biology 27,
DOI 10.1007/978-3-642-19577-8_1,

value. Thus, although developments in detection, isolation and characterization of bacteria mean that we can now better reflect the natural diversity of the microbial world overall, current approaches to classification allow the proposal of single-strain taxa that do not reflect within-species diversity, and our understanding of the roles of these organisms in the natural environment is therefore restricted.

The problem is exacerbated by a huge expansion in the number of valid taxa, and discoveries of strains from novel, and often exotic, environments have helped to fuel this explosion in the recognition of bacterial diversity. There are over 150 species currently allocated to *Bacillus*, and strains of about half of these have been found in soil. Since 1990, 14 further genera have been proposed to accommodate species previously assigned to *Bacillus*, and with the addition of new species these genera now comprise over 220 species. Furthermore, 37 other new genera of aerobic endospore-formers containing organisms not previously assigned to *Bacillus* have also been proposed since 1990. Of these 51 genera, 37 (73%) contain species that have been found in soil or soil-associated environments such as sediments and composts. Notwithstanding the many novel species cultivated from soils in recent years, it is clear that many further taxa await detection and cultivation, and De Vos (2011) gives an account of approaches to studying endospore-former diversity in soil without cultivation.

1.2 Habitats and Dispersal

Most aerobic endospore-forming bacteria are widely distributed in the natural environment as saprophytes, and their most frequent habitats are soils of all kinds, ranging from alkaline through neutral to acid, cold to hot, and desert to fertile, and the water columns and bottom deposits of fresh and marine waters. Many of them will degrade biopolymers, with versatilities varying according to species, and it is therefore assumed that they have important roles in the biological cycling of carbon and nitrogen.

Several factors are known to affect endospore formation (growth temperature, environmental pH, aeration, presence of certain minerals and carbon, nitrogen and phosphorus sources and their concentrations) but two kinds of environmental factors, nutritional deprivation and population density are of particular importance in the induction of sporulation. Endospores are metabolically extremely dormant, and this dormancy is the key to their resistance to many agents, including heat, radiation and chemicals, and so their survival over long periods. These spores can survive distribution in soils, and in dusts and aerosols from soils, to a wide variety of other habitats (such as food, as described in Heyndrickx (2011), and aerobic endospore-formers may dominate the floras of air and airborne dust samples (Harrison et al. 2005; Kellogg and Griffin 2006). The wide distribution of endospore-formers is, of course, in large part owing to the extraordinary longevity of their spores; indeed, there is much interest in the potential for viable endospores to travel between terrestrial planets such as Earth and Mars (Fajardo-Cavazos et al. 2007). It therefore seems remarkable that a metabolically dormant spore can monitor its external environment in order to trigger germination within seconds of finding suitable conditions, and that this triggering mechanism can escape the constraints of dormancy while being resistant to damaging agents. Germination can be induced by exposure to nutrients such as amino acids and sugars, by mixtures of these, by non-nutrients such as dodecylamine and by enzymes and high hydrostatic pressure.

Aerobic endospore-formers are often isolated following heat treatment of specimens in order to select for spores, but of course the presence of spores in a particular soil does not necessarily indicate that the organisms are metabolically active there. However, it is reasonable to assume that large numbers of endospores in a given soil reflect former or current activities of vegetative cells in it (see, for example, Banat and Marchant (2011). Isolations of organisms showing special adaptations to the environments in which they are found, such as acidophily, alkaliphily, halophily, psychrophily and thermophily, also suggest that these organisms must be metabolically active in these niches, but they give us little information about the importance of their roles in the ecosystems, and no information about their interactions with other members of the flora. It is clear, however, that endospore-formers form an important part of the active soil microbiota and so influence soil fertility. Some species, such as *B. cereus* and *B. subtilis*, are very widely distributed worldwide and are often isolated; *B. thuringiensis* has been found growing in soils in all continents, including Antarctica (Forsyth and Logan 2000). Airborne carriage seems the most likely explanation for such wide dissemination, but the Antarctic isolates show some phenotypic and genotypic distinctions from temperate strains, and their arrival in the frozen continent does not appear to have been very recent. Although some other species are apparently quite restricted in their distributions, endospores undoubtedly permit widespread dispersal. For example, *B. fumarioli* was originally isolated from geographically isolated volcanic locations on continental Antarctica, and from Candlemas Island which lies some 5,600 km distant in the South Sandwich archipelago; birds are not known to visit these places, and there are no records of any human travels between the sites, and so airborne dispersal seems likely to have occurred (Logan et al. 2000). Furthermore, strains with similar phenotypic behaviours and substantial genotypic similarities to the Antarctic isolates have been isolated from gelatine production plants in Belgium, France and the USA (De Clerck et al. 2004).

1.3 Species and Ecovars

As Rosselló-Mora and Amann (2001) point out, the prokaryote species concept has evolved with a series of empirical improvements that have followed developments in characterization techniques. Presently, prokaryote taxonomists favour the polyphasic approach, in which a consensus is achieved after extensive investigation of genomic information and phenotype. Comparison of 16S rRNA gene sequences is routinely used to classify bacteria to genus level and to give some guidance of identity at species level, but it has insufficient resolution to identify reliably at species level. Species delineation relies upon DNA–DNA relatedness studies, but these are time-consuming and demanding, and rely on pair-wise comparisons of, usually, small numbers of strains. Unless the reference strains are carefully chosen, and are authentic representatives, erroneous conclusions may easily be reached. Phenotype remains important for routine identification purposes, and phenotypic description is a mandatory element of a valid species proposal. The principal and practical purposes of microbial classification and nomenclature are to facilitate the identification of novel isolates and allow efficient access to information about the taxa to which they belong.

When molecular studies began to influence bacterial classifications in the late 1980s and early 1990s, it was hoped that a unified classification might be constructed that would reflect phylogenetic (natural) relationships and also allow routine identification based upon phenetic (phenotypic and genetic) methods. The second edition of Bergey's Manual of Systematic Bacteriology (publishing in five volumes between 2001 and 2010) arranges the taxa on the basis of perceived evolutionary (i.e., natural) relationships, while providing phenetic information intended to allow routine identification without the necessity for extensive molecular analysis. Phenetic approaches do not attempt to imply evolutionary relationships, and so may be regarded as artificial, but they particularly serve the purpose of facilitating routine identification, and so for practical purposes they are intended for general-purpose use and are based upon many characters.

We are now entering a new era in characterization methods for microbial taxonomy. 16S rRNA gene sequencing analysis is now normally automated and is routine in many laboratories for classification. With nearly two decades of experience, the shortcomings of 16S rRNA gene sequencing analysis for classification at the species level, and therefore for identification as well, are better appreciated. Newer methods, such as multilocus sequence analysis, genome sequencing and microarray analyses, promise useful information for informing classification at the species and subspecies levels, and for assisting identification work. Alongside this increase in discriminatory power, it is often pointed out that the number of recognized bacterial species is very small (around 5,000 species) in comparison with the numbers of plant and animal species recorded – especially given the much longer period of evolution that micro-organisms have enjoyed – and that many soil bacteria have yet to be cultivated or described. It may be argued that the bacterial species definition as presently applied is far broader than the definition applied to animal species, and that if the present 70% DNA relatedness threshold were to be strictly applied, then many thousands of bacterial species might be recognized within a single gram of soil (Staley 2006).

It has been suggested that speciation may occur at the subspecies level within ecological niches (ecovars, Staley 2006; ecotypes, Koeppel et al. 2008) and that it is at such a level that a more natural concept of bacterial species may lie. Staley proposed a genomic–phylogenetic species concept that would recognize the speciation that may occur within ecological niches (leading to the development of ecovars) and become reflected in the geographical distributions of species (or geovars), and encompass an organism's evolution as revealed by genomic analysis such as multilocus sequence analysis. Cohan and Perry (2007) observed that various ecological studies in recent years have shown that a named bacterial species may typically be an assemblage of closely related organisms that represent ecologically distinct populations. They use the term "ecotype", which is defined as "a group of bacteria that are ecologically similar to one another, so similar that genetic diversity within the ecotype is limited by a cohesive force, either periodic selection or genetic drift, or both". Koeppel et al. (2008) introduced the concept of ecotype simulation, which models the ecotype as an ecologically distinct group of organisms whose diversity is limited by a cohesive force such as periodic selection or genetic drift. On this basis, it was suggested that a bacterial ecotype would be closer to the species concept that is widely used outside microbiology – ecologically distinct groups that belong to genetically cohesive and irreversibly separate evolutionary lineages. Ecotype simulation, they argued, would enrich bacterial systematics by acknowledging previously unrecognized bacterial groups, and provide a long-needed natural foundation for microbial ecology and systematics.

These proposals clearly need to be tested extensively in order to investigate their utilities, because if a single system is desired, then a bacterial classification that effectively serves routine identification should not be subservient to one that more accurately reflects ecological and evolutionary relationships. The application of a new bacterial species concept that embraces these ecological and biogeographical considerations, and that is informed by the abovementioned characterization methods of greater discriminatory power, is likely to result in a massive increase in species numbers; yet these species must be identifiable in routine laboratories worldwide if such a classification is to gain acceptance.

As Cohan and Perry (2007) point out, the concept of the ecotype has long been applied in medical bacteriology, as with the distinction of *B. anthracis* from *B. cereus* and also, of course, from *B. thuringiensis*. Horizontal DNA transfer between members of the *B. cereus* group has been demonstrated not only in soil and the rhizosphere (Thomas et al. 2001; Saile and Koehler 2006) but also in insects, mammals and foodstuffs (Andrup et al. 2008). In all three species, virulence factors are borne extra-chromosomally on plasmids, and when such plasmids are lost, rendering a strain avirulent, the species become indistinguishable from one another; their chromosomes are extremely similar (Rasko et al. 2005). The current concepts of the species define *B. anthracis* as having pX01 and pX02 plasmids that, respectively, encode the tripartite toxin and the capsule, and *B. thuringiensis* as having a variety of plasmids that encode isoforms of the δ toxin. Although *B. cereus* has a very variable plasmid profile that lacks well-defined, conserved members that could delineate the species, Rasko et al. (2007) found a group of pX01-like plasmids that may define subgroups associated with different pathogenic presentations among *B. cereus* isolates. In such circumstances, the term pathovar may be regarded as synonymous with ecovar. It would be perfectly acceptable in taxonomic terms to describe both *B. anthracis* and *B. thuringiensis* as varieties of *B. cereus*. Indeed, over 50 years ago Smith et al. (1952) reported that virulent and avirulent strains of *B. anthracis* were indistinguishable from *B. cereus* by routine characterization tests, and they classified *B. anthracis* as *B. cereus* var *anthracis*. Gordon et al. (1973) maintained this view, and subsequently many molecular studies have supported it (Rasko et al. 2005; Tourasse et al. 2006), but it never gained wide acceptance, and it is widely recognized that for medical, economic and social reasons it is still not appropriate to propose any change to the nomenclature of the *B. cereus* group species.

The current taxonomy of the *B. cereus* group thus represents the maintenance of the status quo. That of the *B. subtilis* group is in stark contrast with it, as several taxa have been proposed on the basis of the smallest of differences, with no easily determined characters for species distinction being given; *B. subtilis* has been split into two subspecies and three new species: *B. atrophaeus* (Nakamura 1989), *B. mojavensis* (Roberts et al. 1994), *B. vallismortis* (Roberts et al. 1996), *B. subtilis* subsp. *spizizenii* and *B. subtilis* subsp. *subtilis* (Nakamura et al. 1999). These proposals were based principally upon DNA–DNA relatedness studies, with the 70% relatedness threshold for species having been rigorously applied, and distinctions between these "cryptic" individual taxa were supported by a miscellany of approaches that included small differences in profiles of fatty acid composition, multilocus enzyme electrophoresis, restriction digest analysis of selected genes and transformation resistance. The only distinctive phenotypic character cited among these proposals was the production of brown pigment by *B. atrophaeus* on media containing tyrosine, and so the recognition of the four new taxa is of little practical value. Another example is *B. sonorensis* (Palmisano et al. 2001), a soil ecovar of *B. licheniformis* from the Sonoran Desert, Arizona, and distinguishable from *B. licheniformis* mainly by pigment production on tyrosine agar, certain gene sequences, enzyme electrophoresis and DNA relatedness.

Problems with identification were created by the proposal of *B. vallismortis,* a desert soil ecovar of *B. subtilis;* recently the VITEK2® (bioMérieux Inc. USA), a leading automated bacterial identification system, has been forced to incorporate a note into the identification of *B. vallismortis* that reads: "the taxonomic status of *B. vallismortis* appears to be in flux since it can only be differentiated from *B. subtilis* by molecular methods or by geographic source. As strains of *B. vallismortis* were isolated from Death Valley, California, USA, this should be taken into account for their differentiation". In this example, elevation of an ecovar to species status has hindered and confused many diagnostic bacteriologists that need to identify *B. subtilis* routinely.

Subsequently, *B. axarquiensis* and *B. malacitensis* were proposed (Ruiz-García et al. 2005a) as further members of the *B. subtilis* group, *each upon the basis of a single isolate* showing less than 47% DNA–DNA relatedness with phylogenetic relatives, and yet Wang et al. (2007a) reported homologies of greater than 83% between these two strains and *B. mojavensis,* and found few phenotypic features to separate the three species; they proposed that *B. axarquiensis* and *B. malacitensis* were heterotypic synonyms of *B. mojavensis*. Similarly, *B. velezensis* (Ruiz-García et al. 2005b) was proposed for two river-water isolates that were reported to have less than 20% DNA relatedness with members of the *B. subtilis* group, but

Wang et al. (2008) found 74% relatedness with the type strain of *B. amyloliquefaciens* and declared *B. velezensis* to be a later heterotypic synonym of *B. amyloliquefaciens*.

While DNA–DNA relatedness rightly remains most important in bacterial taxonomy, such data lack transparency because, unlike the 16S rRNA gene sequences that are deposited in public databases and examinable for quality, DNA relatedness values cannot be scrutinized by others and the qualities of the DNAs used in their production cannot be ascertained. Given the danger of error being inherent in the technically demanding methods for DNA–DNA hybridization, the recently proposed minimal standards for describing new taxa of aerobic, endospore-forming bacteria (Logan et al. 2009) "strongly recommended" that the Δthermostability (ΔT_m) of hybrid DNAs be determined and that it is "essential" for researchers to establish the qualities of the DNAs they use and to include reciprocal values and relevant controls.

It has long been recommended (Wayne et al. 1987) that, "a distinct genospecies that cannot be differentiated from another genospecies on the basis of any known phenotypic property not be named until they can be differentiated by some phenotypic property". If microbial classification is to serve a practical diagnostic purpose, it is most important that we continue to adhere to this tenet and that a consensus approach to taxonomy be adopted by scientists. The suggestion of Koeppel et al. (2008), that ecotypes be recognized and named by adding an "ecovar" epithet to the species binomial has much to recommend it. It would allow systematists to feel the way to developing a revised definition of bacterial species that properly utilizes advanced methods of characterization, allows the construction of a practically useful taxonomy, and better reflects the true diversity of bacteria.

1.4 Taxonomic Progress

As mentioned earlier, the genus *Bacillus* and certain of its species hold important places in the history of bacteriology. Christian Ehrenberg described *Vibrio subtilis* in 1835, so that "subtilis" is one of the earliest bacterial species epithets still in use (although "vibrio" may bring a picture of curved rods to a bacteriologist's mind, the name derives from a Latin word meaning "to shake"). Ferdinand Cohn proposed the genus in 1872, for three species of rod-shaped bacteria: *B. subtilis* (the type species), *B. anthracis* and *B. ulna*. The identity of *B. ulna* is obscure but, despite the extensive taxonomic splitting that followed the introduction of 16S rRNA gene sequencing in the 1990s, the other two species remain within the genus, and are probably its best-known representatives. Cohn, who is generally regarded as the founder of bacterial classification, proposed the genus at a time when bacterial taxonomists were still constrained by the small number of differential characters available to them; when assigning the three species to *Bacillus* he was primarily considering cell shape and, as the inclusion of the non-motile *B. anthracis* attests, he did not consider motility to be an important character.

Although Cohn illustrated the spores of *B. subtilis* in a later taxonomic publication of 1875 and was soon to demonstrate their importance as resistant forms in 1876, he did not take sporulation into account as a generic character. Given his interest in bacterial classification this may seem surprising, because Cohn certainly considered spore properties to be of significance in the epidemiology of anthrax. Endospores were discovered independently of Cohn, by Koch and in 1877 by Tyndall, and it is Robert Koch who is usually credited as the founder of medical bacteriology. He recognized the significance of spores in the epidemiology of anthrax and his demonstration of the life history of *B. anthracis* in 1876 confirmed that this bacterium was the agent of anthrax in animals and man; he thus proved the germ theory of disease. It should be noted, however, that others had been investigating anthrax for many years. It was shown to be a contagious disease in 1823, and Casimir Davaine was convinced by 1863 that the "bacteridia" he found in the blood of animals with anthrax were, alone, the cause of this distinct disease, and that anthrax resulted when blood containing them was injected into healthy animals; he later named the organism responsible – *Bacteridium anthracis* – in 1868. We should remember that at this time some workers considered that classification of the "fission fungi" was of no scientific value, and many still believed that all bacteria existed in several morphological and physiological forms – in 1882 Buchner claimed that shaking cultures of *B. subtilis* at different temperatures could yield *B. anthracis*!

Through the 1880s and 1890s, authors of bacterial classifications differed in their opinions of the importance of spores as a key character of the genus. The term *Bacillus* has consequently been employed in two senses – as a genus name and as a general reference to shape – and the latter has, unfortunately, become the most widely accepted definition of the term, especially by medical bacteriologists, and considerable confusion has resulted. Indeed, as early as 1913 Vuillemin considered that the name had become so vulgarized by varied applications that it should lose nomenclatural status; however, he proposed that it should be replaced with the generic name *Serratia*!

The Committee of the Society of American Bacteriologists on Characterization and Classification of Bacterial Types came up with, in the early volumes of the *Journal of Bacteriology* (Winslow et al. 1920), satisfactory and largely uncontested definitions of bacterial groups. The family *Bacillaceae* was defined as "Rods producing endospores, usually Gram-positive. Flagella when present peritrichic. Often decompose protein media actively through the agency of enzymes" and *Bacillus* was described as "Aerobic forms. Mostly saprophytes. Liquefy gelatine. Often occur in long threads and form rhizoid colonies. Form of rod usually not greatly changed at sporulation". The Committee also used the requirement of oxygen and sporangial shape for differentiation between *Bacillus* and *Clostridium*, the other genus in the family *Bacillaceae*, and this description was applied in the first and second editions of *Bergey's Manual of Determinative Bacteriology*.

There existed another problem around this time – confusion had emerged about the identity of the type species of the genus as two very different type strains seemed to exist. One from the University of Marburg bore small spores and germinated equatorially, and the other from the University of Michigan (originating from the laboratory of Koch in 1888) formed much larger spores with germination occurring at the pole. Following extensive studies, Conn suggested in 1930 that the Marburg type should be called *B. subtilis* Cohn, and this strain was finally officially adopted as the generic type in 1936.

Some 70 years on, the phenotypic definition of *Bacillus* as a genus of aerobic endospore-forming rods has been weakened by the introduction of a phylogenetic approach to bacterial classification. There followed the proposals of the strict anaerobes *B. infernus* (Boone et al. 1995), *B. arseniciselenatis* (Switzer Blum et al. 1998) and *B. macyae* (Santini et al. 2004), and the non-sporulating species *B. infernus, Bacillus thermoamylovorans* (Combet-Blanc et al. 1995), *B. selenitireducens* (Switzer Blum et al. 1998), *B. subterraneus* (Kanso et al. 2002), *B. saliphilus* (Romano et al. 2005), *B. foraminis* (Tiago et al. 2006), *B. okhensis* (Nowlan et al. 2006) and *B. qingdaonensis* (Wang et al. 2007b). Furthermore, only coccoid cells have been observed in the single available strain of *B. saliphilus,* so that *Bacillus* is no longer exclusively a genus of rods, and the genus *Sporosarcina*, established in 1936 to accommodate aerobic endospore-forming cocci, now contains 12 species, most of which are rod-shaped.

Nearly 120 *Bacillus* species were proposed after 1990 and many of these were based on the study of a single isolate. The descriptions of a further nine species were based upon only two strains, so that nearly 40% of the species in this genus (and about 50% of those published since 1990) are poorly represented, and we have little or no information about their within-species variation. Also, since 1990, 16 further genera have been established to accommodate species previously assigned to *Bacillus*. With the merger of two genera and the addition of new species these genera now comprise 223 species, while *Bacillus*, with 153 species, has acquired more species than it has lost to new genera. Furthermore, 38 other new genera of aerobic endospore-formers containing organisms not previously assigned to *Bacillus* have also been proposed since 1990; in 35% of these there is only one species in the genus, and, again, most of the species are poorly represented. Lack of knowledge of the between-strain variation of a species will nearly always lead to difficulties in identification, and further isolations regularly result in emendations of descriptions and, not infrequently, in taxonomic and nomenclatural changes. Table 1.1 lists the currently valid genera and shows numbers of species and percentages of poorly represented species.

Overall, therefore, we have 54 new genera (51 following mergers) of aerobic endospore-formers containing 461 new or revived species that have been proposed in the 23 years since the 1986 edition of *Bergey's Manual*, and yet only nine proposals for merging species have been made in that time (*B. galactophilus* as a synonym of *B. agri,* Shida et al. 1994; *Bacillus gordonae* as a synonym of *B. validus*, Heyndrickx et al. 1995; *B. larvae* and *B. pulvifaciens* as subspecies of *Paenibacillus larvae*, Heyndrickx et al. 1996; merger of *P. larvae* and *P. pulvifaciens* as *P. larvae*, Genersch et al. 2006); reclassification of *Paenibacillus durum* [formerly *Clostridium durum*] as a member of the species *P. azotofixans,* Rosado et al. 1997 [but correct name, with gender adjusted, is *P. durus* according to

Table 1.1 The genera of aerobic Endospore-forming bacteria and relatives (as at October 2009)

Family	Genus	Number of species	Percentage of species based on one strain	Proposal of genus	Comment
Bacillaceae	*Bacillus*	153	35	Cohn (1872)	Includes strict aerobes, facultative anaerobes and several strict anaerobes, neutrophiles, alkaliphiles, and acidophiles, mesophiles, psychrophiles and thermophiles, and halophiles from a *wide range of soils* and other habitats
	Amphibacillus	4	75	Niimura et al. (1990)	Alkaliphiles from *composts* and sediments
	Halobacillus	16	94	Spring et al. (1996)	Halophiles from salterns, salt lakes, mangrove and *salt marsh soils*
	*Virgibacillus**	18	50	Heyndrickx et al. (1998)	Halophiles and halotolerant organisms from *soil, saline soil*, salterns, seawater, mural paintings
	*Gracilibacillus**	10	80	Wainø et al. (1999)	Halophiles from salt lakes, *saline and other soils,* fermented fish, iguana
	*“Salibacillus”**	(1)	–	Wainø et al. (1999)	Merged with *Virgibacillus* (Heyrman et al. 2003)
	Anoxybacillus	11	91	Pikuta et al. (2000)	Hot springs, *geothermal soil*, manure, gelatine
	Filobacillus	1	100	Schlesner et al. (2001)	Marine hydrothermal vent
	*Geobacillus**	16	31	Nazina et al. (2001)	Thermophiles from food, milk, water, *soil, geothermal soil, compost*, hot springs, hydrothermal vents, hot springs, sugar beet juice, oilfields
	Oceanobacillus	7	86	Lu et al. (2002)	Obligate and facultative alkaliphiles from deep marine sediments, activated sludge, aquatic animals, shrimp paste, algae, and mural painting

Lentibacillus	11	73	Yoon et al. (2002)	Halophiles from salt lakes, salterns, fermented fish and shrimp products
Paraliobacillus	2	100	Ishikawa et al. (2002)	Halotolerant alkaliphiles from salt lake and marine alga
Cerasibacillus	1	100	Nakamura et al. (2004)	Moderate thermophile and alkaliphile from decomposing kitchen refuse
Pontibacillus	4	75	Lim et al. (2005)	Moderate halophiles from solar salterns and marine animals
Tenuibacilus	1	0	Ren and Zhou (2005)	Moderate halophile from *saline soil*
Halolactibacillus	3	0	Ishikawa et al. (2005)	Halophilic and alkaliphilic marine lactic acid bacteria from marine sponge and algae
Thalassobacillus	2	100	García et al. (2005)	Halophiles from *saline soil* and hypersaline lake
*Alkalibacillus**	5	80	Jeon et al. (2005)	Moderate halophiles from *soil*, *saline soil*, salt, salt lake, mineral pool and camel dung
Salinibacillus	2	50	Ren and Zhou (2006)	Moderate halophiles from a neutral saline lake
Ornithinibacillus	2	100	Mayr et al. (2006)	Halotolerant to moderately halophilic organisms from milk and marine sediment
Vulcanibacillus	1	100	L'Haridon et al. (2006)	Strictly anaerobic thermophile from marine hydrothermal vent
Caldalkalibacillus	2	100	Xue et al. (2006)	Thermophilic alkaliphiles from hot springs
Paucisalibacillus	1	100	Nunes et al. (2006)	From *potting soil*
*Lysinibacillus**	5	40	Ahmed et al. (2007)	*Soil*, water, food, clinical specimens, mosquitoes
Terribacillus	3	67	An et al. (2007)	Moderately halotolerant organisms from *soil* and seawater
Piscibacillus	2	100	Tanasupawat et al. (2007)	Halophiles from fermented fish and hypersaline lake
Salirhabdus	1	100	Albuquerque et al. (2007)	Halotolerant organism from sea salt pond
*Salimicrobium**	4	50	Yoon et al. (2007)	Halophiles from salterns and rotting marine wood

(*continued*)

Table 1.1 (continued)

Family	Genus	Number of species	Percentage of species based on one strain	Proposal of genus	Comment
	Halalkalibacillus	1	100	Echigo et al. (2007)	Moderate halophile and alkaliphile from *soil*
	Salsuginibacillus	2	100	Carrasco et al. (2007)	Moderate halophiles from soda lake sediments
	"Pelagibacillus"	(1)	–	Kim et al. (2007)	Merged with *Terribacillus* (Krishnamurthi and Chakrabarti 2008)
	Aquisalibacillus	1	100	Márquez et al. (2008)	Moderate halophile from saline lake. Spores not observed
	Sediminibacillus	2	100	Carrasco et al. (2008)	Moderate halophiles from saline lake sediments
	Aeribacillus	1	0	Miñana-Galbis et al. (2010)	Sewage and hot spring
	Falsibacillus	1	100	Zhou et al. (2009)	*Soil*
	Marinococcus, Saccharococcus				Related, non-spore-forming genera from salterns, *saline soils*, and sugar beet extraction
"Alicyclobacillaceae"	*Alicyclobacillus**	20	70	Wisotzkey et al. (1992)	Thermoacidophiles from hot springs, *geothermal soils*, *soils*, fruit juices, waste water sludge, metal ore
	Tumebacillus	1	100	Steven et al. (2008)	From *permafrost soil*. Shows some phylogenetic relationship with *Alicyclobacillus*
"Paenibacillaceae"	*Paenibacillus**	110	52	Ash et al. (1993)	*Soil*, *rhizosphere*, *composts*, plant material, foods, warm springs, waste water, insects, nematodes, clinical specimens, mural painting
	Oxalophagus	1	0	Collins et al. (1994)	Oxalotroph from freshwater mud
	*Aneurinibacillus**	5	20	Shida et al. (1996)	*Soil*, *geothermal soil*, faeces, fermentation plant, sugar beet juice

	*Brevibacillus**	16	25	Shida et al. (1996)	*Soil, geothermal soil, compost,* faeces, foods including milk and cheese, pharmaceutical fermentation plant, water, salterns, estuarine sediment, vegetation, insects, clinical specimens
	Ammoniphilus	2	50	Zaitsev et al. (1998)	Ammonium dependent, oxalotropic organisms from *rhizosphere*
	Thermobacillus	2	100	Touzel et al. (2000)	Thermophiles from *soil* and *compost*
	Cohnella	8	100	Kämpfer et al. (2006)	Thermotolerant organisms from water, *soil, volcanic soil, root nodules,* starch, blood
	Saccharibacillus	2	100	Rivas et al. (2008)	Sugar cane, *desert soil*
Planococcaceae	*Sporosarcina**	12	58	Kluyver and van Niel (1936)	Spherical spore-formers from *soil, Antarctic soil* and cyanobacterial mat, brackish and marine waters, blood
	"*Marinibacillus*"*	(2)	–	Yoon et al. (2001)	Merged with *Jeotgalibacillus* (Yoon et al. 2010)
	*Ureibacillus**	5	40	Fortina et al. (2001)	Thermophiles from *soil*, *compost* and air
	Jeotgalibacillus	4	75	Yoon et al. (2001)	Moderate halophiles from marine sediments, salterns, and fermented seafood
	*Viridibacillus**	3	0	Albert et al. (2007)	Spherical spore-formers from *soil*
	*Rummeliibacillus**	2	0	Vaishampayan et al. (2009)	Spherical spore-formers from space craft assembly room, composter and *soil*
	*Solibacillus**	1	100	Krishnamurthi et al. (2009a)	Spherical spore-former from *forest soil*
	Paenisporosarcina	2	100	Krishnamurthi et al. (2009b)	*Landfill soil*, Antarctic cynaobacterial mat
	Planococcus, Caryophanon, Filibacter, Kurthia, Planomicrobium				Related, non-spore-forming genera from cyanobacterial mats, seawater and marine sediment, freshwater sediment, fish and fermented fish, meat, *soil*, mammoth gut, animal faeces

(*continued*)

Table 1.1 (continued)

Family	Genus	Number of species	Percentage of species based on one strain	Proposal of genus	Comment
Pasteuriaceae	*Pasteuria*	4	0	Metchnikoff (1888)	Obligate endoparasites of phytopathogenic nematodes and crustaceans. Not grown axenically
"*Sporolactobacillaceae*"	*Sporolactobacillus**	8	12	Kitahara and Suzuki (1963)	Endospore-forming lactic acid bacteria from *rhizosphere*, *soil*, fruit juice, chicken feed and fermentation starters
	*Pullulanibacillus**	1	100	Hatayama et al. (2006)	Moderate acidophile from *soil*
	Tuberibacillus	1	0	Hatayama et al. (2006)	Thermophile from *compost*
"*Thermooactinomycetaceae*"	*Thermoactinomyces*	2	50	Tsilinsky (1899)	Thermophiles with mycelial growth, from *soil*, *compost*, hay, fodder, dung
	Laceyella	2	0	Yoon et al. (2005)	Thermophiles with mycelial growth, from *soil*, *muds*, sugar cane and bagasse, clinical specimen
	Thermoflavimicrobium	1	0	Yoon et al. (2005)	*Thermophile with mycelial growth, from soil and compost*
	Planifilum	3	67	Hatayama et al. (2005)	Thermophiles with mycelial growth, from *compost* and hot spring
	Mechercharimyces	2	50	Matsuo et al. (2006)	Mesophiles with mycelial growth, from marine lake
	Seinonella	1	0	Yoon et al. (2005)	Mesophile with mycelial growth, from *soil*
	Shimazuella	1	100	Park et al. (2007)	Mesophile with mycelial growth, from *soil*

*Genera containing one or more species that were originally allocated to *Bacillus* priority]; *B. kaustophilus* and *B. thermocatenulatus* as members of *B. thermoleovorans,* Sunna et al. 1997, but not validated; reclassification of *B. axarquiensis* and *B. malacitensis* as later heterotypic synonyms of *B. mojavensis*, Wang et al. 2007a, and three proposals for merging genera (*Virgibacillus* and *Salibacillus,* Heyrman et al. 2003; *Terribacillus* and *Pelagibacillus*, Krishnamurthi and Chakrabarti 2008; *Jeotgalibacillus* and *Marinibacillus*, Yoon et al. 2010).

16S rRNA gene sequence analysis has also allowed the proposal of a clearer phylogenetic structure at higher taxonomic levels, although it is important to appreciate that changes will occur as new data become available and as methods for analysis improve. This "road map" and its revisions were used to guide the arrangement of the Second Edition of *Bergey's Manual of Systematic Bacteriology* (5 volumes, with the first published in 2001; Volume 3 (De Vos et al. 2009) includes the aerobic endospore-formers). The revision used for Volume 3, The *Firmicutes*, was based upon the integrated small-subunit rRNA database of the SILVA project (Pruesse et al. 2007). It has been proposed that the phylum *Firmicutes* contains three classes: *"Bacilli"*, *"Clostridia"* and *"Erysipelotrichia"*, and *Clostridium* clearly no longer lies within the family *Bacillaceae*. Within the class "*Bacilli*" the order *Bacillales* contains nine families, seven of which include genera of aerobic endospore-formers: *Bacillaceae,* "*Alicycobacillaceae*", "*Paenibacillaceae*", *Pasteuriaceae*, *Planococcaceae*, "*Sporolactobacillaceae*" and "*Thermoactinomycetaceae*"; the other two families are "*Listeriaceae*" and "*Staphylococcaceae*". The relationships that led to these proposals are not all clear-cut, however; for example: *Bacillaceae* accommodates misassigned genera, within *Bacillaceae* a large genus group that includes *Amphibacillus, Gracilibacillus, Halobacillus, Virgibacillus* and 12 other recently proposed genera may warrant elevation to family status (Ludwig et al. 2007), and some subclusters of *Bacillus* may warrant elevation to generic status.

1.5 Identification Schemes

However, the taxonomic progress outlined above has not usually revealed readily determinable features characteristic of each of the new genera, and they show wide ranges of sporangial morphologies and phenotypic test patterns. Identification can be very challenging.

The commonest spore-forming bacteria were accurately described in papers published early in the twentieth century, and the number of species recognized at the time was small, but even then identification of fresh isolates was problematic. Ford and coworkers had difficulty in identifying their collection of milk isolates, and so in 1916 they published one of the first practical schemes for *Bacillus* identification, based on an extensive investigation of many strains from various environments. With the resolution of the type strain controversy, the discovery that many pathogenic "*B. subtilis*" strains were in fact *B. cereus* stimulated N. R. Smith and colleagues to conduct a large taxonomic study of the genus followed by a grouping of the mesophilic species in the form of a diagnostic key in the 1930s.

Smith's team meticulously characterized their cultures and then emphasized the similarities rather than the differences between their strains, so that they "lumped" their taxa rather than splitting them. In a report published a decade after the study began, and later revised (Smith et al. 1952), they recognized three groups of species on the basis of spore morphologies, and the first truly workable diagnostic key for *Bacillus* emerged; its effectiveness was soon confirmed by Knight and Proom – in their study of 296 strains all but 51 could be allocated to the species or groups previously described. The studies from Smith's laboratory shaped the future of *Bacillus* taxonomy, and various research groups began applying existing techniques and new methods to the taxonomy of the genus *Bacillus*.

Numerical taxonomic methods were firstly applied to *Bacillus* by Sneath in the early 1960s, using the data of Smith et al. (1952), and a phenogram was constructed that "largely agreed" with the 1952 classification. One of the most influential and significant schemes for the genus was published by Gordon et al. (1973), after protracted study of 1,134 strains. It is perhaps surprising that these authors did not employ numerical analysis, but the scheme was widely and successfully used for many years. During the 1970s several new approaches to characterization such as serology, enzyme and other molecular studies, and pyrolysis gas–liquid chromatography began to emerge as potentially useful taxonomic tools in bacterial taxonomy as a whole, but despite this the taxonomy of *Bacillus* remained relatively untouched, with few new species or subspecies being described and validated. The emphasis at that time was on identification.

A decade later, increasing incidence of *Bacillus* isolations from clinical and industrial environments emphasized the need for a rapid identification scheme, and that could only follow an improved taxonomy of the genus. Consequently, Logan and Berkeley (1984) developed a *Bacillus* identification scheme with a large database for 37 clearly defined taxa (species) based upon miniaturized tests in the API 20E and 50CHB Systems (bioMérieux, Marcy l'Etoile, France).

It was also the classification of Gordon et al. (1973) with support from the work of Logan and Berkeley (1981) that formed the basis of the list of *Bacillus* included in the Approved Lists of Bacterial Names published by the International Committee of Systematic Bacteriology (Skerman et al. 1980). These lists marked a new starting date for bacterial nomenclature, the previous date being that of Linnaeus' monumental classification work, *Species Plantarum,* which was published in 1753. Since that time, many synonyms had inadvertently been proposed and, as mentioned earlier, the number of *Bacillus* species described fluctuated greatly through successive editions of *Bergey's Manual*.

However, since the development of the Gordon et al. and Logan and Berkeley schemes, mainly for neutrophilic and mesophilic species, *Bacillus* has been divided, and the genus and new genera derived from it also encompass many acidophiles, alkaliphiles, halophiles, psychrophiles and thermophiles. The task of identification became more complicated, owing to:

1. The proposal of many new species (frequently from exotic habitats, and often primarily on the basis of molecular analyses).

2. The allocation of strict anaerobes (*B. arseniciselenatis, B. infernus*) and organisms in which spores have not been observed (*B. infernus, B. thermoamylovorans*, etc.) to *Bacillus*.
3. The transfer of many species to new genera, without revealing readily determinable features characteristic of each genus.
4. Many recently described species being proposed on the basis of very few strains so that the within-species diversities of such taxa, and so their true boundaries, remain unknown.
5. Many recently described species being based on genomic groups disclosed by DNA–DNA pairing experiments, with routine phenotypic characters for their distinction often being very few and of unproven value.

The identification schemes of Gordon et al. (1973) and Logan and Berkeley (1984) embraced 18 and 37 species, respectively, and in both cases only about half of the species remain in *Bacillus*; such methods can no longer be expected, even with substantial modifications and expanded databases, to allow recognition of all or even most species of aerobic endospore-formers. Identification with routine phenotypic tests must therefore call upon a variety of characterization methods, and a unified approach is no longer possible. Despite this, the traditional characterization tests used by Gordon et al. retain their place in *Bacillus* identification, because the most commonly encountered species are still distinguishable by these methods. Fritze (2002) recommended a stepwise approach to identification of the aerobic endospore-formers, using the characterization tests of Gordon et al., and many of the same characters are included in the Proposed minimal standards for describing new taxa of aerobic, endospore-forming bacteria (Logan et al. 2009).

However, further problems have emerged with the splitting of well-established (but not necessarily homogeneous) species or groups into large numbers of new taxa over a short period. For example, *Bacillus circulans* was long referred to as a complex rather than a species, but the revision of the taxonomy of this group, and consequent proposals for several new species to be derived from it, have led to difficulties in identification. Although the proposals were mostly based upon polyphasic taxonomic studies, initial recognition of the new taxa depended largely upon DNA relatedness data. A DNA:DNA re-association study of *B. circulans* strains yielded *B. circulans sensu stricto, B. amylolyticus, B. lautus, B. pabuli* and *B. validus* and evidence for the existence of five other species (Nakamura and Swezey 1983; Nakamura 1984). *Bacillus amylolyticus, B. lautus, B. pabuli* and *B. validus* are now accommodated in *Paenibacillus*, and these species and *B. circulans* are difficult to distinguish using routine phenotypic tests. Such radical taxonomic revisions have left many culture collections worldwide with few representatives of *B. circulans sensu stricto*, but with numerous misnamed strains of this species, which may or may not belong to one of the newly proposed taxa. The curators will normally not be able to know which are which without considerable expenditure in scholarship and experimental work, and in many cases a collection will hold only one authentic strain, the type strain, of a species – be it an old or new species. If microbial classification does not serve a practical diagnostic purpose, it is of little interest, and it is essential that if a new species is to be proposed it should be identifiable by readily available phenotypic and/or genomic methods.

The written description of a species or genus must therefore be seen as a practically essential, integral part of any taxonomic proposal, and care needs to be taken to ensure that it is as comprehensive and practically useful as possible; the diagnostic characters should not be seen merely as adjuncts to the delineation of a taxon that has been recognized primarily on the basis of genomic data. Phenotype, along with genotype, "continues to play a salient role in the decision about cut-off points of genomic data for species delineation ... [and] ... description of species... should be based on the use of well-documented criteria, laboratory protocols and reagents which are reproducible" (Stackebrandt et al. 2002). Unfortunately, comparisons of related species are often made by reference to descriptions given in the literature, rather than by laboratory study of reference cultures. It needs to be remembered that characterization methods and their interpretation vary, and typographic errors in the compilation of descriptions are bound to occur (many were encountered by NAL in the preparation of sections on *Bacillus* and other genera for the second edition of *Bergey's Manual of Systematic Bacteriology*), so compilations as well as original descriptions should never be relied upon entirely.

Current schemes for identifying aerobic endospore-formers may be roughly divided into four categories effectively based upon (1) traditional biochemical, morphological and physiological characters, (2) miniaturized versions of traditional biochemical tests (e.g., API kits, VITEK cards and Biolog plates), (3) chemotaxonomic characters (such as PAGE of whole cell proteins, and fatty acid methyl ester [FAME] profiles) and (4) genomic characters (16S rRNA gene sequencing, DNA–DNA relatedness, etc.). As early as the work of Smith et al. (1952), it was becoming clear that no one phenotypic technique would be suitable for identifying all species of aerobic endospore-former. Identification problems have mounted up as further species from all kinds of niches have subsequently been proposed. Chemotaxonomic analyses and studies of nucleic acids have resolved only some of these problems, as it is impossible to devise standardized conditions to accommodate the growth of all strains of all species for chemotaxonomic work, and it remains unknown to the taxonomist if differences between taxa are consequences of genetic or, as discussed earlier, environmental factors. The need to substantiate each characterization method by other techniques (be they phenotypic or genotypic) has become increasingly important as new techniques emerge. This need is satisfied by the polyphasic approach now usual for serious classification studies, and the same approach may sometimes be necessary in order to identify strains from some of the less familiar species.

The International Committee on Systematics of Prokaryotes (ICSP) is an international committee within the International Union of Microbiological Societies, and it is responsible for matters relating to prokaryote nomenclature and taxonomy. The ICSP has a number of subcommittees that deal with matters relating to the nomenclature and taxonomy of specific groups of prokaryotes. One of the main remits of these subcommittees is the proposal of minimal standards for describing new taxa within their fields of expertise (Recommendation 30b of the

Bacteriological Code (1990 Revision), Lapage et al. 1992), and the Subcommittee on the Taxonomy of the Genus *Bacillus* and Related Organisms has recently proposed its own such standards. These give guidance on the preparation of proposals of new genera and species of aerobic, endospore-forming bacteria; they recommend that such proposals be based upon more than one isolate, that cultivation conditions be given in detail, and that microscopic characters be clearly described, and the phenotypic, chemotaxonomic and genotypic characters that are essential or highly desirable for the establishment of a new species or genus, and its subsequent recognition, are listed (Logan et al. 2009).

1.6 Current Taxonomy

Table 1.1 summarizes the present classification of the aerobic endospore-formers, and highlights the genera that contain species isolated from soil; Fig. 1.1 shows the relationships of many of the aerobic endospore-forming genera, based upon comparison of 16S rRNA gene sequences. Although *Bacillus* species have long been regarded as primarily soil organisms, it seems that soils are not the main sources for many of the newer (albeit poorly represented) taxa. The following paragraphs give a little more information about some of the larger genera.

In the family *Bacillaceae*, *Bacillus* continues to accommodate the best-known species such as *B. subtilis* (the type species), *B. anthracis, B. cereus, B. licheniformis, B. megaterium, B. pumilus* and *B. thuringiensis*. It still remains a large genus, with over 150 species, as transfers of species to other genera have been balanced by proposals for new *Bacillus* species (and several species await transfer to *Geobacillus* and *Paenibacillus*). The pathogenic members of the *B. cereus* group, *B. anthracis, B. cereus* and *B. thuringiensis*, are really pathovars of a single species, and along with *B. mycoides, B. pseudomycoides* and *B. weihenstephanensis*, these species form the *B. cereus* group, as this phylogenetic subcluster is informally known. Another such subcluster contains *B. amyloliquefaciens*, *B. licheniformis*, and other close relatives of *B. subtilis* – the *B. subtilis* group. Most of the obligate thermophiles originally allocated to *Bacillus* have been transferred to *Geobacillus*, with the former "*Bacillus stearothermophilus*" as the type species; the genus now comprises 16 species including *G. kaustophilus, G. thermodenitrificans, G. thermoglucosidasius* and *G. thermoleovorans*, but several species probably warrant merging. The distinctiveness of the spherical-spored species *B. sphaericus* and *B. fusiformis* had long been recognized, as they are unreactive in routine characterization tests and possess distinctive petidoglycans containing lysine. They have accordingly been transferred to a new genus, *Lysinibacillus*, with a new species *L. boronitolerans* as the type. Most of the remaining genera contain small numbers of species that were not originally accommodated in *Bacillus*, and many of these species are halophilic or halotolerant, and often alkaliphilic.

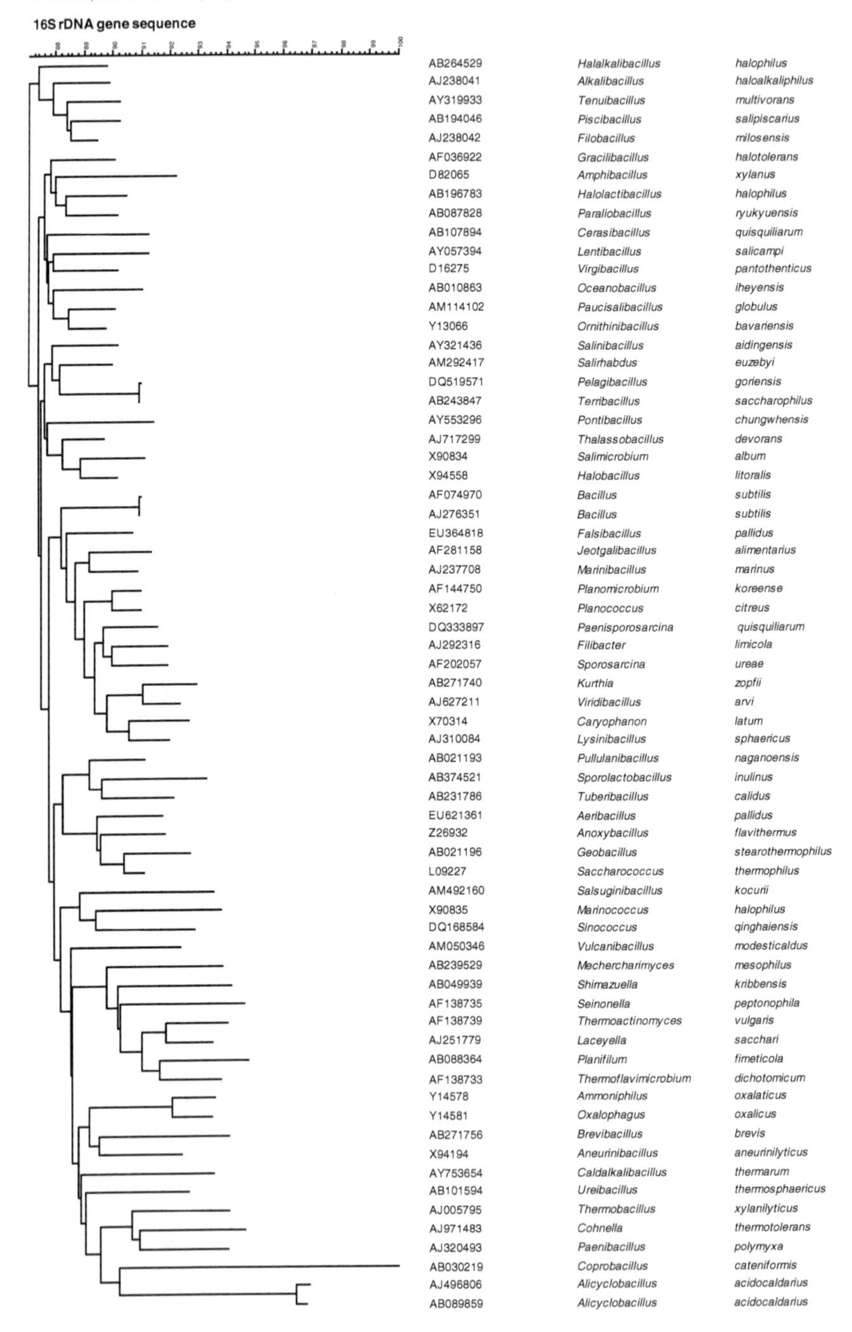

Fig. 1.1 Neighbour-joining phylogenetic tree of aerobic endospore-forming bacteria based on 16S rDNA gene sequences, showing accession numbers, genera and species. Cluster analysis was performed using BioNumerics 5.10 (Applied Maths, Sint-Martens-Latem, Belgium)

At present only one genus of thermoacidophiles, *Alicyclobacillus*, most of whose species possess ω-alicyclic fatty acids, is allocated to the proposed family "*Alicyclobacillaceae*". *Tumebacillus permanentifrigoris*, a facultatively chemolithoautotrophic and mesophilic neutrophile isolated from permafrost soil, shows some phylogenetic relationship with *Alicyclobacillus*.

In the family "*Paenibacillaceae*", *Paenibacillus*, with 110 species, is the largest genus of aerobic endospore-formers after *Bacillus*. It was proposed to accommodate species formerly allocated to *Bacillus*, such as *B. polymyxa* (the type species), *B. macerans*, *B. alvei*, and the honeybee pathogen *B. larvae*. Many species are highly active in tests for acid production from carbohydrates, and share other phenotypic characters with *B. circulans*; 16S rRNA gene sequence studies revealed that many strains formerly allocated to that species actually represent several taxa within *Paenibacillus*. Members of this genus are widely distributed in soil, often in association with plants, and several species can fix nitrogen. *Aneurinibacillus* and *Brevibacillus* species are also particularly associated with soil, but members of both genera are unreactive in routine laboratory characterization tests, unlike *Paenibacillus* species. The other genera in this family also show associations with soil and plants.

Pasteuriaceae contains the single genus *Pasteuria*, which contains four species of endospore-forming, obligate endoparasites of phytopathogenic nematodes and crustaceans, with *P. ramosa* as type species; none of the species have been grown axenically, and several further species have been described and await validation (see Bishop 2011).

Planococcaceae contains genera of endospore-formers and non-endospore-formers. *Sporosarcina* was proposed in 1936 to accommodate the motile, spore-forming coccus *S. ureae*, and its relationship to *Bacillus pasteurii* was recognized in the early days of 16S rRNA gene sequence comparisons. It remained a single-species genus for 48 years. In 1984 *Sporosarcina halophila* was added, but 12 years later it was removed to the new genus *Halobacillus* (in *Bacillaceae*); subsequently 11 other, rod-shaped, round-spored species, including *B. pasteurii*, *B. globisporus* and *B. psychrophilus* were placed in the genus.

"*Sporolactobacillaceae*" is another single-genus family, although some *Bacillus* species show a relationship to it, as do the recently proposed genera *Pullulanibacillus* and *Tuberibacillus*. Following its proposal in 1963, *Sporolactobacillus* comprised a single species for over 30 years, until four new species and two subspecies were proposed in 1997. The type species, *Sporolactobacillus inulinus*, was at one time regarded as a "*Bacillus–Lactobacillus*" intermediate on account of its homolactic fermentation and microaerophily, and some more recent attention has been stimulated by current interest in probiotics. The sharing of a lineage with *Bacillus laevolacticus* had been recognized for some years (Stackebrandt and Swiderski 2002), and Hatayama et al. (2006) proposed the transfer of *B. laevolacticus* to *Sporolactobacillus*.

Thermoactinomyces was proposed in 1899 to harbour a thermophilic organism with *Actinomyces*-like mycelial growth, but its spores are true endospores, it lies within the evolutionary radiation of *Bacillus* and related genera, and shows a close relationship with *Pasteuria*. The family "*Thermoactinomycetaceae*" comprises seven genera of endospore-formers: *Laceyella* and *Thermoflavimicrobium*, which contain species formerly within *Thermoactinomyces*, and *Mechercharimyces*, *Planifilum*, *Seinonella* and *Shimazuella*.

References

Ahmed I, Yokota A, Yamazoe A, Fujiwara T (2007) Proposal of *Lysinibacillus boronitolerans* gen. nov. sp. nov., and transfer of *Bacillus fusiformis* to *Lysinibacillus fusiformis* comb. nov. and *Bacillus sphaericus* to *Lysinibacillus sphaericus* comb. nov. Int J Syst Evol Microbiol 57:1117–1125

Albert RA, Archambault J, Lempa M, Hurst B, Richardson C, Gruenloh S, Duran M, Worliczek HL, Huber BE, Rosselló-Mora R, Schumann P, Busse H-J (2007) Proposal of *Viridibacillus* gen. nov. and reclassification of *Bacillus arvi*, *Bacillus arenosi* and *Bacillus neidei* as *Viridibacillus arvi* gen. nov., comb. nov., *Viridibacillus arenosi* comb. nov. and *Viridibacillus neidei* comb. nov. Int J Syst Evol Microbiol 57:2729–2737

Albuquerque L, Tiago I, Rainey FA, Taborda M, Nobre MF, Veríssimo A, da Costa MS (2007) *Salirhabdus euzebyi* gen. nov., sp. nov., a Gram-positive, halotolerant bacterium isolated from a sea salt evaporation pond. Int J Syst Evol Microbiol 57:1566–1571

An S-Y, Asahara M, Goto K, Kasai H, Yokota A (2007) *Terribacillus saccharophilus* gen. nov., sp. nov. and *Terribacillus halophilus* sp. nov., spore-forming bacteria isolated from field soil in Japan. Int J Syst Evol Microbiol 57:51–55

Andrup L, Barfod KK, Jensen GB, Smidt L (2008) Detection of large plasmids from the *Bacillus cereus* group. Plasmid 59:139–143

Ash C, Priest FG, Collins MD (1993) Molecular identification of rRNA group 3 bacilli (Ash, Farrow, Wallbanks and Collins) using a PCR probe test. Ant van Leeuwenhoek 64:253–260

Banat IM, Marchant R (2011) *Geobacillus* activities in soil and oil contamination remediation. In: Logan NA, De Vos P (eds) Endospore forming soil bacteria (Soil Biology 27). Springer, Heidelberg, doi: 10.1007/978-3-642-19577-13

Bishop AH (2011) *Pasteuria penetrans* and its parasitic interaction with plant parasitic nematodes. In: Logan NA, De Vos P (eds) Endospore forming soil bacteria (Soil Biology 27). Springer, Heidelberg, doi: 10.1007/978-3-642-19577-9

Boone DR, Liu Y, Zhao Z-J, Balkwill DL, Drake GR, Stevens TO, Aldrich HC (1995) *Bacillus infernus* sp. nov., an Fe(III)- and Mn(IV)-reducing anaerobe from the deep terrestrial subsurface. Int J Syst Bacteriol 45:441–448

Carrasco IJ, Márquez MC, Xue Y, Ma Y, Cowan DA, Jones BE, Grant WD, Ventosa A (2007) *Salsuginibacillus kocurii* gen. nov., sp. nov., a moderately halophilic bacterium from soda-lake sediment. Int J Syst Evol Microbiol 57:2381–2386

Carrasco IJ, Márquez MC, Xue Y, Ma Y, Cowan DA, Jones BE, Grant WD, Ventosa A (2008) *Sediminibacillus halophilus* gen. nov., sp. nov., a moderately halophilic, Gram-positive bacterium from a hypersaline lake. Int J Syst Evol Microbiol 58:1961–1967

Cohan FM, Perry EB (2007) A systematics for discovering the fundamental units of bacterial diversity. Curr Biol 17:R373–R386

Cohn F (1872) Untersuchungen über Bakterien. Beiträge zur Biologie der Pflanzen 1. Heft II:127–224

Collins MD, Lawson PA, Willems A, Cordoba JJ, Fernandez-Garayzabal J, Garcia P, Cai J, Hippe H, Farrow JAE (1994) The phylogeny of the genus *Clostridium*: proposal of five new genera and eleven species combinations. Int J Syst Bacteriol 44:812–826

Combet-Blanc Y, Ollivier B, Streicher C, Patel BKC, Dwivedi PP, Pot B, Prensier G, Garcia J-L (1995) *Bacillus thermoamylovorans* sp. nov., a moderately thermophilic and amylolytic bacterium. Int J Syst Bacteriol 45:9–16

De Clerck E, Gevers D, Sergeant K, Rodríguez-Díaz M, Herman L, Logan NA, Van Beeumen J, De Vos P (2004) Genomic and phenotypic comparison of *Bacillus fumarioli* isolates from geothermal Antarctic soil and gelatine. Res Microbiol 155:483–490

De Vos P (2011) Studying the bacterial diversity of the soil by culture-independent approaches. In: Logan NA, De Vos P (eds) Endospore forming soil bacteria (Soil Biology 27). Springer, Heidelberg, doi:10.1007/978-3-642-19577-3

De Vos P, Garrity G, Jones D, Krieg NR, Ludwig W, Rainey FA, Schleifer K-H, Whitman WB (eds) (2009) Bergey's Manual of Systematic Bacteriology, 2nd Edition, Vol. 3, The *Firmicutes*. Springer, New York, 1450 pp

Echigo A, Fukushima T, Mizuki T, Kamekura M, Usami R (2007) *Halalkalibacillus halophilus* gen. nov., sp. nov., a novel moderately halophilic and alkaliphilic bacterium isolated from a non-saline soil sample in Japan. Int J Syst Evol Microbiol 57:1081–1085

Fajardo-Cavazos P, Schuerger AC, Nicholson WL (2007) Testing interplanetary transfer of bacteria between Earth and Mars as a result of natural impact phenomena and human space-flight activities. Acta Astronaut 60:534–540

Forsyth G, Logan NA (2000) Isolation of *Bacillus thuringiensis* from northern Victoria Land, Antarctica. Lett Appl Microbiol 30:263–266

Fortina MG, Pukall R, Schumann P, Mora D, Parini C, Manachini PL, Stackebrandt E (2001) *Ureibacillus* gen. nov., a new genus to accommodate *Bacillus thermosphaericus* (Andersson *et al.* 1995), emendation of *Ureibacillus thermosphaericus* and description of *Ureibacillus terrenus* sp. nov. Int J Syst Evol Microbiol 51:447–455

Fritze D (2002) *Bacillus* identification – traditional approaches. In: Berkeley RCW, Heyndrickx M, Logan NA, De Vos P (eds) Applications and systematics of *Bacillus* and relatives. Blackwell Science, Oxford, pp 100–122

García MT, Gallego V, Ventosa A, Mellado E (2005) *Thalassobacillus devorans* gen. nov., sp. nov., a moderately halophilic, phenol-degrading, Gram-positive bacterium. Int J Syst Evol Microbiol 55:1789–1795

Genersch E, Forsgren E, Pentikäinen J, Ashiralieva A, Rauch S, Kilwinski J, Fries I (2006) Reclassification of *Paenibacillus larvae* subsp. *pulvifaciens* and *Paenibacillus larvae* subsp. *larvae* as *Paenibacillus larvae* without subspecies differentiation. Int J Syst Evol Microbiol 56:501–511

Gordon RE, Haynes WC, Pang CH-N (1973) The genus *Bacillus*. U.S. Department of Agriculture, Agricultural Handbook No. 427. U.S. Department of Agriculture, Washington, DC

Harrison RM, Jones AM, Biggins PDE, Pomeroy N, Cox CS, Kidd SP, Hobman JL, Brown NL, Beswick A (2005) Climate factors influencing bacterial count in background air samples. Int J Biometeorol 49:167–178

Hatayama K, Shoun H, Ueda Y, Nakamura A (2005) *Planifilum fimeticola* gen. nov., sp. nov. and *Planifilum fulgidum* sp. nov., novel members of the family '*Thermoactinomycetaceae*' isolated from compost. Int J Syst Evol Microbiol 55:2101–2104

Hatayama K, Shoun H, Ueda Y, Nakamura A (2006) *Tuberibacillus calidus* gen. nov., sp. nov., isolated from a compost pile and reclassification of *Bacillus naganoensis* Tomimura *et al.* 1990 as *Pullulanibacillus naganoensis* gen. nov., comb. nov. and *Bacillus laevolacticus* Andersch *et al.* 1994 as *Sporolactobacillus laevolacticus* comb. nov. Int J Syst Evol Microbiol 56:2545–2551

Heyndrickx M (2011) Dispersal of aerobic endospore-forming bacteria from soil and agricultural activities to food and feed. In: Logan NA, De Vos P (eds) Endospore forming soil bacteria (Soil Biology 27). Springer, Heidelberg, doi: 10.1007/978-3-642-19577-7

Heyndrickx M, Logan NA, Ali N, Vandemeulebroecke K, Scheldemann P, Hoste B, Kersters K, Aziz AM, Berkeley RCW, De Vos P (1995) *Paenibacillus* (formerly *Bacillus*) *gordonae*, is a later subjective synonym of *Paenibacillus* (formerly *Bacillus*) *validus* (ex Bredemann and Heigener, 1935) Nakamura 1984. Emended description of *P. validus*. Int J Syst Bacteriol 45:661–669

Heyndrickx M, Vandemeulebroecke K, Hoste B, Janssen P, Kersters K, De Vos P, Logan NA, Ali N, Berkeley RCW (1996) Reclassification of *Paenibacillus* (formerly *Bacillus*) *pulvifaciens*

(Nakamura 1984) Ash et al. 1994, a later subjective synonym of *Paenibacillus* (formerly *Bacillus*) *larvae* (White 1906) Ash et al. 1994, as a subspecies of *P. larvae,* with emended descriptions of *P. larvae* as *P. larvae* subsp. *larvae* and *P. larvae* subsp. *pulvifaciens*. Int J Syst Bacteriol 46:270–279

Heyndrickx M, Lebbe L, Vancanneyt M, Kersters K, De Vos P, Forsyth G, Logan NA (1998) *Virgibacillus:* a new genus to accommodate *Bacillus pantothenticus* (Proom and Knight 1950). Emended description of *Virgibacillus pantothenticus*. Int J Syst Bacteriol 48:99–106

Heyrman J, Balcaen A, Lebbe L, Rodríguez-Díaz M, Logan NA, Swings J, De Vos P (2003) *Virgibacillus carmonensis* sp. nov., *Virgibacillus necropolis* sp. nov. and *Virgibacillus picturae* sp. nov., three new species isolated from deteriorated mural paintings, transfer of the species of the genus *Salibacillus* to *Virgibacillus*, as *Virgibacillus marismortui* comb. nov. and *Virgibacillus salexigens* comb. nov., and emended description of the genus *Virgibacillus*. Int J Syst Evol Microbiol 53:501–511

Ishikawa M, Ishizaki S, Yamamoto Y, Yamasato K (2002) *Paraliobacillus ryukyuensis* gen. nov., sp. nov., a new Gram-positive, slightly halophilic, extremely halotolerant, facultative anaerobe isolated from a decomposing marine alga. J Gen Appl Microbiol 48:269–279

Ishikawa M, Nakajima K, Itamiya Y, Furukawa S, Yamamoto Y, Yamasato K (2005) *Halolactibacillus halophilus* gen. nov., sp. nov. and *Halolactibacillus miurensis* sp. nov., halophilic and alkaliphilic marine lactic acid bacteria constituting a phylogenetic lineage in *Bacillus* rRNA group 1. Int J Syst Evol Microbiol 55:2427–2439

Jeon CO, Lim J-M, Lee J-M, Xu L-H, Jiang C-L, Kim C-J (2005) Reclassification of *Bacillus haloalkaliphilus* Fritze 1996 as *Alkalibacillus haloalkaliphilus* gen. nov., comb. nov. and the description of *Alkalibacillus salilacus* sp. nov., a novel halophilic bacterium isolated from a salt lake in China. Int J Syst Evol Microbiol 55:1891–1896

Kämpfer P, Rosselló-Mora R, Falsen E, Busse H-J, Tindall BJ (2006) *Cohnella thermotolerans* gen. nov., sp. nov., and classification of '*Paenibacillus hongkongensis*' as *Cohnella hongkongensis* sp. nov. Int J Syst Evol Microbiol 56:781–786

Kanso S, Greene AC, Patel BKC (2002) *Bacillus subterraneus* sp. nov., an iron- and manganese-reducing bacterium from a deep subsurface Australian thermal aquifer. Int J Syst Evol Microbiol 52:869–874

Kellogg CA, Griffin DW (2006) Aerobiology and the global transport of desert dust. Trends Ecol Evol 21:638–644

Kim YG, Hwang CY, Yoo KW, Moon HT, Yoon J-H, Cho BC (2007) *Pelagibacillus goriensis* gen. nov., sp. nov., a moderately halotolerant bacterium isolated from coastal water off the east coast of Korea. Int J Syst Evol Microbiol 57:1554–1560

Kitahara K, Suzuki J (1963) *Sporolactobacillus* nov. subgen. J Gen Appl Microbiol 9:59–71

Kluyver AJ, van Niel CB (1936) Prospects for a natural classification of bacteria. Zentbl Bakteriol Parasitenkd Infektkrankh Hyg Abt II 94:369–403

Koeppel A, Perry EB, Sikorski J, Krizanc D, Warner A, Ward DM, Rooney AP, Brambilla E, Connor N, Ratcliff RM, Nevo E, Cohan FM (2008) Identifying the fundamental units of bacterial diversity: a paradigm shift to incorporate ecology into bacterial systematics. Proc Natl Acad Sci USA 105:2504–2509

Krishnamurthi S, Chakrabarti T (2008) Proposal for transfer of *Pelagibacillus goriensis* Kim *et al.* 2007 to the genus *Terribacillus* as *Terribacillus goriensis* comb. nov. Int J Syst Evol Microbiol 58:2287–2291

Krishnamurthi S, Chakrabarti T, Stackebrandt E (2009a) Re-examination of the taxonomic position of *Bacillus silvestris* Rheims *et al.* 1999 and proposal to transfer it to *Solibacillus* gen. nov. as *Solibacillus silvestris* comb. nov. Int J Syst Evol Microbiol 59:1054–1058

Krishnamurthi S, Bhattacharya A, Mayilraj S, Saha P, Schumann P, Chakrabarti T (2009b) Description of *Paenisporosarcina quisquiliarum* gen. nov., sp. nov., and reclassification of *Sporosarcina macmurdoensis* Reddy et al. 2003 as *Paenisporosarcina macmurdoensis* comb. nov. Int J Syst Evol Microbiol 59:1364–1370

Lapage SP, Sneath PHA, Lessel EF, Skerman VBD, Seeliger HPR, Clark WA (eds) (1992) International code of nomenclature of bacteria (1990 Revision). Bacteriological code. American Society for Microbiology, Washington, DC

L'Haridon S, Miroshnichenko ML, Kostrikina NA, Tindall BJ, Spring S, Schumann P, Stackebrandt E, Bonch-Osmolovskaya EA, Jeanthon C (2006) *Vulcanibacillus modesticaldus* gen. nov., sp. nov., a strictly anaerobic, nitrate-reducing bacterium from deep-sea hydrothermal vents. Int J Syst Evol Microbiol 56:1047–1053

Lim J-M, Jeon C-O, Song SM, Kim CJ (2005) *Pontibacillus chungwhensis* gen. nov., sp. nov., a moderately halophilic Gram-positive bacterium from a solar saltern in Korea. Int J Syst Evol Microbiol 55:165–170

Logan NA, Berkeley RCW (1981) Classification and identification of members of the genus *Bacillus*. In: Berkeley RCW, Goodfellow M (eds) The aerobic endospore-forming bacteria. Academic, London, pp 105–140

Logan NA, Berkeley RCW (1984) Identification of *Bacillus* strains using the API system. J Gen Microbiol 130:1871–1882

Logan NA, Lebbe L, Hoste B, Goris J, Forsyth G, Heyndrickx M, Murray BL, Syme N, Wynn-Williams DD, De Vos P (2000) Aerobic endospore-forming bacteria from geothermal environments in northern Victoria Land, Antarctica, and Candlemas Island, South Sandwich archipelago, with the proposal of *Bacillus fumarioli* sp. nov. Int J Syst Evol Microbiol 50:1741–1753

Logan NA, Berge O, Bishop AH, Busse H-J, De Vos P, Fritze D, Heyndrickx M, Kämpfer P, Salkinoja-Salonen MS, Seldin L, Rabinovitch L, Ventosa A (2009) Proposed minimal standards for describing new taxa of aerobic, endospore-forming bacteria. Int J Syst Evol Microbiol 59:2114–2121

Lu J, Nogi Y, Takami H (2002) *Oceanobacillus iheyensis* gen. nov., sp. nov., a deep-sea extremely halotolerant and alkaliphilic species isolated from a depth of 1,050 m on the Iheya Ridge. FEMS Microbiol Lett 205:291–297

Ludwig W, Schleifer K-H, Whitman WB (2007) Revised roadmap to the phylum *Firmicutes*. Bergey's Manual Trust. http://www.bergeys.org/outlines/Bergeys_Vol_3_Outline.pdf

Márquez MC, Carrasco IJ, Xue Y, Ma Y, Cowan DA, Jones BE, Grant WD, Ventosa A (2008) *Aquisalibacillus elongatus* gen. nov., sp. nov., a moderately halophilic bacterium of the family *Bacillaceae* isolated from a saline lake. Int J Syst Evol Microbiol 58:1922–1926

Matsuo Y, Katsuta A, Matsuda S, Shizuri Y, Yokota A, Kasai H (2006) *Mechercharimyces mesophilus* gen. nov., sp. nov. and *Mechercharimyces asporophorigenens* sp. nov., antitumour substance-producing marine bacteria, and description of *Thermoactinomycetaceae* fam. nov. Int J Syst Evol Microbiol 56:2837–2842

Mayr R, Busse H-J, Worliczek HL, Ehling-Schulz M, Scherer S (2006) *Ornithinibacillus* gen. nov., with the species *Ornithinibacillus bavariensis* sp. nov. and *Ornithinibacillus californiensis* sp. nov. Int J Syst Evol Microbiol 56:1383–1389

Metchnikoff ME (1888) *Pasteuria ramosa* un représentant des bactéries à division longitudinale. Ann Inst Pasteur 2:165–170

Miñana-Galbis D, Pinzón DL, Lorén JG, Manresa A, Oliart-Ros RM (2010) Reclassification of *Geobacillus pallidus* (Scholz et al. 1988) Banat et al. 2004 as *Aeribacillus pallidus* gen nov., comb. nov. Int J Syst Evol Microbiol 60:1600–1604

Nakamura LK (1984) *Bacillus pulvifaciens* sp. nov., nom. rev. Int J Syst Bacteriol 34:410–413

Nakamura LK (1989) Taxonomic relationship of black-pigmented *Bacillus subtilis* strains and a proposal for *Bacillus atrophaeus* sp. nov. Int J Syst Bacteriol 39:295–300

Nakamura LK, Swezey J (1983) Taxonomy of *Bacillus circulans* Jordan 1890: base composition and reassociation of deoxyribonucleic acid. Int J Syst Bacteriol 33:46–52

Nakamura LK, Roberts MS, Cohan FM (1999) Relationship of *Bacillus subtilis* clades associated with strains 168 and W23: a proposal for *Bacillus subtilis* subsp. *subtilis* subsp. nov. and *Bacillus subtilis* subsp. *spizizenii* subsp. nov. Int J Syst Bacteriol 49:1211–1215

Nakamura K, Haruta S, Ueno S, Ishii M, Yolota A, Igarashi Y (2004) *Cerasibacillus quisquiliarum* gen. nov., sp. nov., isolated from a semi-continuous decomposing system of kitchen refuse. Int J Syst Evol Microbiol 54:1063–1069

Nazina TN, Tourova TP, Poltaraus AB, Novikova EV, Grigoryan AA, Ivanova AE, Lysenko AM, Petrunyaka VV, Osipov GA, Belyaev SS, Ivanov MV (2001) Taxonomic study of aerobic thermophilic bacilli: descriptions of *Geobacillus subterraneus* gen nov, sp. nov. and *Geobacillus uzenensis* sp. nov. from petroleum reservoirs and transfer of *Bacillus stearothermophilus, Bacillus thermocatenulatus, Bacillus thermoleovorans Bacillus kaustophilus, Bacillus thermoglucosidasius, Bacillus thermodenitrificans* to *Geobacillus* as *Geobacillus stearothermophilus, Geobacillus thermocatenulatus, Geobacillus thermoleovorans Geobacillus kaustophilus, Geobacillus thermoglucosidasius, Geobacillus thermodenitrificans.* Int J Syst Evol Microbiol 51:433–446

Niimura Y, Koh E, Yanagida F, Suzuki K-I, Komagata K, Kozaki M (1990) *Amphibacillus xylanus* gen. nov., sp. nov., a facultatively anaerobic sporeforming xylan-digesting bacterium which lacks cytochrome, quinone, and catalase. Int J Syst Bacteriol 40:297–301

Nowlan B, Dodia MS, Singh SP, Patel BKC (2006) *Bacillus okhensis* sp. nov., a halotolerant and alkalitolerant bacterium from an Indian saltpan. Int J Syst Evol Microbiol 56:1073–1077

Nunes I, Tiago I, Pires AL, da Costa MS, Veríssimo A (2006) *Paucisalibacillus globulus* gen. nov., sp. nov., a Gram-positive bacterium isolated from potting soil. Int J Syst Evol Microbiol 56:1841–1845

Palmisano MM, Nakamura LK, Duncan KE, Istock CA, Cohan FM (2001) *Bacillus sonorensis* sp. nov., a close relative of *Bacillus licheniformis*, isolated from soil in the Sonoran Desert, Arizona. Int J Syst Evol Microbiol 51:1671–1679

Park D-J, Dastager SG, Lee J-C, Yeo S-H, Yoon J-H, Kim C-J (2007) *Shimazuella kribbensis* gen. nov., sp. nov., a mesophilic representative of the family *Thermoactinomycetaceae*. Int J Syst Evol Microbiol 57:2660–2664

Pikuta E, Lysenko A, Chuvilskaya N, Mendrock U, Hippe H, Suzina N, Nikitin D, Osipov G, Laurinavichius K (2000) *Anoxybacillus pushchinensis* gen. nov., sp. nov., a novel anaerobic, alkaliphilic, moderately thermophilic bacterium from manure, and description of *Anoxybacillus flavithermus* comb. nov. Int J Syst Evol Microbiol 50:2109–2117

Pruesse E, Quast C, Knittel K, Fuchs B, Ludwig W, Peplies J, Glöckner FO (2007) SILVA: a comprehensive online resource for quality checked and aligned rRNA sequence data compatible with ARB. Nucleic Acids Res 35:7188–7196

Rasko DA, Altherr MR, Han CS, Ravel J (2005) Genomics of the *Bacillus cereus* group of organisms. FEMS Microbiol Rev 29:303–329

Rasko DA, Rosovitz MJ, Økstad OA, Fouts DE, Jiang L, Cer RZ, Kolstø AB, Gill SR, Ravel J (2007) Complete sequence analysis of novel plasmids from emetic and periodontal *Bacillus cereus* isolates reveals a common evolutionary history among the *B. cereus*-group plasmids, including *Bacillus anthracis* pXO1. J Bacteriol 189:52–64

Ren P-G, Zhou P-J (2005) *Tenuibacillus multivorans* gen. nov., sp. nov., a new moderately halophilic bacterium isolated from saline soil in Xin-Jiang, China. Int J Syst Evol Microbiol 55:95–99

Ren P-G, Zhou P-J (2006) *Salinibacillus aidingensis* gen. nov., sp. nov. and *Salinibacillus kushneri* sp. nov., moderately halophilic bacteria isolated from a neutral saline lake in Xin-Jiang, China. Int J Syst Evol Microbiol 56:949–953

Rivas R, García-Fraile P, Zurdo-Piñeiro JL, Mateos PF, Martínez-Molina E, Bedmar EJ, Sánchez-Raya J, Velázquez E (2008) *Saccharibacillus sacchari* gen. nov., sp. nov., isolated from sugar cane. Int J Syst Evol Microbiol 58:1850–1854

Roberts MS, Nakamura LK, Cohan FM (1994) *Bacillus mojavensis* sp. nov., distinguishable by sexual isolation, divergence in DNA sequence, and differences in fatty acid composition. Int J Syst Bacteriol 44:256–264

Roberts MS, Nakamura LK, Cohan FM (1996) *Bacillus vallismortis* sp. nov., a close relative of *Bacillus subtilis*, isolated from soil in Death Valley, California. Int J Syst Bacteriol 46:470–475

Romano I, Lama L, Nicolaus B, Gambacorta A, Giordano A (2005) *Bacillus saliphilus* sp. nov., isolated from a mineral pool in Campania, Italy. Int J Syst Evol Microbiol 55:159–163

Rosado AS, van Elsas JD, Seldin L (1997) Reclassification of *Paenibacillus durum* (formerly *Clostridium durum* Smith and Cato 1974) Collins et al. 1994 as a member of the species *P. azotofixans* (formerly *Bacillus azotofixans* Seldin et al. 1984) Ash et al. 1994. Int J Syst Bacteriol 47:569–572

Rosselló-Mora R, Amann R (2001) The species concept for prokaryotes. FEMS Microbiol Rev 25:39–67

Ruiz-García C, Béjar V, Martínez-Checa F, Llamas I, Quesada E (2005a) *Bacillus velezensis* sp. nov., a surfactant-producing bacterium isolated from the river Vélez in Málaga, southern Spain. Int J Syst Evol Microbiol 55:191–195

Ruiz-García C, Quesada E, Martínez-Checa F, Llamas I, Urdaci MC, Béjar V (2005b) *Bacillus axarquiensis* sp. nov. and *Bacillus malacitensis* sp. nov., isolated from river-mouth sediments in southern Spain. Int J Syst Evol Microbiol 55:1279–1285

Saile E, Koehler TM (2006) *Bacillus anthracis* multiplication, persistence, and genetic exchange in the rhizosphere of grass plants. Appl Environ Microbiol 72:3168–3174

Santini JM, Streimann ICA, vanden Hoven RN (2004) *Bacillus macyae* sp. nov., an arsenate-respiring bacterium isolated from an Australian gold mine. Int J Syst Evol Microbiol 54:2241–2244

Schlesner H, Lawson PA, Collins MD, Weiss N, Wehmeyer U, Volker H, Thomm M (2001) *Filobacillus milensis* gen. nov., sp. nov., a new halophilic spore-forming bacterium with Orn-D-Glu-type peptidoglycan. Int J Syst Evol Microbiol 51:425–431

Shida O, Takagi H, Kadowaki K, Udaka S, Komagata K (1994) *Bacillus galactophilus* is a later subjective synonym of *Bacillus agri*. Int J Syst Bacteriol 44:172–173

Shida O, Takagi H, Kadowaki K, Komagata K (1996) Proposal for two new genera. *Brevibacillus* gen. nov. and *Aneurinibacillus* gen. nov. Int J Syst Bacteriol 46:939–946

Skerman VBD, McGowan V, Sneath PHA (1980) Approved lists of bacterial names. Int J Syst Bacteriol 30:225–420

Smith NR, Gordon RE, Clark FE (1952) Aerobic spore-forming bacteria. Monograph No. 16. United States Department of Agriculture, Washington, DC, 148 pp

Spring S, Ludwig W, Marquez MC, Ventosa A, Schleifer K-H (1996) *Halobacillus* gen. nov., with descriptions of *Halobacillus litoralis* sp. nov. and *Halobacillus trueperi* sp. nov., and transfer of *Sporosarcina halophila* to *Halobacillus halophilus* comb. nov. Int J Syst Bacteriol 46:492–496

Stackebrandt E, Swiderski J (2002) From phylogeny to systematics: the dissection of the genus *Bacillus*. In: Berkeley RCW, Heyndrickx M, Logan NA, De Vos P (eds) Applications and systematics of *Bacillus* and relatives. Blackwell Science, Oxford, pp 8–22

Stackebrandt E, Frederiksen W, Garrity GM, Grimont PAD, Kämpfer P, Maiden MCJ, Nesme X, Rosselló-Mora R, Swings J, Trüper HG, Vauterin L, Ward AC, Whitman WB (2002) Report of the ad hoc committee for the re-evaluation of the species definition in bacteriology. Int J Syst Evol Microbiol 52:1043–1047

Staley JT (2006) The bacterial species dilemma and the genomic–phylogenetic species concept. Philos Trans Roy Soc B Biol Sci 361:1899–1909

Steven B, Chen MQ, Greer CW, Whyte LG, Niederberger TD (2008) *Tumebacillus permanentifrigoris* gen. nov., sp. nov., an aerobic, spore-forming bacterium isolated from Canadian high Arctic permafrost. Int J Syst Evol Microbiol 58:1497–1501

Sunna A, Tokajian S, Burghardt J, Rainey F, Antranikian G, Hashwa F (1997) Identification of *Bacillus kaustophilus, Bacillus thermocatenulatus* and *Bacillus* strain HSR as members of *Bacillus thermoleovorans*. Syst Appl Microbiol 20:232–237

Switzer Blum J, Burns Bindi A, Buzzelli J, Stolz JF, Oremland RS (1998) *Bacillus arsenicoselenatis*, sp. nov., and *Bacillus selenitireducens*, sp. nov.: two haloalkaliphiles from Mono Lake, California that respire oxyanions of selenium and arsenic. Arch Microbiol 171:19–30

Tanasupawat S, Namwong S, Kudo T, Itoh T (2007) *Piscibacillus salipiscarius* gen. nov., sp. nov., a moderately halophilic bacterium from fermented fish (pla-ra) in Thailand. Int J Syst Evol Microbiol 57:1413–1417

Thomas DJ, Morgan JA, Whipps JM, Saunders JR (2001) Plasmid transfer between *Bacillus thuringiensis* subsp. *israelensis* strains in laboratory culture, river water, and dipteran larvae. Appl Environ Microbiol 67:330–338

Tiago I, Pires C, Mendes V, Morais PV, da Costa MS, Veríssimo A (2006) *Bacillus foraminis* sp. nov., isolated from a non-saline alkaline groundwater. Int J Syst Evol Microbiol 56:2571–2574

Tourasse NJ, Helgason E, Økstad OA, Hegna IK, Kolstø A-B (2006) The *Bacillus cereus* group: novel aspects of population structure and genome dynamics. J Appl Microbiol 101:579–593

Touzel JP, O'Donohue M, Debeire P, Samain E, Breton C (2000) *Thermobacillus xylanilyticus* gen. nov., sp. nov., a new aerobic thermophilic xylan-degrading bacterium isolated from soil. Int J Syst Evol Microbiol 50:315–320

Tsilinsky P (1899) On the thermophilic moulds. Ann Inst Pasteur 13:500–505 (in French)

Vaishampayan P, Miyashita M, Ohnishi A, Satomi M, Rooney A, La Duc MT, Venkateswaran K (2009) Description of *Rummeliibacillus stabekisii* gen. nov., sp. nov. and reclassification of *Bacillus pycnus* Nakamura *et al.* 2002 as *Rummeliibacillus pycnus* comb. nov. Int J Syst Evol Microbiol 59:1094–1099

Wainø M, Tindall BJ, Schumann P, Ingvorsen K (1999) *Gracilibacillus* gen. nov., with description of *Gracilibacillus halotolerans* gen. nov., sp. nov.: transfer of *Bacillus dipsosauri* to *Gracilibacillus dipsosauri* comb. nov., and *Bacillus salexigens* to the genus *Salibacillus* gen. nov., as *Salibacillus salexigens* comb. nov. Int J Syst Bacteriol 49:821–831

Wang L-T, Lee F-L, Tai C-J, Yokota A, Kuo H-P (2007a) Reclassification of *Bacillus axarquiensis* Ruiz-Garcia et al. 2005 and *Bacillus malacitensis* Ruiz-Garcia et al. 2005 as later heterotypic synonyms of *Bacillus mojavensis* Roberts et al. 1994. Int J Syst Evol Microbiol 57:1663–1667

Wang QF, Li W, Liu YL, Cao HH, Li Z, Guo GQ (2007b) *Bacillus qingdaonensis* sp. nov., a moderately haloalaliphilic bacterium isolated form a crude sea-salt sample collected near Qindao in eastern China. Int J Syst Evol Microbiol 57:1143–1147

Wang L-T, Lee F-L, Tai C-J, Kuo H-P (2008) *Bacillus velezensis* is a later heterotypic synonym of *Bacillus amyloliquefaciens*. Int J Syst Evol Microbiol 58:671–675

Wayne LG, Brenner DJ, Colwell RR, Grimont PAD, Kandler O, Kritchevsky MI, Moore LH, Murray RGE, Stackebrandt E, Starr MP, Trüper HG (1987) Report of the ad hoc committee on reconciliation of approaches to bacterial systematics. Int J Syst Bacteriol 37:463–464

Winslow C-EA, Broadhurst J, Buchanan RE, Krumwiede C, Rogers LA, Smith GH (1920) The families and genera of the bacteria. Final report of the committee of the society of American bacteriologists on characterization and classification of bacterial types. J Bacteriol 5:191–229

Wisotzkey JD, Jurtshuk P Jr, Fox GE, Deinhard G, Poralla K (1992) Comparative sequences analyses on the 16S rRNA (rDNA) of *Bacillus acidocaldarius*, *Bacillus acidoterrestris*, and *Bacillus cycloheptanicus* and proposal for creation of a new genus *Alicyclobacillus* gen. nov. Int J Syst Bacteriol 42:263–269

Xue Y, Zhang X, Zhou C, Zhao Y, Cowan DA, Heaphy S, Grant WD, Jones BE, Ventosa A, Ma Y (2006) *Caldalkalibacillus thermarum* gen. nov., sp. nov., a novel alkalithermophilic bacterium from a hot spring in China. Int J Syst Evol Microbiol 56:1217–1221

Yoon J-H, Weiss N, Lee K-C, Kho I-S, Kang KH, Park Y-H (2001) *Jeotgalibacillus alimentarius* gen. nov., sp. nov., a novel bacterium isolated from jeotgal with L-lysine in the cell wall, and reclassification of *Bacillus marinus* Rüger 1983 as *Marinibacillus marinus,* gen. nov., comb. nov. Int J Syst Evol Microbiol 51:2087–2093

Yoon J-H, Kang KH, Park Y-H (2002) *Lentibacillus salicampi* gen. nov., sp. nov., a moderately halophilic bacterium isolated from a salt field in Korea. Int J Syst Evol Microbiol 52:2043–2048

Yoon J-H, Kim IG, Shin YK, Park YH (2005) Proposal of the genus *Thermoactinomyces sensu stricto* and three new genera *Laceyella*, *Thermoflavimicrobium* and *Seinonella*, on the basis of phenotypic, phylogenetic and chemotaxonomic analyses. Int J Syst Evol Microbiol 55:395–400

Yoon J-H, Kang S-J, Oh TK (2007) Reclassification of *Marinococcus albus* Hao *et al.* 1985 as *Salimicrobium album* gen. nov., comb. nov. and *Bacillus halophilus* Ventosa *et al.* 1990 as *Salimicrobium halophilum* comb. nov., and description of *Salimicrobium luteum* sp. nov. Int J Syst Evol Microbiol 57:2406–2411

Yoon J-H, Kang S-J, Schumann P, Oh T-K (2010) *Jeotgalibacillus salarius* sp. nov., isolated from a marine saltern, and reclassification of *Marinibacillus marinus* and *Marinibacillus campisalis* into the genus *Jeotgalibacillus* as *Jeotgalibacillus marinus* comb. nov. and *Jeotgalibacillus campisalis* comb. nov., respectively. Int J Syst Evol Microbiol 60:15–20

Zaitsev G, Tsitko IV, Rainey FA, Trotsenko YA, Uotila JS, Stackebrandt E, Salkinoja-Salonen MS (1998) New aerobic ammonium-dependent obligately oxalotrophic bacteria: description of *Ammoniphilus oxalaticus* gen. nov., sp. nov. and *Ammoniphilus oxalivorans* gen. nov., sp. nov. Int J Syst Bacteriol 48:151–163

Zhou Y, Xu J, Xu L, Tindall BJ (2009) *Falsibacillus pallidus* to replace the homonym *Bacillus pallidus* Zhou et al. 2008. Int J Syst Evol Microbiol 59:3176–3180

Chapter 2
Diversity of Endospore-forming Bacteria in Soil: Characterization and Driving Mechanisms

Ines Mandic-Mulec and James I. Prosser

2.1 Introduction

Aerobic endospore-formers have long been considered to be important components of the soil bacterial community. Inoculation of standard nutrient media with soil that has been heat-treated, to kill vegetative cells, leads to growth of high numbers of endospore-formers and the availability of isolates has facilitated detailed taxonomic and physiological studies. Both taxonomic and physiological diversity of soil isolates is high and extrapolation of this metabolic capability and diversity implies important roles in a wide range of soil ecosystem functions and processes. Their heterotrophic life style suggests an obvious role in the carbon cycle but, as a group, aerobic endospore-formers are also important in the soil nitrogen cycle, as denitrifiers, nitrogen fixers and degraders of organic nitrogen; in the sulphur cycle as sulphur oxidizers; and in transformation of other soil nutrients, e.g., through manganese reduction. Their abilities to break down cellulose, hemicelluloses and pectins suggest major roles in mineralization of plant material and humic material, while chitinase activity facilitates degradation of fungal cell walls and insect exoskeletons. Thermophilic bacilli dominate the high-temperature stages of composting and they produce a wide range of commercially valuable extracellular enzymes, including thermostable enzymes. Endospore-formers are important in soil bioremediation, through their ability to degrade BTEX (benzene, toluene, ethylbenzene and xylene) compounds and to methylate mercury. They produce a wide range of antiviral, antibacterial and antifungal compounds, which may be important in interactions with other soil microorganisms and have significant commercial potentials in agriculture and medicine. Many have phosphatase

I. Mandic-Mulec
Biotechnical Faculty, Department of Food Science and Technology, University of Ljubljana, Vecna pot 111, Ljubljana, Slovenia

J.I. Prosser (✉)
Institute of Biological and Environmental Sciences, University of Aberdeen, Cruickshank Building, St. Machar Drive, Aberdeen AB24 3UU, UK
e-mail: j.prosser@abdn.ac.uk

N.A. Logan and P. De Vos (eds.), *Endospore-forming Soil Bacteria*, Soil Biology 27,
DOI 10.1007/978-3-642-19577-8_2, © Springer-Verlag Berlin Heidelberg 2011

activity, releasing phosphate for plant growth, and they possess a range of other plant-beneficial properties that have led to commercial applications as plant-growth-promoting bacteria.

Specific aspects of the ecology of aerobic endospore-formers and their roles in remediation, plant-growth promotion, biological control and other applications are discussed in later chapters. Here we focus on their diversity and community composition in soil, the implications of recent and future methodological developments on diversity studies, and their contributions to our understanding of the mechanisms driving microbial diversity in soil. Most of the studies we discuss preceded recent taxonomic revisions, and aerobic endospore-formers are frequently referred to using the collective term bacilli; this term will be used here, except where studies focus on particular genera or species.

2.1.1 Methods for Determining Diversity of Soil Bacilli

Soil bacilli have been investigated since the first isolation of bacteria from soil in the nineteenth century and methods to study their diversity have followed those used for all soil bacteria. These began with phenotypic characterization of soil isolates, the application of numerical taxonomy and the subsequent introduction of molecular phylogenetic approaches more than 20 years ago. The past 15 years have seen considerable development of cultivation-independent methods for characterizing bacterial communities, based on molecular analysis of nucleic acids extracted directly from the soil, rather than from isolates. These developments will be discussed in this section, considering specific factors related to endospore-formers, examples of their use and their benefits and limitations. Results arising from use of these techniques are discussed in Sect. 2.3.3, while detailed protocols are described in other chapters.

2.1.2 Cultivation-Based Methods

Traditional approaches to determination of soil bacterial community composition and diversity relied largely on cultivation-based methods. Typically, a solid, complex organic growth medium inoculated with dilutions of a soil suspension is incubated and colonies are removed, subcultured and purified. Identification of isolates provides information on the relative importance of endospore-formers within total bacterial communities. More detailed studies exploit the heat-resistance of endospores and select for bacilli by pasteurization of soil, or soil dilutions, to kill vegetative cells. Selection for specific taxonomic or functional groups of bacilli (e.g., nitrogen fixers) is also possible through the use of selective media, based on nutritional or physiological characteristics of the target group. For example, Seldin et al. (1998) describe a method for selection of *Paenibacillus azotofixans* from soil.

2.1.2.1 Phenotypic Characterization

Isolates may be characterized by a range of methods. Traditionally, morphological and physiological characteristics were used for classification and identification. For the large numbers of isolates generated by ecological studies, this process is facilitated by miniaturized methods, such as API strips. While such techniques provided the basis for bacterial, and *Bacillus* taxonomy, additional tests were necessary for confirmation and for fine-scale resolution. For example, analysis of protein or enzyme composition, using multiple-locus enzyme electrophoresis (MLEE), has been used to analyse intraspecific variation in *Bacillus cereus* and *B. thuringiensis* (Helgason et al. 1998; Vilas-Boas et al. 2002) and serotyping, against flagellar antigens, is well-established for distinguishing strains of *B. thuringiensis* (Helgason et al. 1998). Bacterial soil isolates can also be characterized by fatty acid methyl ester (FAME) analysis, which can distinguish bacilli within total bacterial communities, as well as identifying isolates.

2.1.2.2 Identification Using Molecular Techniques

Bacterial taxonomy was revolutionized by the application of molecular techniques and, in particular, analysis of 16S rRNA gene sequences to determine phylogenetic relationships. This approach is much more robust and is now routinely used for classification and identification of soil isolates, including bacilli. Other genes have been employed for phylogenetic analysis and identification. Some increase discriminatory power beyond that of 16S rRNA genes while others have potential ecological relevance and importance and can be used to determine relationships between bacilli and other bacteria sharing particular functions.

16S rRNA and functional gene-based phylogenies are generally useful for delineation to the genus- and, occasionally, species-levels but analysis of chromosome-wide differences increases taxonomic resolution and enable intraspecies discrimination. For example, Meintanis et al. (2008) compared 11 *Geobacillus* and *Bacillus* strains isolated from a volcanic region by *rpoB* sequence analysis, repetitive extragenic palindromic-PCR (REP-PCR) and BOX-PCR; da Mota et al. (2002) compared *Paenibacillus polymyxa* isolates from maize rhizosphere using REP-PCR and randomly amplified polymorphic DNA (RAPD) analysis; and Ryu et al. (2005) used amplified fragment length polymorphism (AFLP) and multilocus variable-number tandem repeat analysis (MLVA) to characterize soil and clinical strains of *Bacillus anthracis*. REP-PCR, BOX-PCR and MLVA are alternative techniques for analysis of interspersed repeated sequences within the genome, while AFLP and RAPD, respectively, amplify genomic DNA using random primers or non-specific primers that are complementary to a number of sites within the genome.

Taxonomic resolution can also be increased using multilocus sequence typing (MLST) and analysis (MLSA). This involves sequencing of short regions of several (typically seven) housekeeping genes distributed throughout the chromosome.

Relatedness of strains is determined by comparison of sequence types of all seven genes. MLST was used by Sorokin et al. (2006) to distinguish 115 *B. cereus* group soil isolates and by Bizzarri et al. (2008) to determine relatedness of 22 phylloplane isolates, comparing results with analysis of plasmid profiles and *cry* gene sequences.

2.1.2.3 Benefits and Limitations

Each of these approaches has its advantages and limitations, with regard to taxonomic resolution, ease of use, cost, and availability, and their values will be evident elsewhere in this volume. An important factor is the amount and type of information they provide. Several of the techniques are valuable in discriminating and grouping strains, but give no useful ecological information. For example, 16S rRNA gene-based methods provide good phylogenetic information to the genus level, but, in themselves, give little information on function. Importantly, traditional approaches that group isolates on the basis of common metabolic properties may be limited in terms of phylogenetic power, but provide clues to environmental factors favouring and selecting for particular groups and can be strong indicators of potential ecosystem function. Classification often requires a multiphasic approach and the different approaches may not agree. For example, groupings indicated by FAME and 16S rRNA gene analysis of rhizosphere communities can differ (Kim et al. 2003). In addition, while phylogenetic analysis may group isolates with similar physiological characteristics, many important ecological traits are borne on plasmids.

Cultivation-based analysis of diversity of endospore-formers is facilitated, in one sense, by the ease with which they can be selected, i.e., through pasteurization of samples before cultivation. This is appropriate for qualitative studies of diversity, where the aim is merely to determine "who is there". However, selection against vegetative cells will introduce bias towards those bacilli that produce the greatest numbers of spores, those producing spores that germinate most rapidly on laboratory media, and those that may have produced large numbers of spores through stress, which makes them inactive in the soil. This approach is therefore severely restricted for studies in which relative abundances of total numbers of bacilli (vegetative cells and spores) are required, or where information is required on active, rather than potentially active organisms. Consequently, it is limited in its ability to link community structure and composition to soil ecosystem processes carried out by bacilli.

Cultivation studies also suffer from the major disadvantages associated with the inability of the majority of soil microorganisms to grow on standard enrichment growth media and under standard laboratory incubation conditions. Selection for particular functional groups (e.g., nitrogen fixers, denitrifiers) is possible by careful design of media and growth conditions, but the broad physiological and metabolic diversity within the bacilli makes it difficult to design media that are selective for particular phylogenetic groups.

2.1.3 Cultivation-Independent Analysis of Diversity

Molecular approaches for phylogenetic analysis of cultivated organisms quickly led to the development and application of similar approaches for analysis of natural soil communities that avoid prior enrichment or isolation of pure cultures. They have transformed our view of soil microbial communities and have uncovered vast and previously unsuspected diversity within groups, such as bacilli, that are well represented among soil isolates. They have also revealed abundant, novel groups of bacteria and archaea performing important soil ecosystem functions.

2.1.3.1 16S rRNA Gene-Based Analysis of Prokaryote Communities

Soil prokaryote community composition is now routinely determined by the amplification of 16S rRNA genes from DNA extracted from soil using primers that are specific for a particular target group. The presence of regions within the 16S rRNA gene with differing degrees of sequence variability allows the design of primers with different resolution. For example, primers are available for amplification of all bacteria or for specific genera and, occasionally, for different functional groups. Amplification products can be cloned and sequenced, for phylogenetic analysis or for identification, by comparison with database sequences obtained from cultivated organisms and other environmental studies. Sequences can also be used to design probes for in situ detection using fluorescence in situ hybridization (FISH) or for probing of nucleic acids. Increasingly, high-throughput sequencing techniques are being applied for analysis of extracted soil DNA (Roesch et al. 2007). These avoid the cloning step, and potential cloning bias, and enable sequencing of hundreds of thousands of amplicons in a single run.

Amplification products can also be analysed using fingerprinting techniques, such as denaturing gradient gel electrophoresis (DGGE), temperature gradient gel electrophoresis (TGGE), terminal restriction fragment length polymorphism (T-RFLP) and single-strand conformation polymorphism (SSCP). DGGE and TGGE separate amplicons on the basis of heterogeneities in GC content and sequence, and consequent differences in melting properties, when run on a gel containing a gradient of a denaturant or temperature, respectively. For T-RFLP, amplified DNA is digested with restriction enzymes, distinguishing amplicons with sequence polymorphisms. SSCP involves electrophoretic separation of single-stranded nucleic acids with differences in secondary structure. All are less expensive than sequencing methods and allow rapid analysis of many samples and assessment of relative abundances of different phylotypes, but provide less information on identity of organisms present. Felske et al. (2003) adopted an alternative approach, multiplex PCR, to amplify simultaneously 16S rRNA genes using primers targeting several groups: *Acidobacteria*, *Verrucomicrobia*, *Bacillus megaterium*, *Paenibacillus* and *Bacillus* RNA groups 1 and 3. Simultaneous detection of many thousands of organisms can now be achieved using microarrays with probes for functional genes and 16S rRNA genes, including many *Bacillus* representatives (Andersen et al. 2010).

2.1.3.2 Community Composition of Specific Groups

The relative abundance of bacilli in total bacterial communities can be determined by identification of *Bacillus* sequences in 16S rRNA gene clone libraries constructed using universal bacterial primers and using DNA fingerprinting methods. If high numbers of *Bacillus* sequences are obtained, within-group diversity can also be studied. Alternatively, primers targeting specific groups can be used, e.g., for *Bacillus* (Garbeva et al. 2003) and *Paenibacillus* (da Silva et al. 2003), although few have been designed.

Functional genes may also be used to determine the potential contribution of bacilli to specific ecosystem functions. Nitrite reductase (*nirK*) genes can be used to characterize bacteria with potential roles in denitrification (Philippot et al. 2007), and comparison of sequences with those in databases will indicate which *nirK* genes are from bacilli. The value of this approach depends on the number of functional gene sequences in databases and the extent to which they can be linked to taxonomic groups. The latter relies on the ability to link 16S rRNA and functional genes, e.g., by sequencing both genes in cultivated organisms, and on the extent of gene transfer, which is significant for *nirK* genes. Table 2.1 provides

Table 2.1 Examples of functional genes used to classify and identify bacilli isolated from soil and/or cultivated and uncultured bacilli, by amplification from extracted soil DNA

Gene	Function or target group	Reference
*16S rRNA	*Bacillus*	Garbeva et al. (2003)
*16S rRNA	*Paenibacillus*	da Silva et al. (2003)
**apr*	Alkaline metalloprotease	Sakurai et al. (2007)
cheA	Histidine kinase	Reva et al. (2004)
**cbbL*	Ribulose-1,5-bisphosphate carboxylase/oxygenase (RubisCO)	Selesi et al. (2005)
cry	Bt endotoxin	Collier et al. (2005)
cspA	Cold shock protein A	von Stetten et al. (1999)
gyrA	Gyrase A	Reva et al. (2004)
merA	Mercury reductase	Hart et al. (1998)
**nifH*	Dinitrogen reductase	Rosado et al. (1998)
**nirK*	Nitrite reductase	Philippot et al. (2007)
**nosZ*	Nitrous oxide reductase	Philippot et al. (2007), Kraigher et al. (2008), Stres et al. (2008)
**npr*	Neutral metalloprotease	Sakurai et al. (2007)
panC	Pantothenate synthesis	Guinebretière et al. (2008)
rpoB	β-Subunit of RNA polymerase	da Mota et al. (2004)

*Indicate genes used for cultivation-independent studies

examples of functional genes that have been used in cultivation-dependent and -independent studies.

2.1.3.3 Diversity of Active Organisms

DNA-targeted methods demonstrate the presence of particular groups but not, necessarily, their activity in soil ecosystem processes, which is better achieved by targeting RNA. Extracted RNA is reverse-transcribed to DNA and RNA genes are then amplified as described above. Targeting of 16S rRNA gene sequences in this manner is more sensitive, as cells contain more ribosomes than rRNA genes, but is also believed to characterize active communities. This is based on the assumption that active and growing cells will contain more ribosomes than dormant or inactive cells. This approach was used to target the active soil bacterial community in acidic peat grassland soils (Felske et al. 2000).

Activity associated with specific processes performed by bacteria can be determined by transcriptional analysis of genes controlling specific functions. For example, quantification of *nirK* expression will indicate denitrifier transcriptional activity, and sequence analysis of expressed genes may indicate which denitrifiers are active. Putative activity of uncultured organisms can be assessed by amplification of large sections of chromosomal DNA (typically >100 kb), containing many functional genes and, potentially, several genes encoding a metabolic pathway. This is achieved by constructing bacterial artificial chromosome (BAC) or fosmid libraries, screening to target clones containing the target 16S rRNA gene sequence and full sequencing to determine which functional genes are associated with the target group. Activity can also be assessed by techniques such as stable isotope probing, to determine which organisms are utilizing specific ^{13}C- or ^{15}N-labelled compounds, or BrdU capture, which separates organisms incorporating the thymidine analogue bromodeoxyuridine (Prosser et al. 2010).

2.1.3.4 Benefits and Limitations

The major advantage of nucleic-acid-based techniques is their lack of dependence on laboratory cultivation of soil bacilli. All growth media and cultivation conditions are selective, leading to inevitable bias, and only a small fraction of soil bacterial communities grow on enrichment and isolation plates. However, molecular methods also have potential biases. (See Prosser et al. (2010) for a discussion of benefits and limitations of molecular techniques and of developing methods for assessment of bacterial community diversity and activity.) Nevertheless, there is now a suite of techniques available to identify soil bacilli, measure changes in their diversities and community structures and determine their abundance, using quantitative PCR methods.

Although molecular techniques eliminate the major restrictions of cultivation-based techniques for analysis of diversity of soil bacilli, spore production will introduce bias which may be important for some studies. Molecular techniques are based on extraction of nucleic acids, which requires lysis of vegetative cells. Lysis of spores requires more severe conditions and most studies achieve this through physical disruption by bead-beating. All methods for cell lysis and nucleic acid extraction require a balance between conditions and lengths of treatment that are sufficiently rigorous to optimize lysis of cells and spores and minimization of DNA degradation, that will be increased by length of treatment. This balance is particularly difficult to achieve for bacilli, because of significant differences in the conditions required to lyse vegetative cells and spores. Consequently, many of the problems associated with cultivation-based analysis of soil bacilli diversity apply equally to molecular methods.

2.2 Diversity of Endospore-Formers Within Soil Bacterial Communities

Investigations of the diversity of soil bacilli fall into two classes. In the first, bacilli are studied as members of the total soil bacterial community. In the second, diversity within specific groups of bacilli is determined. This distinction is followed here, combining both cultivation-dependent and -independent approaches. Again, the term "bacilli" will be used to represent aerobic endospore-formers, particularly for older studies; in addition, some molecular studies do not distinguish bacilli from other bacteria within the *Firmicutes*. Although characterization of isolates provides clues to the roles of bacilli, and the studies described below provide evidence for the influence of a range of factors on their diversities, it is rarely easy to link isolates to soil ecosystem function. For example, *Geobacillus* is typically isolated from thermophilic environments, but is commonly found in soils from temperate environments (Marchant et al. 2002). Isolates from these environments can grow at temperatures up to 80°C, but not below 40°C, calling into question their role in these environments (see Banat and Marchant 2011).

2.2.1 Bacilli Within Soil Bacterial Communities

The introduction of molecular techniques led to a reassessment of soil bacterial diversity, mainly through analysis of 16S rRNA gene clone libraries. Libraries invariably contain sequences closely related to bacilli, but the proportion varies between studies – not only because of differences in environmental conditions, but also through the use of different techniques for isolation of nucleic acids, use of different primers and analysis methods, and differences in timing; sequence databases have developed considerably as molecular methods have generated new sequence data over the past 15–20 years. Early studies were limited by the resources (cost and time) required for sequencing, and involved libraries containing only tens of clones. The frequency of *Bacillus* sequences in these libraries was often low. For example one *Bacillus globisporus*-related sequence was found in a grassland soil library with a total of 275 clones (McCaig et al. 1999), but 8 of 52 clones from an oilseed rape rhizosphere fell within a *Bacillus* group related to *B. megaterium* (Macrae et al. 2000). DGGE analysis also shows the presence of bacilli, for example, a *Bacillus* sp. in bulk soil and *B. megaterium* in barley rhizosphere and rhizoplane (Normander and Prosser 2000). More recently, DGGE analysis with four primers targeting the V6 region indicated that *Firmicutes* constituted between 19% and 32% of sequences in a grassland soil, and the majority (76%–86%) of these were bacilli (Brons and Van Elsas 2008).

Liles et al. (2003) amplified 16S rRNA gene sequences from DNA extracted from soil using a bead-beating method and from BAC libraries. The latter were constructed by extracting high-molecular weight genomic DNA using successive freeze–thaw cycles and cloning into an *E. coli* vector. Traditional clone libraries from soil samples indicated that bacilli comprised 3–15% of total clone sequences, while one of the 28 BAC clones harbouring 16S rRNA genes contained a *Bacillus*-related sequence, closely related to *Bacillus licheniformis*. Thus, the two different DNA extraction processes led to significant differences in recovery of *Bacillus* sequences, presumably because the gentler freeze–thaw method (required for recovery of high-molecular weight DNA) was less efficient at lysing spores and, possibly, vegetative cells of bacilli. Kraigher et al. (2006) also extracted DNA without bead-beating and found that 2.6% of 114 partial 16S rRNA sequences from a high organic grassland fen soil belonged to *Firmicutes* with only two, closely related sequences from the *Bacillus* group, possibly through poor extraction from spores.

High-throughput sequencing methods greatly increase the depth of coverage of soil bacterial diversity. Roesch et al. (2007) used pyrosequencing to obtain 26,140–53,533 16S rRNA gene sequences from each of four soils. This approach removes potential cloning bias and *Firmicutes* comprised 2–5% of sequences from the four soils. Microarray systems also provide the potential to obtain detailed information on soil diversity and have been used to determine differences in bacterial community structure between a soil that was suppressive against a plant-parasitic nematode and an adjacent non-suppressive soil (Valinsky et al. 2002). Bacilli were tenfold more abundant in the non-suppressive soil, but reliable quantification was difficult.

Most diversity studies have targeted DNA, providing information on "total" communities, i.e., assessing active and dormant growth forms, including spores. Felske et al. (1998) characterized the active soil bacterial community in acidic peat grassland soils by targeting RNA, rather than DNA. TGGE analysis and sequencing of clone library representatives indicated dominance of the active community by bacilli. More than 50% of sequenced clones were related to bacilli, and 20% were closely related to a previously uncultivated strain, *Bacillus benzoevorans*, including one clone which gave the strongest band on TGGE gels, implying high relative abundance.

A meta-analysis of 32 soil clone libraries (Janssen 2006) indicated that bacilli encompassed less than 1% of soil bacterial 16S rRNA gene sequences. In contrast, bacilli comprised 5–45% of isolates from traditional cultivation-based studies. Similar patterns were found for other bacterial groups that were traditionally considered to be "dominant" soil organisms. This may reflect difficulties in cultivating truly dominant soil organisms but, for bacilli, this analysis is confounded by lack of information on contributions by spores and vegetative cells.

2.2.2 Diversity of Plant-Associated Bacilli

Root exudates are the major source of organic matter input to soil and lead to high microbial biomass in the rhizoplane and rhizosphere, where their compositions are likely to influence the composition and diversity of the root-associated bacteria, including bacilli. Several studies have therefore compared bulk soil bacterial communities and rhizosphere communities of different plants, to assess selection and the influence of plant root exudates as a driver of bacterial diversity. An understanding of rhizosphere bacterial diversity is also important for the commercial development of plant-growth-promoting bacterial inocula and biocontrol agents.

The reported importance of bacilli in the rhizosphere varies significantly between studies. For example, Chin et al. (1999) characterized nine isolates from anoxic rice paddy soils using different isolation media with xylan, pectin or a mixture of seven mono- and disaccharides as the growth substrates. Isolates were obtained from terminal dilutions of most probable number counts, to obtain the most abundant organisms; *Bacillus* was only obtained on the sugar mixture and had 16S rRNA gene sequences similar to sequences from other rice paddy studies. Garbeva et al. (2001) also found that bacilli were not major components in endophytic bacterial populations of potato plants, and the only isolate was *Paenibacillus pabuli*.

In contrast, Smalla et al. (2001) compared bulk soil and rhizosphere communities of field-grown strawberry (*Fragaria ananassa* Duch.), oilseed rape (*Brassica napus* L.), and potato (*Solanum tuberosum* L.). DGGE profiles from bulk soil and potato rhizosphere were dominated by two *B. megaterium*-related bands, which were also found in strawberry and oilseed rape rhizospheres. Pankhurst et al. (2002) found higher *Bacillus* populations in roots growing within macropores in subsoil than in bulk soil and Duineveld et al. (2001) found several *Bacillus*-related DGGE bands in the *Chrysanthemum* rhizosphere, with little difference with growth stage or between rhizosphere and bulk soil. The number of bands amplified when targeting RNA was less than that from DNA, suggesting that not all strains present were active and that active organisms were phylogenetically diverse. Garbeva et al. (2008) found differences in bacilli under maize (*Bacillus* sp. and *B. thuringiensis*), a commercial grass mix (*B. benzoevorans* and *B. pumilus*) and oats and barley (*Bacillus* sp. and *B. fumarioli*). *Bacillus*-specific primers generated most phylotypes from grass and maize rhizosphere, while diversity and abundance were greatest in permanent grassland and arable land originating from grassland. Reva et al. (2004) classified 17 *Bacillus* root isolates on the basis of 16S rRNA, gyraseA (*gyrA*) and the *cheA* histidine kinase sequences. The isolates were closely related to two *Bacillus subtilis* strains and *B. mojavensis*. They were also related to, but distinct from, *Bacillus amyloliquefaciens* – which showed greatest rhizosphere colonization with oilseed rape (*Brassica napus*), barley (*Hordeum vulgare*) and thale cress (*Arabidopsis thaliana*).

Risk assessment of genetically modified crops requires assessment of impact on rhizosphere communities and a number of studies have found differences between *Bacillus* communities colonizing rhizospheres of wild-type and genetically modified plants. For example, Tesfaye et al. (2003) found fewer *Bacillus* clones in alfalfa when it was over-expressing a nodule-enhanced malate dehydrogenase, and Siciliano and Germida (1999) found differences in root-associated *Bacillus* communities of three Canola (rapeseed cultivar) varieties, one of which was genetically engineered to tolerate the herbicide glyphosate. The significance of these and other reported differences is unclear, however; both in terms of the mechanisms driving community composition, and the impact on the colonized plant.

2.2.3 Effects of Fertilizer Application and Removal on Diversity of Bacilli

Changes in soil nutrients through fertilization, or changes in plant communities associated with different management strategies, are likely to influence soil bacterial communities. Soil fumigation appears to select for endospore-formers, with high proportions (81%) of bacilli among isolates (Mocali et al. 2008). In many studies, however, changes in aerobic endospore-former communities are not great. Smit et al. (2001) investigated seasonal changes in isolates from a wheat field soil and only found *Bacillus* isolates in July, but *Bacillus* sequences were not detected in clone libraries. Chu et al. (2007) found some evidence for selection of a *Bacillus*-related strain following treatment with organic manure, but not inorganic fertilizer, and Sturz et al. (2004) found increased diversity in rhizosphere isolates following sulphate fertilization to control potato common scab, with increased antibiosis of isolates against *Streptomyces scabies*.

Studies of soil bacterial diversity focus on changes in community structure; i.e., changes in the relative abundances of different groups, but not in actual abundance. Therefore there is often a need for quantification, using qPCR or alternative techniques, to understand the effects of environmental change on bacterial communities. Most studies also give no information on bacterial activity, with the exception of a small number of studies targeting rRNA. Felske et al. (2000) addressed these issues when investigating the effects of grassland succession on bacterial communities. Bacterial RNA, quantified by RNA probing (dot-blot hybridization), doubled for several years after cessation of fertilization of a grassland soil, probably owing to increased levels of root residues following replacement of *Lolium perenne* by other grasses. rRNA levels then decreased, but no changes were detected in major phylogenetic groups. Multiplex PCR, however, coupled with TGGE showed changes in the 20 dominant bacterial ribotypes, including several representing bacilli. Thus, four *Bacillus* types increased during the final stage of succession, one increased during the early stage, three showed no clear pattern with succession but fell to low relative abundances during the later stages, and two decreased with grassland succession. These results demonstrate an ability to follow dynamics of soil *Bacillus* communities in soil. In this study, changes were accompanied by changes in plant communities, earthworm activity and soil processes, but establishment of links between these changes, and of links to physiological characteristics of different ribotypes, was not possible and remain difficult to discern in most studies.

2.2.4 Links Between Diversity and Ecosystem Function

Several studies have investigated bacterial diversity by the use of growth conditions selective for particular metabolic processes, or by amplifying functional genes associated with specific processes. The ability to identify functional genes derived from bacilli depends on the size and reliability of sequence databases, which are much smaller than those for 16S rRNA genes. Also, phylogenies of 16S rRNA and functional genes may not be congruent, making identification difficult, and important functional genes may also be plasmid-encoded and subject to gene transfer.

Bacilli have roles in two soil nitrogen cycle processes: denitrification and nitrogen fixation. Denitrifying bacilli have been targeted either by enrichment on selective media or by targeting functional genes, rather than by using 16S rRNA genes. For example, 138 denitrifying isolates from three soils, classified using ARDRA and 16S rRNA gene sequence analysis, fell within five groups, one of which comprised bacilli (Chèneby et al. 2000). *Bacillus* and *Paenibacillus* strains have been found among diverse communities of nitrogen fixers associated with *Drosera villosa*, a Brazilian carnivorous plant (Albino et al. 2006), wheat rhizosphere (Beneduzi et al. 2008a) and rice rhizosphere and bulk soil (Beneduzi et al. 2008b), where there was evidence of a relationship between different RFLP groups and soil pH. Most isolates were members of *Bacillus* or *Paenibacillus*. Wheat rhizosphere and bulk soil strains were dominated by *Paenibacillus*, particularly *P. borealis* and *P. graminis*, the remainder being identified as *Bacillus* sp.

Communities involved in degradation of proteins were characterized by Sakurai et al. (2007) in fertilized soils planted with lettuce. Proteolytic activity was greater following addition of organic, rather than inorganic, fertilizer and was greater in rhizosphere than bulk soil. Sequences of alkaline (*apr*) and neutral metalloprotease (*npr*) genes associated with DGGE bands were closely related to those of *Pseudomonas fluorescens* and *B. megaterium*, respectively, with homology of one band to *Bacillus vietnamensis*. Fertilizer type affected *apr* community composition in the rhizosphere, but not bulk soil, and affected *npr* composition in both. Rhizosphere and bulk soil communities were different for both genes. Multiple regression analysis of protease activity and DGGE profiles of both genes showed significant relationships: the *apr* community was affected by fertilizer treatment and the rhizosphere, while the *npr* community was affected mainly by fertilizer. The results were interpreted as indicating different roles for pseudomonads and bacilli in the rhizosphere and bulk soil, and showing that community structure has an important influence on soil protease activity.

Chitinolytic activity contributes to the role of bacilli in mineralization of soil organic matter and Hallmann et al. (1999) found *Bacillus* and *Arthrobacter* to be the dominant genera in the soil and rhizosphere of cotton; but they were not detected as endophytes. Addition of chitin to soil, to reduce fungal pathogens, decreased the frequency of *Bacillus* among isolates suggesting a role for other organisms in chitin degradation in this habitat. Degradation of cellulose, hemicellulose and aromatic compounds by bacilli is also believed to be important in the guts of soil invertebrates, which contain a diverse range of bacilli (see König 2011).

2.2.5 The Influence of Environmental Factors on Diversity of Bacilli

One obvious factor favouring selection of endospore-formers is their ability to survive temperatures that kill vegetative cells. This is of little advantage under normal soil conditions, but can be important following forest fires. For example, *Bacillus aminovorans*-related sequences were detected in a spruce-dominated boreal forest ecosystem one year after a large wildfire (Smith et al. 2008). Incineration of municipal solid waste material, before disposal to landfill soil, also leads to selection (Mizuno et al. 2008). *B. cereus* and *B. megaterium* were the most common isolates from forest and cultivated soils, while buried ash contained *B. licheniformis*, *Bacillus firmus*, *Bacillus thioparus* and *Bacillus krulwichiae*, the first two species also being found in the overlying forest soil. Incinerated ash has high concentrations of Na, Ca, K and Cl, and a high pH, and phenotypic analysis indicated that selection was due to ability to grow under anaerobic conditions and at high pH, in addition to selection of *Bacillus* during incineration.

Flooding reduces oxygen availability in soil and might be expected to select for groups carrying out processes such as denitrification, ferric iron reduction, sulphate reduction or methanogenesis. Graff and Conrad (2005) determined the effects of flooding on bacterial communities both in bulk soil and soil associated with roots of poplar trees, grown in microcosms. *Bacillus*-related sequences represented 16% of the bacterial community in unflooded bulk soil and the rhizosphere, but this proportion was reduced following flooding, although a *Paenibacillus*-related sequence increased in relative abundance. Rhizosphere soil contained 42% *Bacillus* sequences, but sequence types were different to those in bulk soil. Although some of these findings are consistent with other studies, no mechanistic basis for the changes was proposed. Kim et al. (2005) also found significant effects of flooding in rice fields, with reduction in biodiversity of bacilli from $>10\%$ to undetectable, and domination by *Arthrobacter*. Clone sequences were less diverse and contained a lower proportion of bacilli than isolates.

Lear et al. (2004) investigated the effect of applying an electrical field to soil, to increase biodegradation of pollutants. Although this had no detectable effect on the majority of the members of the bacterial community, some effects were seen in the immediate vicinity of the anode. In control soil, *Bacillus mycoides* and *Bacillus sphaericus* dominated isolates. *B. sphaericus* was not detected in the electrokinetic cells and *B. megaterium*, which was not found in control soil isolates, was the most abundant *Bacillus* isolate. This was explained in terms of protection afforded by production of a cell capsule and greater stress at the anode arising from changes in soil characteristics, particularly increased acidity.

2.2.6 Effects of Soil Contaminants on Diversity of Bacilli

Mercury resistance in soil bacilli is widespread and is encoded through a chromosomally borne *mer* operon. Hart et al. (1998) found that 5% of bacilli in a mercury-contaminated soil were resistant to mercury. RFLP analysis of mercuric reductase (*merA*) genes gave 14 RFLP groups from soil isolates and 11 from directly extracted soil DNA, with three common to both, reflecting difficulties in cultivation of dominant organisms. Phenotypic analysis indicated significant physiological diversity among isolates, suggesting high levels of genetic exchange. There is also evidence for selection of a *Bacillus* strain in auriferous soils (Reith and Rogers 2008), supporting earlier observation of a 1,000-fold increase in *B. cereus* spores (Reith et al. 2005). In contrast, there was no evidence of selection of members of the *Firmicutes* following spiking of soil with mercury (Nazaret et al. 2003), in mine tailings, or in cultures grown on media containing cadmium (Zhang et al. 2007). A lead-enriched community contained both *Paenibacillus* and *Bacillus* sequences, but there are no reports of metal resistance in *Paenibacillus* (Zhang et al. 2007).

Bodour et al. (2003) screened 1,305 isolates from soil contaminated with metals and/or hydrocarbons and found greater numbers of Gram-positive, biosurfactant-producing isolates in metal-contaminated or uncontaminated soils and more Gram-negative isolates in hydrocarbon-contaminated or co-contaminated soils. Of the 45 isolates, eight were *B. subtilis* and three *B. licheniformis*, both of which produce a range of biosurfactants. A decrease in the proportion of bacilli among isolates was seen following contamination of soil with toluene (Chao and Hsu 2004) or long-term contamination with the organophosphate pesticide methylparathion (Zhang et al. 2006).

2.2.7 *Intraspecific Diversity Within Soil Bacilli*

In general, studies of diversity of bacilli within "total" prokaryote soil communities are descriptive, giving information on which groups are present and, in some studies, estimates of the relative abundance of bacilli. The physiological characteristics of bacilli, notably spore formation but also degradative capacities, provide the potential for an understanding of mechanisms determining diversity and ecosystem function. This potential is, however, severely limited by inability to distinguish the relative contributions of spores and vegetative cells and ignorance of which bacilli are active, and the nature and extent of their activity. In contrast, studies of within-species diversity of soil bacilli have given clues to mechanisms that control diversity within this group, and provide model systems with which to investigate the mechanisms controlling general prokaryotic diversity and community structure. This section therefore provides descriptions of diversity within species of soil endospore-formers. For several species, these descriptions link closely with the mechanistic studies that are described in the following section.

2.2.8 *Bacillus benzoevorans* and *Bacillus niacini*

Molecular techniques uncover and demonstrate the potential importance of groups with no cultivated representatives, and can thereby direct work towards the isolation of these groups. Two examples of this are detection of *B. benzoevorans* in a Dutch soil by FISH and qPCR and the presence in soil clone libraries of sequences associated with *B. niacini*, constituting 15% of the uncultured *Bacillus* community in soil. Screening of 4,224 soil isolates by multiplex PCR generated several hundred novel *B. benzoevorans* relatives (Felske et al. 2003). A similar search, using several enrichment media, led to isolation of 64 isolates of the *B. niacini* group from different Dutch soils (Felske et al. 2004). The isolates grew best on acetate and were very diverse, but their metabolic diversities gave no clues to their likely ecosystem functions. Metabolic potentials of isolates on laboratory media give good information on their genetic potentials, but it is difficult to predict the ecosystem function of soil isolates because metabolic genes may be responsive to environmental cues that may significantly change the gene expression patterns from those observed in the laboratory.

2.2.9 *Paenibacillus polymyxa*

Mavingui et al. (1992) isolated 130 *Bacillus* (now *Paenibacillus*) *polymyxa* strains from rhizosphere, non-rhizosphere and rhizoplane soils by immunotrapping and characterized them phenotypically, serologically, by RFLP analysis of total DNA, and by hybridization with a rRNA probe. Phenotypic analysis placed isolates in four groups. Two contained isolates from non-rhizosphere soil, one contained isolates from the rhizosphere soil and the fourth contained only rhizoplane isolates. Serological and molecular methods indicated greater diversity among isolates, with lowest diversity in rhizoplane isolates. Von der Weid et al. (2000) isolated 67 *B. polymyxa* strains from maize rhizosphere and analysed them phenotypically and by hybridization with a nifKDH probe and by BOX-PCR. The different methods gave different numbers of major clusters, with BOX-PCR showing greatest discrimination, and communities changed during plant development. Finer scale molecular methods (RADP-PCR and BOX-PCR) indicated high diversity of *P. polymyxa* strains colonizing rhizospheres of four maize cultivars, with evidence of selection by different plants for different strains, and implications for choice of inocula to improve plant growth (da Mota et al. 2002).

2.2.10 Bacillus cereus–Bacillus thuringiensis Group

This group contains a wide diversity of strains, including six closely related species: *B. anthracis, B. thuringiensis*, *B. mycoides*, *Bacillus pseudomycoides*, *Bacillus weihenstephanensis* and *B. cereus* (*sensu stricto*). They inhabit diverse soil habitats and have many economically important representatives, notably *B. thuringiensis*, but also many plant-growth-promoting bacteria. Diversity is greatest in *B. cereus* and *B. thuringiensis*, but low in *B. anthracis*. For example, Helgason et al. (1998) found high diversity in soil isolates collected from five geographic regions in Norway ranging from coastal to the Arctic, while the diversity of *B. anthracis*, but not *B. cereus* or *B. thuringiensis*, isolates collected worldwide was low (Keim et al. 2000). *B. cereus* or *B. thuringiensis* strains appeared to co-exist in a French forest soil (Vilas-Boas et al. 2002), but were genetically distinct and diverged to a greater extent than strains of the same species isolated from geographically different locations. MLST (rather than MLEE) analysis of these strains and 19 strains from elsewhere separated them into three clusters, one exhibiting frequent exchange between strains, while the other two were clonal (Sorokin et al. 2006). These differences may reflect differences in life style between strains. *B. weihenstepahniensis* isolates are psychrotolerant and the role of temperature in determining their distribution is discussed in Sect. 2.3.5.

2.2.11 Bacillus simplex and *Bacillus subtilis*

Both *B. simplex* and *B. subtilis* are common members of the soil endospore-former community and are important model organisms for mechanistic studies. Their diversity is described in greater detail in Sect. 2.4.

2.2.12 *Paenibacillus*

Both cultivation-based and molecular methods have been used to investigate associations between *Paenibacillus* strains and plant roots. *Paenibacillus durus* (previously *P. azotofixans*) is a common member of the rhizospheres of maize, sorghum, sugarcane, wheat and forage grasses, where it fixes nitrogen and produces antimicrobial compounds. Rosado et al. (1998) isolated 53 *P. azotofixans* strains from the rhizoplane and rhizosphere of different grasses and from soil, and characterized them using a *nifKDH* probe, RAPD, BOX-PCR and API. There was little evidence of specific plant associations, although some clusters were isolated more frequently from wheat and sugarcane. Subsequent work (Rosado et al. 1998; de Albuquerque et al. 2006) placed *P. durus* isolates in two clusters, the first associated with wheat, maize and sugarcane rhizospheres, and the second with all plant species investigated and bulk soil. The different clusters showed different patterns of carbohydrate metabolism: wheat isolates could metabolize sorbitol, and sugarcane isolates could metabolize starch and glycogen, suggesting a selective role for root exudates.

Paenibacillus communities in the rhizospheres of four maize cultivars have been investigated in two soils. Cultivation-based methods (Rosado et al. 1998; da Mota et al. 2002) indicated that *Paenibacillus* diversity was determined by soil type, rather than by cultivar. This was confirmed by molecular methods based on 16S rRNA genes (da Silva et al. 2003) and *rpoB* genes (da Mota et al. 2005). A clone library constructed using *Paenibacillus*-specific 16S rRNA gene primers indicated high diversity and clustering into 12 groups, with *P. azotofixans* the most abundant (19% of clones). DGGE analysis showed clear differences in rhizosphere communities from the different soils.

Vollú et al. (2003) classified cyclodextrin-producing *P. graminis* root-associated isolates from wheat, maize and sorghum sown in Australia, Brazil and France using *rpo*B-RFLP and *gyr*B-RFLP and *rpo*B gene sequencing. Brazilian isolates clustered separately from Australian and French isolates and strains fell within four clusters. For the Brazilian isolates, soil type was more important than plant host in determining the rhizosphere community.

2.3 Mechanisms Driving Diversity

2.3.1 *General Mechanisms Driving Microbial Diversity in Soil*

A combination of speciation, extinction, dispersal and microbial interactions is responsible for the creation and maintenance of diversity (Horner-Devine et al. 2004a; Ramette and Tiedje 2007). Speciation is driven by natural selection, acting on a pool of genotypes in a specific environment. The rate of speciation in bacteria is high owing to large population size and high reproduction rates. Diversity in bacteria may also be high because of low extinction rates, and the ability to form highly resistant life forms such as the spores that are typical of bacilli. Dispersal rates of spore-formers are also high (Roberts and Cohan 1995), reducing extinction rates further by giving spore-formers opportunities to colonize new habitats and escape harsh conditions.

Prokaryotic species have been discriminated by divergence in 16S rRNA gene sequences of 2.5–3%. Isolates that diverged more at the 16S rRNA gene level also show less than 70% DNA–DNA re-association values and are thus placed in different species (Stackebrandt and Goebel 1994) but ecologically distinct groups, termed ecotypes, can be discerned within such species from protein coding sequences (Cohan and Perry 2007) and phenotypic characteristics. An ecotype is defined as an ecologically distinct group of organisms that fall into distinct sequence clusters (lineages), sharing a common evolutionary path, with diversity limited by periodic selection and genetic drift. Ecotype formation is purged by frequent recombination (Cohan and Perry 2007), which may occur between closely related bacilli. Genetic exchange among closely related bacilli occurs at high frequency in laboratory soil microcosms (Graham and Istock 1978; Duncan et al. 1995). In addition, linkage disequilibrium and genetic recombination among wild isolates of *B. subtilis* obtained from a microsite (200 cm^3) of surface desert soil was higher than in *E. coli* populations, based on MLEE, phage and antibiotic resistance and RFLP analysis (Istock et al. 1992). The potential for genetic exchange is further increased by transformation, which has been observed within natural populations of *B. subtilis* and closely related species isolated from various desert soils, whose transformation rates range over three orders of magnitude (Cohan et al. 1991). To determine actual, rather than potential, rates of recombination Roberts and Cohan (1995) analysed restriction patterns of three housekeeping genes among natural closely related *Bacillus* isolates. Recombination within *B. subtilis* or *B. mojavensis* was too low to prevent adaptive divergence between ecotypes (Cohan 2002; Maynard Smith and Szathmary 1993). Diversification between ecologically distinct populations increases, due to sequence divergence and differences in restriction modification systems between donor and recipient (Dubnau et al. 1965; Roberts et al. 1994).

2.3.2 Environmental Factors Driving Diversification in Bacilli

Bacteria are highly adaptable and can exploit a wide range of environmental opportunities for diversification (Horner-Devine et al. 2004b; Ramette and Tiedje 2007), but little is known of the influence of environmental factors on diversity and community composition of soil bacilli. However, an ecotype simulation algorithm (Koeppel et al. 2008) has been used to model evolutionary dynamics of bacterial populations and was tested using *Bacillus* ecotypes belonging to two clades isolated from two “Evolutionary Canyons” in Israel. These sites show interesting topographies, with three major habitats: north facing (European) slopes, south facing, more stressful (African) slopes and a canyon bottom with greater access to water (Nevo 1995). Nine and 13 ecotypes were identified among isolates belonging to the *B. simplex* and *B. subtilis–B. licheniformis* clades, respectively (Fig. 2.1). *B. simplex* strains within each ecotype were exclusively or sometimes predominantly isolated from one of the habitats (Sikorski and Nevo 2007). Similar, but less strong, associations between habitat and ecotype were also detected within the *B. subtilis–B. licheniformis* clade (Roberts and Cohan 1995; Koeppel et al. 2008). This suggests that specialization for environmental conditions associated with one of the three habitats can be discerned at the level of sequence clustering and provides strong evidence that ecotype clustering and ecological distinctness may correlate. However, ecotypes did not correlate well with physiological characteristics, determined using Biolog, suggesting that energy metabolism is not determined by soil characteristics or by solar radiation, temperature and drought, which are considered to be the major abiotic features of this environment (Sikorski et al. 2008).

The study by Koeppel et al. (2008) demonstrates a significant impact of environment on *Bacillus* diversity, but more ecological and physiological data are

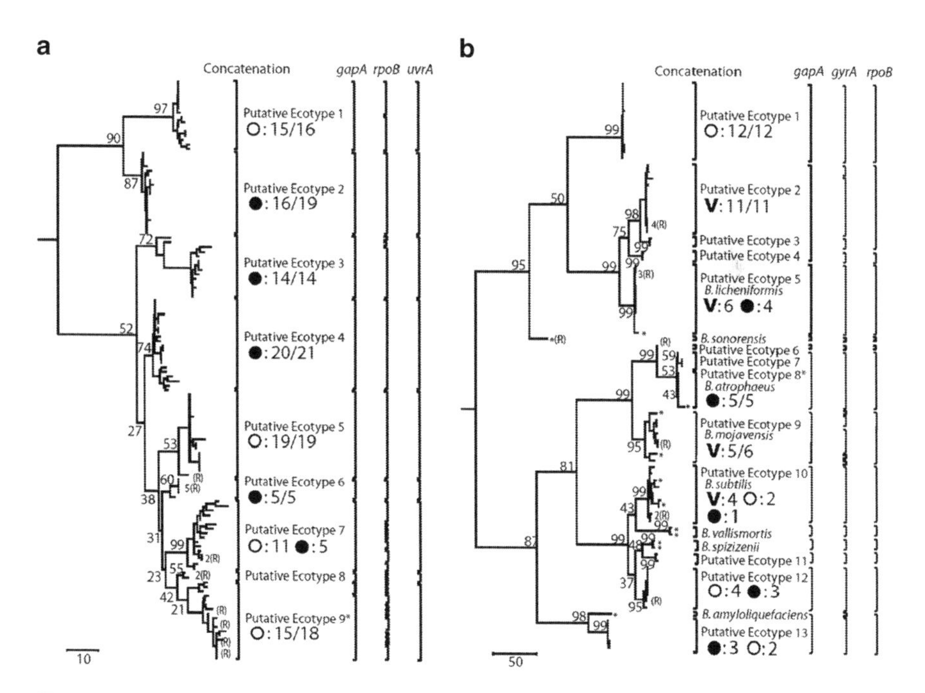

Fig. 2.1 Phylogeny and ecotype segregation of the *B. simplex* and *B. subtilis–B. licheniformis* clades. Analysis was (**a**) *gapA*, *rpoB* and *uvrA* genes for the *B. simplex* clade, resulting in nine putative ecotypes, and (**b**) *gapA*, *gyrA* and *rpoB* genes for the *B. subtilis–B. licheniformis* clade, giving 13 putative ecotypes. Bacilli were isolated from the south facing slope (*open circle*), the north facing slope (*filled circle*) and the canyon bottom (V) at Evolutionary Canyon (Israel). Habitat source is indicated for ecotypes represented by at least four isolates and only one habitat is depicted if >80% of isolates originated from one habitat. All habitat sources are indicated for clusters not dominated by isolates from one habitat (From Koeppel et al. (2008), with permission)

required to understand the specific factors leading to ecological distinctness of each putative ecotype (Koeppel et al. 2008). More information is also required to determine whether variation between ecotypes is correlated with ecosystem function and whether ecotypes identified by sequence analysis correspond to ecologically distinct groups.

Temperature has been shown to determine the relative distributions of the psychrotolerant *B. weihenstepahniensis*, which can grow below 7°C and up to 38°C, and the mesophilic *B. cereus sensu stricto*, which grows in the range 7°C–46°C. The psychrotolerant and mesophilic phenotypes are reflected in genotypic differences in the cold shock protein A (*cspA*) gene. Von Stetten et al. (1999) studied the distribution of 1,060 mesophilic and psychrotolerant isolates obtained from a tropical soil, a temperate soil and two alpine habitats, with average annual temperatures of 28°C, 7°C, 4°C and 1°C, respectively. Isolates were characterized phenotypically, in terms of their growth–temperature responses and psychrotolerance, and genotypically (16S rRNA and *cspA* gene sequences). The proportions of psychrotolerant isolates in these four habitats were 0%, 45%, 86% and 98%, respectively, indicating strong temperature selection. Psychrotolerant strains were able to grow at temperatures below 7°C and up to 38°C, while mesophilic strains could grow at temperatures above 7°C and possessed psychrotolerant or mesophilic *cspA* genotypes.

Only *B. cereus* isolates were obtained from the tropical habitat and isolates from the alpine habitats were heavily dominated by *B. weihenstephanensis*. These isolates also contained the corresponding psychrotolerant or mesophilic *cspA* genotype. Both groups were found in the temperate habitat together with isolates named "intermediate thermal types". The latter carried the psychrotolerant *cspA* gene, but showed mesophilic phenotype or carried the mesophilic *cspA* gene and had the psychrophilic phenotype, and sometimes even had mesophilic and psychrotolerant 16S rRNA operon copies within a single isolate. These intermediate thermal types may represent ongoing adaptation to prevalent temperatures. In addition, diversity was greater in temperate soils, potentially reflecting greater variation in temperature around annual means, leading to co-existence of psychrotolerant and mesophilic organisms.

2.3.3 Diversification Within Species of Aerobic Endospore-Formers

B. subtilis is one of the best-studied bacteria at the molecular level, but relatively little is known of its ecology and diversity. Strains within the *B. subtilis* clade form two subclusters, 168, delineating *B. subtilis subsp. subtilis*, and W23, representing *B. subtilis subsp. spizizenii* (Roberts and Cohan 1995; Nakamura et al. 1999). The level of genetic diversity within W23 is considerably higher than within 168 or the closely related *B. mojavensis* cluster. The ecological significance of this diversity is not understood, but microarray-based comparative genomic hybridization (M-CGH) (Earl et al. 2008) confirmed closer relations within than between subspecies, with 30% divergence of genes within species. Diversity was highest for genes involved in the synthesis of secondary metabolites, teichoic acid and the adaptive response to alkylation DNA damage, but there was variation in all functional groups for genes of potential ecological importance in adaptation to different environments, including environmental sensing and carbohydrate or amino acid metabolism. In contrast, genes belonging to the core genome, previously identified as essential under laboratory conditions in *B. subtilis* 168, were highly conserved. Divergence was greater in germination than sporulation genes, suggesting that environmental cues for outgrowth might vary between strains. In addition, genes involved in the ability of *B. subtilis* to become naturally competent and take up DNA from the environment were highly conserved, except for the first three genes in the *comQXPA* operon that have been previously indicated as highly polymorphic (Tran et al. 2000; Tortosa et al. 2001) (see below).

2.3.4 Cell–Cell Signalling Driving Diversification

Soil bacilli provide an interesting system with which to investigate the influence of microbial interactions on diversification. Competence in *B. subtilis* is controlled through a population-density dependent, quorum-sensing system, encoded by the *comQXPA* operon for enzymes involved in the synthesis, processing and recognition of the extracellular pheromone ComX. The genomic diversity of comQXP genes results in functional diversity, so that strains producing similar pheromones are able to induce competence in each other, while divergent strains do not (Ansaldi et al. 2002; Mandic-Mulec et al. 2003; Stefanic and Mandic-Mulec 2009). This therefore provides an example of functional diversification with a potential ecological role. Polymorphism of competence-signalling may act as a sexual isolation mechanism (Tortosa et al. 2001; Ansaldi et al. 2002), lowering the frequency of recombination even among members of the same species, and so increasing diversification among strains of different pherotypes. It would be interesting to examine the diversity of members of different pherotypes at other loci that may be adaptive in certain environments.

Stefanic and Mandic-Mulec (2009) examined polymorphism of highly related *B. subtilis* isolates from soil aggregates that were exposed to the same environmental conditions. All four pherotypes previously found among strains isolated from distant geographical locations were present in microscale samples taken from within aggregates, but only three were specific for *B. subtilis* (Fig. 2.2). The fourth, previously associated with *B. subtilis* and the closely related *B. mojavensis*, was detected in *B. amyloliquefaciens* isolates, suggesting a role for horizontal gene transfer in interspecies distribution of the pherotypes.

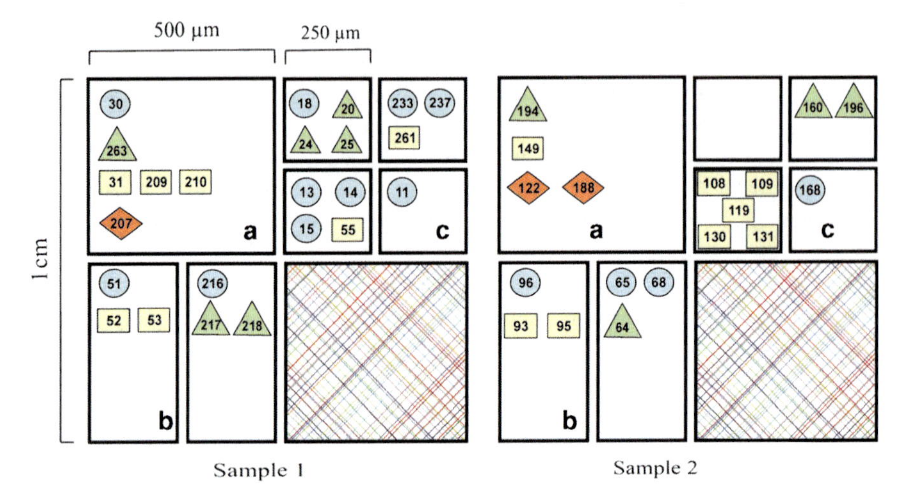

Fig. 2.2 The distribution of *B. subtilis* and *B. amyloliquefaciens* pherotypes in two 1-cm^3 samples of river bank soil that were sectioned to progressively smaller subsamples, comprising 1/4, 1/8 and 1/16 of the initial soil sample. From each sub-sample 30 spore-formers were isolated but, among those, only two, three and sometimes more were identified as *B. subtilis* and were analysed further. Each pherotype is depicted by colour and shape: the 168, RS-D-2/NAF4, RO-B-2/RO-H-1 and RO-E-1 pherotype with *blue circles*, *green triangles*, *yellow rectangles* and *orange diamonds*, respectively (From Stefanic and Mandic-Mulec (2009), with permission)

2.3.5 Diversification and Biogeography of Bacilli

Similarities in macroscale and microscale diversity of *B. subtilis* pherotypes are consistent with the cosmopolitan nature of bacilli. Global distribution of spore-forming bacteria such as *B. mojavensis* and *B. subtilis* is indicated by analysis of protein coding genes, which show similar linkage disequilibrium ($D = 0.50 - 0.87$) in local populations and global ($D = 0.71$) populations. A similar linkage disequilibrium, which is the non-random association of genetic loci, suggests high migration rates and lack of geographical isolation (Roberts and Cohan 1995). Migration rate increased with geographical scale, but even populations separated by the greatest distances were not sufficiently isolated to demonstrate genetic drift. This suggests that, for the genes investigated, evolutionary processes that require continued geographical isolation are unlikely to occur, although it would be interesting to see whether diversification due to geographical isolation would be detectable in faster-evolving genes. Diversification of the fast-evolving *comQXP* loci has been observed even within soil aggregates, where spatial isolation might not be expected, and the data indicate that diversity decreases with aggregate size (Fig. 2.3) (Stefanic and Mandic-Mulec 2009). The spatial heterogeneity of the soil environment may therefore lead to geographic isolation and niche differentiation even at the 250-µm scale. Evolutionary principles and competition between bacteria for natural resources acting at this scale are poorly understood and warrant further studies (Grundmann 2004).

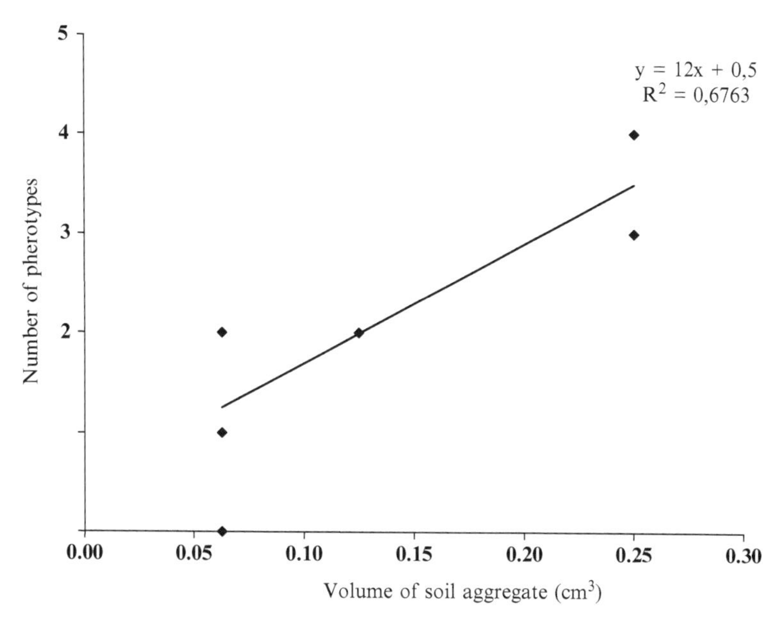

Fig. 2.3 The correlation between number of pherotypes and sample size in the study of Stefanic and Mandic-Mulec (2009) (see legend to Fig. 2.2). The number of pherotypes decreased with soil sample size

2.4 Conclusions

Aerobic endospore-forming bacteria are taxonomically and physiologically diverse, and are ubiquitous members of soil microbial communities. Understanding diversity patterns in soil is of particular interest, given their cosmopolitan nature and high biotechnological potentials that range from the biological degradation of pollutants, through the production of industrially interesting enzymes and bioactive chemicals, to their roles as biopesticides. The nature of these patterns and the importance of this group for soil ecosystem function are, however, difficult to assess. Few studies attempt to determine quantitative activity of bacilli in soil, or the specific processes which they carry out, in relation to the activities of the total microbial community. To some extent this reflects methodological limitations. Cultivation-based approaches are highly selective, and will underestimate abundance, and they give information on potential, rather than actual activity. Molecular techniques avoid cultivation bias, but very few studies have utilized primers that target bacilli or specific genera, species or ecotypes. None of these has been used to investigate in situ activity. Design and application of group-specific primers is required therefore, combined with molecular techniques that assess activity rather than just presence. The next generation of molecular techniques, based on high-throughput sequencing, genomics, metatranscriptomics and proteomics, should also be exploited. More importantly, there is a need for studies that aim to assess the links between diversity, community structure, physiological diversity and ecosystem function, rather than merely characterizing the presence, absence and identity of strains present.

A major conceptual and technical issue is the distinction between vegetative cells and spores. For cultivation-based methods, the presence of spores is likely to lead to overestimation of the importance of bacilli in soil processes. This could be addressed by comparisons with and without heat-treatment of samples, but is rarely attempted. Consequently, although the ability to eliminate non-spore-formers makes it easy to study spore-formers, there is little attempt to address the relative sensitivities of spores of different organisms, the relative importances of spores and vegetative cells for cultivation, and the relative culturabilities both of spores (from spore-formers) and vegetative cells of non-spore-formers.

Molecular techniques cannot currently distinguish spores and vegetative cells, and nucleic acid extraction techniques represent a compromise between lysis efficiency and nucleic acid degradation. Therefore, differences between cultivation-based and molecular studies may be exaggerated for spore-formers, in comparison with other microbial groups. These problems and biases can actually be used to advantage, depending on the ecological question being addressed, but in general these issues are not considered. While these issues limit our knowledge of the ecology of soil bacilli, the study of specific genera and species has generated important and generic advances in our understanding of the mechanisms driving microbial diversity. These include studies of biogeography and the impact of environmental factors, signalling, spatial scale and horizontal gene transfer on evolution of bacterial species, ecotype formation and speciation. These studies have been based mainly on the analysis of laboratory isolates and it will be interesting to see how the results obtained correlate with diversity patterns detected directly in soil and how communities respond spatially and temporally to changing environmental parameters.

References

Albino U, Saridakis DP, Ferreira MC, Hungria M, Vinuesa P, Andrade G (2006) High diversity of diazotrophic bacteria associated with the carnivorous plant *Drosera villosa var. villosa* growing in oligotrophic habitats in Brazil. Plant Soil 287:199–207

Andersen GL, He Z, DeSantis TZ, Brodie EL, Zhou J (2010) The use of microarrays in microbial ecology. In: Liu W-T, Jansson JK (eds) Environmental molecular microbiology. Horizon Scientific, Norfolk, UK

Ansaldi M, Marolt D, Stebe T, Mandic-Mulec I, Dubnau D (2002) Specific activation of the *Bacillus* quorum-sensing systems by isoprenylated pheromone variants. Mol Microbiol 44:1561–1573

Banat IM and Marchant R (2011) *Geobacillus* activities in soil and oil contamination remediation. In: Logan NA (ed) Endospore forming soil bacteria (Soil Biology 27). Springer, Heidelberg, doi: 10.1007/978-3-642-19577-13

Beneduzi A, Peres D, Vargas LK, Bodanese-Zanettini MH, Passaglia LMP (2008a) Evaluation of genetic diversity and plant growth promoting activities of nitrogen-fixing bacilli isolated from rice fields in South Brazil. Appl Soil Ecol 39:311–320

Beneduzi A, Peres D, da Costa PB, Bodanese Zanettini MH, Passaglia LMP (2008b) Genetic and phenotypic diversity of plant-growth-promoting bacilli isolated from wheat fields in southern Brazil. Res Microbiol 159:244–250

Bizzarri MF, Prabhakar A, Bishop AH (2008) Multiple-locus sequence typing analysis of *Bacillus thuringiensis* recovered from the phylloplane of clover (*Trifolium hybridum*) in vegetative form. Microb Ecol 55:619–625

Bodour AA, Drees KP, Maier RM (2003) Distribution of biosurfactant-producing bacteria in undisturbed and contaminated arid southwestern soils. Appl Environ Microbiol 69:3280–3287

Brons JK, Van Elsas JD (2008) Analysis of bacterial communities in soil by use of denaturing gradient gel electrophoresis and clone libraries, as influenced by different reverse primers. Appl Environ Microbiol 74:2717–2727

Chao WL, Hsu SF (2004) Response of the soil bacterial community to the addition of toluene and toluene-degrading bacteria. Soil Biol Biochem 36:479–487

Chèneby D, Philippot L, Hartmann A, Hénault C, Germon J-C (2000) 16S rDNA analysis for characterization of denitrifying bacteria isolated from three agricultural soils. FEMS Microbiol Ecol 34:121–128

Chin K, Hahn D, Hengstmann U, Liesack W, Janssen PH (1999) Characterization and identification of numerically abundant culturable bacteria from the anoxic bulk soil of rice paddy microcosms. Appl Environ Microbiol 65:5042–5049

Chu H, Lin X, Fujii T, Morimoto S, Yagi K, Hu J, Zhang J (2007) Soil microbial biomass, dehydrogenase activity, bacterial community structure in response to long-term fertilizer management. Soil Biol Biochem 39:2971–2976

Cohan FM (2002) What are bacterial species? Annu Rev Microbiol 56:457–487

Cohan FM, Perry EB (2007) A systematics for discovering the fundamental units of bacterial diversity. Curr Biol 17:R373–R386

Cohan FM, Roberts MS, King EC (1991) The potential for genetic exchange by transformation within a natural population of *Bacillus subtilis*. Evolution 45:1393–1421

Collier FA, Elliot SL, Ellis RF (2005) Spatial variation in *Bacillus thuringiensis*/*cereus* populations within the phyllosphere of broad-leaved dock (*Rumex obtusifolius*) and surrounding habitats. FEMS Microbiol Ecol 54:417–425

da Mota FF, Nóbrega A, Evódio Marriel I, Paiva E, Seldin L (2002) Genetic diversity of *Paenibacillus polymyxa* populations isolated from the rhizosphere of four cultivars of maize (*Zea mays*) planted in Cerrado soil. Appl Soil Ecol 20:119–132

da Mota FF, Gomes EA, Paiva E, Rosado AS, Seldin L (2004) Use of *rpoB* gene analysis for identification of nitrogen-fixing *Paenibacillus* species as an alternative to the 16S rRNA gene. Lett Appl Microbiol 39:34–40

da Mota FF, Gomes EA, Paiva E, Seldin L (2005) Assessment of the diversity of *Paenibacillus* species in environmental samples by a novel *rpoB*-based PCR-DGGE method. FEMS Microbiol Ecol 53:317–328

da Silva KRA, Salles JF, Seldin L, Van Elsas JD (2003) Application of a novel *Paenibacillus*-specific PCR-DGGE method and sequence analysis to assess the diversity of *Paenibacillus* spp. in the maize rhizosphere. J Microbiol Methods 54:213–231

de Albuquerque PJ, da Mota FF, von der Weid I, Seldin L (2006) Diversity of *Paenibacillus durus* strains isolated from soil and different plant rhizospheres evaluated by ARDRA and *gyrB*-RFLP analysis. Eur J Soil Biol 42:200–207

Dubnau D, Smith I, Morell P, Marmur J (1965) Gene conservation in *Bacillus* species. I. Conserved genetic and nucleic acid base sequence homologies. Proc Natl Acad Sci USA 54:491–498

Duineveld BM, Kowalchuk GA, Keijzer A, van Elsas JD, van Veen JA (2001) Analysis of bacterial communities in the rhizosphere of chrysanthemum via denaturing gradient gel electrophoresis of PCR-amplified 16S rRNA as well as DNA fragments coding for 16S rRNA. Appl Environ Microbiol 67:172–178

Duncan KE, Ferguson N, Istock CA (1995) Fitnesses of a conjugative plasmid and its host bacteria in soil microcosms. Mol Biol Evol 12:1012–1021

Earl AM, Losick R, Kolter R (2008) Ecology and genomics of *Bacillus subtilis*. Trends Microbiol 16:269–275

Felske A, Akkermans ADL, De Vos WM (1998) Quantification of 16S rRNAs in complex bacterial communities by multiple competitive reverse transcription-PCR in temperature gradient gel electrophoresis fingerprints. Appl Environ Microbiol 64:4581–4587

Felske A, Wolterink A, Van Lis R, De Vos WM, Akkermans ADL (2000) Response of a soil bacterial community to grassland succession as monitored by 16S rRNA levels of the predominant ribotypes. Appl Environ Microbiol 66:3998–4003

Felske ADM, Heyrman J, Balcaen A, De Vos P (2003) Multiplex PCR screening of soil isolates for novel *Bacillus*-related lineages. J Microbiol Methods 55:447–458

Felske ADM, Tzeneva V, Heyrman J, Langeveld MA, Akkermans ADL, De Vos P (2004) Isolation and biodiversity of hitherto undescribed soil bacteria related to *Bacillus niacini*. Microb Ecol 48:111–119

Garbeva P, Van Overbeek LS, Van Vuurde JWL, Van Elsas JD (2001) Analysis of endophytic bacterial communities of potato by plating and denaturing gradient gel electrophoresis (DGGE) of 16S rDNA based PCR fragments. Microb Ecol 41:369–383

Garbeva P, Van Veen JA, Van Elsas JD (2003) Predominant *Bacillus* spp. in agricultural soil under different management regimes detected via PCR-DGGE. Microb Ecol 45:302–316

Garbeva P, Van Elsas JD, Van Veen JA (2008) Rhizosphere microbial community and its response to plant species and soil history. Plant Soil 302:19–32

Graff A, Conrad R (2005) Impact of flooding on soil bacterial communities associated with poplar (*Populus* sp.) trees. FEMS Microbiol Ecol 53:401–415

Graham JB, Istock CA (1978) Genetic exchange in *Bacillus subtilis* in soil. Mol Gen Genet 166:287–290

Grundmann GL (2004) Spatial scales of soil bacterial diversity – the size of a clone. FEMS Microbiol Ecol 48:119–127

Guinebretière M-H, Thompson FL, Sorokin A, Normand P, Dawyndt P, Ehling-Schulz M, Svensson B, Sanchis V, Nguyen-The C, Heyndrickx M, De Vos P (2008) Ecological diversification in the *Bacillus cereus* Group. Environ Microbiol 10:851–865

Hallmann J, Rodríguez-Kábana R, Kloepper JW (1999) Chitin-mediated changes in bacterial communities of the soil, rhizosphere and within roots of cotton in relation to nematode control. Soil Biol Biochem 31:551–560

Hart MC, Elliott GN, Osborn AM, Ritchie DA, Strike P (1998) Diversity amongst *Bacillus merA* genes amplified from mercury resistant isolates and directly from mercury polluted soil. FEMS Microbiol Ecol 27:73–84

Helgason E, Caugant DA, Lecadet M, Chen Y, Mahillon J, Lövgren A, Hegna I, Kvaløy K, Kolstø AB (1998) Genetic diversity of *Bacillus cereus/B. thuringiensis* isolates from natural sources. Curr Microbiol 37:80–87

Horner-Devine MC, Carney KM, Bohannan BJM (2004a) An ecological perspective on bacterial biodiversity. Proc R Soc Lond Ser B Biol Sci 271:113–122

Horner-Devine MC, Lage M, Hughes JB, Bohannan BJM (2004b) A taxa-area relationship for bacteria. Nature 432:750–753

Istock CA, Duncan KE, Ferguson N, Zhou X (1992) Sexuality in a natural population of bacteria – *Bacillus subtilis* challenges the clonal paradigm. Mol Ecol 1:95–103

Janssen PH (2006) Identifying the dominant soil bacterial taxa in libraries of 16S rRNA and 16S rRNA genes. Appl Environ Microbiol 72:1719–1728

Keim P, Price LB, Klevytska AM, Smith KL, Schupp JM, Okinaka R, Jackson PJ, Hugh-Jones ME (2000) Multiple-locus variable-number tandem repeat analysis reveals genetic relationships within *Bacillus anthracis*. J Bacteriol 182:2928–2936

Kim J-S, Kwon S-W, Jordan F, Ryu J-C (2003) Analysis of bacterial community structure in bulk soil, rhizosphere soil, and root samples of hot pepper plants using FAME and 16S rDNA clone libraries. J Microbiol Biotechnol 13:236–242

Kim M-S, Ahn J-H, Jung M-K, Yu J-H, Joo D, Kim M-C, Shin H-Y, Kim T, Ryu T-H, Kweon S-J, Kim T, Kim D-H, Ka J-O (2005) Molecular and cultivation-based characterization of bacterial community structure in rice field soil. J Microbiol Biotechnol 15:1087–1093

Koeppel A, Perry EB, Sikorski J, Krizanc D, Warner A, Ward DM, Rooney AP, Brambilla E, Connor N, Ratcliff RM, Nevo E, Cohan FM (2008) Identifying the fundamental units of bacterial diversity: a paradigm shift to incorporate ecology into bacterial systematics. Proc Natl Acad Sci USA 105:2504–2509

König H (2011) Aerobic endospore-forming bacteria and soil invertebrates. In: Logan NA (ed) Endospore forming soil bacteria (Soil Biology 27). Springer, Heidelberg, doi: 10.1007/978-3-642-19577-10

Kraigher B, Stres B, Hacin J, Ausec L, Mahne I, Van Elsas JD, Mandic-Mulec I (2006) Microbial activity and community structure in two drained fen soils in the Ljubljana Marsh. Soil Biol Biochem 38:2762–2771

Kraigher B, Kosjek T, Heath E, Kompare B, Mandic-Mulec I (2008) Influence of pharmaceutical residues on the structure of activated sludge bacterial communities in wastewater treatment bioreactors. Water Res 42:4578–4588

Lear G, Harbottle MJ, van der Gast CJ, Jackman SA, Knowles CJ, Sills G, Thompson IP (2004) The effect of electrokinetics on soil microbial communities. Soil Biol Biochem 36:1751–1760

Liles MR, Manske BF, Bintrim SB, Handelsman J, Goodman RM (2003) A census of rRNA genes and linked genomic sequences within a soil metagenomic library. Appl Environ Microbiol 69:2684–2691

Macrae A, Rimmer DL, O'Donnell AG (2000) Novel bacterial diversity recovered from the rhizosphere of oilseed rape (*Brassica napus*) determined by the analysis of 16S ribosomal DNA. Antonie van Leeuwenhoek Int J Gen Mol Microbiol 78:13–21

Mandic-Mulec I, Kraigher B, Cepon U, Mahne I (2003) Variability of the quorum sensing system in natural isolates of *Bacillus* sp. Food Technol Biotechnol 41:23–28

Marchant R, Banat IM, Rahman TJ, Berzano M (2002) The frequency and characteristics of highly thermophilic bacteria in cool soil environments. Environ Microbiol 4:595–602

Mavingui P, Laguerre G, Berge O, Heulin T (1992) Genetic and phenotypic diversity of *Bacillus polymyxa* in soil and in the wheat rhizosphere. Appl Environ Microbiol 58:1894–1903

Maynard Smith J, Szathmary E (1993) The origin of chromosomes I. Selection for linkage. J Theor Biol 164:437–446

McCaig AE, Glover LA, Prosser JI (1999) Molecular analysis of bacterial community structure and diversity in unimproved and improved upland grass pastures. Appl Environ Microbiol 65:1721–1730

Meintanis C, Chalkou KI, Kormas KA, Lymperopoulou DS, Katsifas EA, Hatzinikolaou DG, Karagouni AD (2008) Application of *rpoB* sequence similarity analysis, REP-PCR and BOX-PCR for the differentiation of species within the genus *Geobacillus*. Lett Appl Microbiol 46:395–401

Mizuno K, Fukuda K, Fujii A, Shiraishi A, Takahashi K, Taniguchi H (2008) *Bacillus* species predominated in an incineration ash layer at a landfill. Biosci Biotechnol Biochem 72:531–539

Mocali S, Paffetti D, Emiliani G, Benedetti A, Fani R (2008) Diversity of heterotrophic aerobic cultivable microbial communities of soils treated with fumigants and dynamics of metabolic, microbial, and mineralization quotients. Biol Fertil Soil 44:557–569

Nakamura LK, Roberts MS, Cohan FM (1999) Relationship of *Bacillus subtilis* clades associated with strains 168 and W23: a proposal for *Bacillus subtilis* subsp. *subtilis* subsp. *nov.* and *Bacillus subtilis* subsp. *spizizenii* subsp. *nov.* Int J Syst Bacteriol 49:1211–1215

Nazaret S, Brothier E, Ranjard L (2003) Shifts in diversity and microscale distribution of the adapted bacterial phenotypes due to Hg(II) spiking in soil. Microb Ecol 45:259–269

Nevo E (1995) Asian, African and European biota meet at 'Evolution Canyon' Israel: Local tests of global biodiversity and genetic diversity patterns. Proc R Soc B Biol Sci 262:149–155

Normander B, Prosser JI (2000) Bacterial origin and community composition in the barley phytosphere as a function of habitat and presowing conditions. Appl Environ Microbiol 66:4372–4377

Pankhurst CE, Pierret A, Hawke BG, Kirby JM (2002) Microbiological and chemical properties of soil associated with macropores at different depths in a red-duplex soil in NSW Australian Plant Soil 238:11–20

Philippot L, Hallin S, Schloter M (2007) Ecology of denitrifying prokaryotes in agricultural soil. Adv Agron 96:249–305

Prosser JI, Jansson JK, Liu W-T (2010) Nucleic-acid-based characterisation of community structure and function. In: Liu W-T, Jansson JK (eds) Environmental molecular microbiology. Horizon Scientific, Norfolk, UK, pp 65–88

Ramette A, Tiedje JM (2007) Biogeography: an emerging cornerstone for understanding prokaryotic diversity, ecology, and evolution. Microb Ecol 53:197–207

Reith F, Rogers SL (2008) Assessment of bacterial communities in auriferous and non-auriferous soils using genetic and functional fingerprinting. Geomicrobiol J 25:203–215

Reith F, McPhail DC, Christy AG (2005) *Bacillus cereus*, gold and associated elements in soil and other regolith samples from Tomakin Park Gold Mine in southeastern New South Wales, Australia. J Geochem Explor 85:81–98

Reva ON, Dixelius C, Meijer J, Priest FG (2004) Taxonomic characterization and plant colonizing abilities of some bacteria related to *Bacillus amyloliquefaciens* and *Bacillus subtilis*. FEMS Microbiol Ecol 48:249–259

Roberts MS, Cohan FM (1995) Recombination and migration rates in natural populations of *Bacillus subtilis* and *Bacillus mojavensis*. Evolution 49:1081–1094

Roberts MS, Nakamura LK, Cohan FM (1994) *Bacillus mojavensis* sp. nov., distinguishable from *Bacillus subtilis* by sexual isolation, divergence in DNA sequence, and differences in fatty acid composition. Int J Syst Bacteriol 44:256–264

Roesch LFW, Fulthorpe RR, Riva A, Casella G, Hadwin AKM, Kent AD, Daroub SH, Camargo FA, Farmerie WG, Triplett EW (2007) Pyrosequencing enumerates and contrasts soil microbial diversity. ISME J 1:283–290

Rosado AS, Duarte GF, Seldin L, Van Elsas JD (1998) Genetic diversity of *nifH* gene sequences in *Paenibacillus azotofixans* strains and soil samples analyzed by denaturing gradient gel electrophoresis of PCR-amplified gene fragments. Appl Environ Microbiol 64:2770–2779

Ryu C, Lee K, Hawng H-J, Yoo C-K, Seong W-K, Oh H-B (2005) Molecular characterization of Korean *Bacillus anthracis* isolates by amplified fragment length polymorphism analysis and multilocus variable-number tandem repeat analysis. Appl Environ Microbiol 71:4664–4671

Sakurai M, Suzuki K, Onodera M, Shinano T, Osaki M (2007) Analysis of bacterial communities in soil by PCR-DGGE targeting protease genes. Soil Biol Biochem 39:2777–2784

Seldin L, Rosado AS, Da Cruz DW, Nobrega A, Van Elsas JD, Paiva E (1998) Comparison of *Paenibacillus azotofixans* strains isolated from rhizoplane, rhizosphere, and non-root-associated soil from maize planted in two different Brazilian soils. Appl Environ Microbiol 64:3860–3868

Selesi D, Schmid M, Hartmann A (2005) Diversity of green-like and red-like ribulose-1, 5-bisphosphate carboxylase/oxygenase large-subunit genes (cbbL) in differently managed agricultural soils. Appl Environ Microbiol 71:175–184

Siciliano SD, Germida JJ (1999) Taxonomic diversity of bacteria associated with the roots of field-grown transgenic *Brassica napus* cv. Quest, compared to the non-transgenic *B. napus* cv. Excel and *B. rapa* cv. Parkland. FEMS Microbiol Ecol 29:263–272

Sikorski J, Nevo E (2007) Patterns of thermal adaptation of *Bacillus simplex* to the microclimatically contrasting slopes of 'Evolution Canyons' I and II, Israel. Environ Microbiol 9:716–726

Sikorski J, Pukall R, Stackebrandt E (2008) Carbon source utilization patterns of *Bacillus simplex* ecotypes do not reflect their adaptation to ecologically divergent slopes in 'Evolution Canyon', Israel. FEMS Microbiol Ecol 66:38–44

Smalla K, Wieland G, Buchner A, Zock A, Parzy J, Kaiser S, Roskot N, Heuer H, Berg G (2001) Bulk and rhizosphere soil bacterial communities studied by denaturing gradient gel electrophoresis: plant-dependent enrichment and seasonal shifts revealed. Appl Environ Microbiol 67:4742–4751

Smit E, Leeflang P, Gommans S, Van Den Broek J, Van Mil S, Wernars K (2001) Diversity and seasonal fluctuations of the dominant members of the bacterial soil community in a wheat field as determined by cultivation and molecular methods. Appl Environ Microbiol 67:2284–2291

Smith NR, Kishchuk BE, Mohn WW (2008) Effects of wildfire and harvest disturbances on forest soil bacterial communities. Appl Environ Microbiol 74:216–224

Sorokin A, Candelon B, Guilloux K, Galleron N, Wackerow-Kouzova N, Ehrlich SD, Bourguet D, Sanchis V (2006) Multiple-locus sequence typing analysis of *Bacillus cereus* and *Bacillus thuringiensis* reveals separate clustering and a distinct population structure of psychrotrophic strains. Appl Environ Microbiol 72:1569–1578

Stackebrandt E, Goebel BM (1994) Taxonomic note: A place for DNA-DNA reassociation and 16S rRNA sequence analysis in the present species definition in bacteriology. Int J Syst Bacteriol 44:846–849

Stefanic P, Mandic-Mulec I (2009) Social interactions and distribution of *Bacillus subtilis* pherotypes at microscale. J Bacteriol 191:1756–1764

Stres B, Danevčič T, Pal L, Fuka MM, Resman L, Leskovec S, Hacin J, Stopar D, Mahne I, Mandic-Mulec I (2008) Influence of temperature and soil water content on bacterial, archaeal and denitrifying microbial communities in drained fen grassland soil microcosms. FEMS Microbiol Ecol 66:110–122

Sturz AV, Ryan DAJ, Coffin AD, Matheson BG, Arsenault WJ, Kimpinski J, Christie BR (2004) Stimulating disease suppression in soils: sulphate fertilizers can increase biodiversity and antibiosis ability of root zone bacteria against *Streptomyces scabies*. Soil Biol Biochem 36:343–352

Tesfaye M, Dufault NS, Dornbusch MR, Allan DL, Vance CP, Samac DA (2003) Influence of enhanced malate dehydrogenase expression by alfalfa on diversity of rhizobacteria and soil nutrient availability. Soil Biol Biochem 35:1103–1113

Tortosa P, Logsdon L, Kraigher B, Itoh Y, Mandic-Mulec I, Dubnau D (2001) Specificity and genetic polymorphism of the *Bacillus* competence quorum-sensing system. J Bacteriol 183:451–460

Tran L-P, Nagai T, Itoh Y (2000) Divergent structure of the ComQXPA quorum-sensing components: Molecular basis of strain-specific communication mechanism in *Bacillus subtilis*. Mol Microbiol 37:1159–1171

Valinsky L, Della Vedova G, Scupham AJ, Alvey S, Figueroa A, Yin B, Hartin RJ, Chrobak M, Crowley DE, Jiang T, Borneman J (2002) Analysis of bacterial community composition by oligonucleotide fingerprinting of rRNA genes. Appl Environ Microbiol 68:3243–3250

Vilas-Boas G, Sanchis V, Lereclus D, Lemos MVF, Bourguet D (2002) Genetic differentiation between sympatric populations of *Bacillus cereus* and *Bacillus thuringiensis*. Appl Environ Microbiol 68:1414–1424

Vollú RE, Dos Santos SCC, Seldin L (2003) 16S rDNA targeted PCR for the detection of *Paenibacillus macerans*. Lett Appl Microbiol 37:415–420

von der Weid I, Paiva E, Nóbrega A, Dirk Van Elsas J, Seldin L (2000) Diversity of *Paenibacillus polymyxa* strains isolated from the rhizosphere of maize planted in Cerrado soil. Res Microbiol 151:369–381

von Stetten F, Mayr R, Scherer S (1999) Climatic influence on mesophilic *Bacillus cereus* and psychrotolerant *Bacillus weihenstephanensis* populations in tropical, temperate and alpine soil. Environ Microbiol 1:503–515

Zhang R, Jiang J, Gu J-D, Li S (2006) Long term effect of methylparathion contamination on soil microbial community diversity estimated by 16S rRNA gene cloning. Ecotoxicology 15:523–530

Zhang H-B, Yang M, Shi W, Zheng Y, Sha T, Zhao Z-W (2007) Bacterial diversity in mine tailings compared by cultivation and cultivation-independent methods and their resistance to lead and cadmium. Microb Ecol 54:705–712

Chapter 3
Studying the Bacterial Diversity of the Soil by Culture-Independent Approaches

Paul De Vos

3.1 Introduction

Bacterial diversity in soil is very complex, and all the various micro-organisms involved are far from identified. Approaching this complex microbial ecosystem is therefore not straightforward, not least because of the enormous macro-variety of soil habitats that exist: sandy soils, clay soils, organic soils, non-agricultural soils, etc. that often occur as soil aggregates; these are the basic microhabitats. Furthermore, the environmental macroconditions such as temperature, pH, water availability, salt concentration, anaerobic versus anoxic, etc. vary considerably, and these of course affect endemic microbial populations. Apart from this macrovariety, the micro-ecological aspects of, for example, a soil particle/soil aggregate, in which the O_2 concentration changes from almost complete saturation at the outer border to anoxic in the middle, affect the development of micro-organisms and hence the microbial diversity at the microscopic level in a dramatic way. Consequently, cultivation-based approaches need a huge variety of incubation conditions if one hopes to investigate even a representative part of the cultivable fractions of these ecosystems, and so such investigations are extremely laborious. This has led to a widespread opinion amongst microbiologists that only a small fraction of the bacterial population present in various ecosystems and niches is cultivable. Therefore, with the support of the molecular tools that became available during the last decennium, culture-independent approaches attracted more and more attention in comparison with culture-dependent methods. This interpretation certainly refutes the historical evolution of our knowledge on microbial diversity, as stated by Leadbetter (2003), that breakthroughs in environmental physiology based on isolation since the 1970s are perhaps even more impressive than those based on

P. De Vos
Laboratory for Microbiology, University of Gent, K.L. Ledeganckstraat, 35, 9000 Gent, Belgium
e-mail: Paul.DeVos@UGent.be

N.A. Logan and P. De Vos (eds.), *Endospore-forming Soil Bacteria*, Soil Biology 27,
DOI 10.1007/978-3-642-19577-8_3, © Springer-Verlag Berlin Heidelberg 2011

molecular approaches alone. Furthermore, it came to the fore in the ongoing debate (Nichols 2007) that molecular data alone are not sufficient to understand functionalities. Indeed, achieving whole genome sequences is not straightforward when the micro-organism is not available (Tringe and Rubin 2005), and this often leaves the microbial heterogeneity unresolved. In addition, functional interpretation of the proteins derived from the generated DNA sequences remains for the most part unresolved, and molecular ecological studies must go hand-in-hand with sequence analyses.

It is, however, clear that knowing the bacterial composition of different ecosystems such as soils covers only one aspect; the question is not only Who is out there? but also What are they doing? and What are they doing with whom?, so as to address the functionalities that are of increasing interest. It is now realized that mixed cultures and natural consortia may be the keys to our better understanding of microbial population ecology. This is clearly an underexplored research field.

Despite this renewed interest in culture-dependent approaches, culture-independent studies have revealed an overwhelming amount of data concerning individual bacteria, specific bacterial taxa and complex bacterial communities, and they have clearly improved our understanding of microbial ecology. It should be appreciated, therefore, that culture-independent approaches may address questions relating both to individual/specific taxa and complex communities.

In the case of *Bacillus*, or in a more broad sense the aerobic endospore-formers, it must be remembered that their survival strategies may result in temporal and spatial resting or dormant stages. The process of switching from dormant spore to active vegetative cell in the natural environment is not well understood (Setlow 2003), but it depends upon changes in environmental conditions such as nutritional components, the so-called germinants, and also – of course – the existence of suitable macro-environmental conditions (pH, temperature, water availability, etc.). When these conditions are unfavourable for such organisms to be in their vegetative states at the time of sampling, molecular approaches that involve cell lysis or cell penetration give misleading results. Therefore, it can easily be appreciated that the effective contributions of aerobic endospore-formers to soil ecology, be they biogeochemical cycling or beneficial effects on plants via the rhizosphere, etc., can be underestimated. On the other hand, preceding such studies with enrichment or cultivation steps to encourage spore germination will bias the final results should only a limited range of culture conditions be applied. For example, the contributions of *Bacillus* and relatives to denitrification processes in activated sludge and water have been underestimated, and, judging by their observed abundances when appropriate media and cultivation conditions have been provided, these organisms most probably make important contributions to the denitrification process.

Finally, in case of direct culture-independent approaches, the same organisms risk being underestimated because they belong to the so-called *Firmicutes*. The cell lysis of such organisms is often difficult, and so subsequent DNA purification and/or PCR-based applications may also be biased.

Researchers must take these considerations into account when designing culture-independent as well as culture-dependent experiments, and the interested reader is also referred to Mandic-Mulec and Prosser (2011).

Investigations into biodiversities of the so-called bulk soils have mainly been made using microscopic or DNA-based molecular methodologies, or combinations of these. The overview below focuses upon those studies that address both methodologies and their application to the aerobic endospore-formers.

3.2 Microscopy

Microscopic investigations of soil often concentrate on the interactions between the natural soil bacteria and the rhizosphere, in order to reveal the development of bacterium/plant relationships and the entry of bacteria into the root system. Although the aerobic endospore-formers are important for certain activities, such as nitrogen fixation or biocontrol of plant pathogens, most studies have concentrated on Gram-negative bacteria. Nevertheless, these methodologies can be applied, after certain adjustments, to the study of Gram-positive bacteria such as the aerobic endospore-formers in their vegetative states. Spores of course are less easy to address, and will demand special procedures. A flow chart for microscopic investigations is shown in Fig. 3.1.

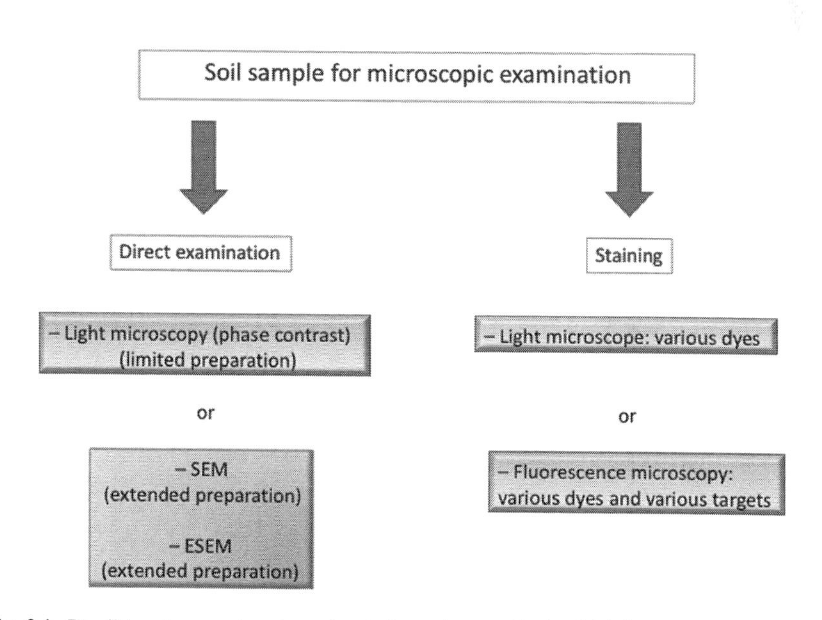

Fig. 3.1 Possible microscopic approaches to investigate soil microbial flora

3.2.1 *Direct Microscopic Examination Without Staining*

Although of little specific value, direct light microscopic analysis can be of some importance, so as to have an idea of the variation of morphological forms of the microbial flora in certain biotopes such as soil.

Scanning electron microscopy (SEM) in its classical application needs an extensive sample preparation procedure that for direct soil studies includes embedding into a resin matrix with subsequent sectioning of the specimen. Although classical SEM also requires the specimen to undergo a dehydration process (because the specimen chamber is under vacuum), environmental scanning electron microscopy (ESEM; Danilatos 1997) does not. In ESEM an oversaturated water vapour pressure in the specimen chamber is maintained, and this allows the study of hydrated biological samples. Although the technique has mainly been used in environmental mineralogical studies, the growth and development of biofilms have been followed in at least one application (Wolgelius et al. 2007).

3.2.2 *Direct Microscopic Examination After Staining*

3.2.2.1 General Staining

Like unstained, direct examination, direct microscopic examination after cell staining allows us to study members of the indigenous microbial flora and their spatial relationships in habitats such as soil. It is clear that, apart from the general staining methods that visualize the overall microbiotic component of the habitat, more specific staining approaches are needed to reveal the abundance of certain bacterial groups such as the aerobic endospore-formers at given times and under particular environmental conditions.

Light microscopy. Light microscopic examination after staining involves Gram and methylene blue staining procedures and may provide a general impression of the variability of the microbiota present in a soil sample. Furthermore, traditional malachite green staining specifically demonstrates the presence of spores. Again, this staining can only provide a general impression of the abundance of endospore-formers in a complex biotope.

Fluorescence microscopy and confocal laser scanning microscopy. An overview of the use of fluorescence microscopy for the visualization of soil micro-organisms is given by Li et al. (2004). Two types of fluorescent microscopy can be distinguished according to the optical paths employed, namely, the transmitted-light fluorescence microscopes and the epi-illumination fluorescence microscopes. Confocal microscopy is based on the same principle as the epifluorescence microscopy with the addition of two pinhole apertures that are placed behind the light source and in front of the detector to eliminate the out-of-focus light so that only in-focus light is detected; as these apertures have the same focus, the arrangement is described as confocal. The image obtained by confocal microscopy is not a real 2D image but a "translated 2D image" of higher quality than one obtained from epifluorescence microscopy. As the focused spot of light scans across the specimen, an image is constructed from the output of a photomultiplier. A 3D image is then built up with the aid of a focus tool that allows the specimen to be scanned at different depths.

Fluorescent dyes and staining. Various dyes are available and different staining procedures have been developed for a wide range of applications: general or specific staining (phylogenetic and physiological groups), the target for staining (DNA, cell wall components, etc.), composition of the soil (clay, sandy, etc.), pH of the soil and so on.

DNA as target for fluorescent microscopy. Acridine orange is probably the best known and most widely used nucleic acid stain for micro-organisms. It is mostly employed for direct total counting. Cell counts obtained by flow cytometry and acridine orange-based direct counting may be in very good agreement, though the latter is much more time-consuming and has the restriction of a small sample size. Furthermore, with the flow cytometric approach, cell sizes and forms can be linked to specific bacteria, such as members of *Bacillus* (DeLeo and Baveye 1996). To minimize the non-specific staining of soil particles by acridine orange, resin embedding is recommended (Altemüller and Vliet-Lanoe 1990). Because acridine orange is positively charged, it binds to and colours the negatively charged clay particles. Ethidium bromide and DAPI (4′,6-diamidino-2-phenylindole) are the two other fluorescent nucleic acid stains that are commonly used for enumeration or biomass measurements of soil preparations.

3.2.2.2 Specific Cell Stains

FISH (*Fluorescent in situ hybridization*). The development of fluorescent probes allows the direct staining of specific phylogenetic groups of bacteria in a soil sample. The methodology that was, and still is, most successful is based upon specific domains in the 16S RNA gene in its transcripted rRNA complement. This is present in a vast number of copies in the cell, and it can bind the selective fluorescent probe so as to give a detectable signal. The method only allows the detection of actively growing cells, and the penetration of the reagents through the cell membranes of *Firmicutes* may be less effective than that for Gram-negative organisms, and so this deserves special attention. FISH has often been used to study the relationships between rhizosphere bacteria and plants; examples can be found for rhizobia (Santaella et al. 2008), *Pseudomonas* (Watt et al. 2006) and *Paenibacillus* (Timmusk et al. 2005). More recently, FISH has been used to demonstrate the expression of functional genes by targeting mRNA, for example, the toluene monooxygenase gene (*tom*) of *Pseudomonas putida* (Wu et al. 2008) and the nitrite reductase gene (*nirK*) of dentrifiers (Pratscher et al. 2009).

GFP (*green fluorescent protein gene*) *LUX* (lux *gene*) *tagging.* Tracking of single bacterial strains in the soil can be achieved by the insertion of fluorescence marker genes. Design and insertion of plasmids or transposons controlled by promoters have allowed in bioluminescence (e.g., *lux* gene) or green fluorescent protein (e.g., *gfp* gene) tagging. These so-called reporter genes may be regarded as non-specific or specific. The non-specific reporters have a lux reporter system that is under the control of a constitutive promoter, and bacteria will emit constant light when supplied with oxygen and energy. They have been used to monitor the development of the metabolic activities of bacterial inoculants in drying soils (Meikle et al. 1995). Specific reporters have been used to observe responses to specific stress factors, compounds or elements (Jaeger et al. 1999).

3.3 Molecular Approaches

Despite the efforts to improve the cultivation approaches (Janssen et al. 2002; Heylen 2007), only a fraction of a soil's microbial diversity is cultivable. Recent developments of molecular tools might appear to minimize the need for cultivation. A range of approaches, collectively known as "metagenomics", has been proposed in various forms. Despite important progress and understanding to overcome various biases, major hurdles still exist:

1. The correct relative coverage of the microbial diversity in the DNA after its extraction from the crude sample
2. Selection of appropriate gene sequences and their integration into functionalities to unravel the existing functional networks and
3. Computational aspects in order to move curated DNA sequences of gene fragments to the databases for further interpretation

The strategy to be chosen depends on the target of interest: is it the complete bacterial community, or specific parts of it such as individual phylogenetic groups, or organisms with particular physiological traits? In all cases, treatment of the samples must be followed by DNA or RNA extraction procedures as the analytical starting points. The next question to be answered concerns functionality aspects of the bacterial community.

An overview of the various strategies is shown in Fig. 3.2.

3.3.1 Sampling

Depending on the nature of the biotope, water or soil samples need to be treated differently. The treatment of water samples will depend on the volume of the samples; up to 20 l of samples can be directly filtered using, for example, a 500-kDa filtration disc, while larger samples may require tangential flow filtration in order to separate solid particles before filtration. In both cases the loaded filters can be cut into pieces for DNA/RNA extraction. In the case of soil samples, the efficiency of the metagenomic approach is directly linked to the method of DNA purification used. When DNA is

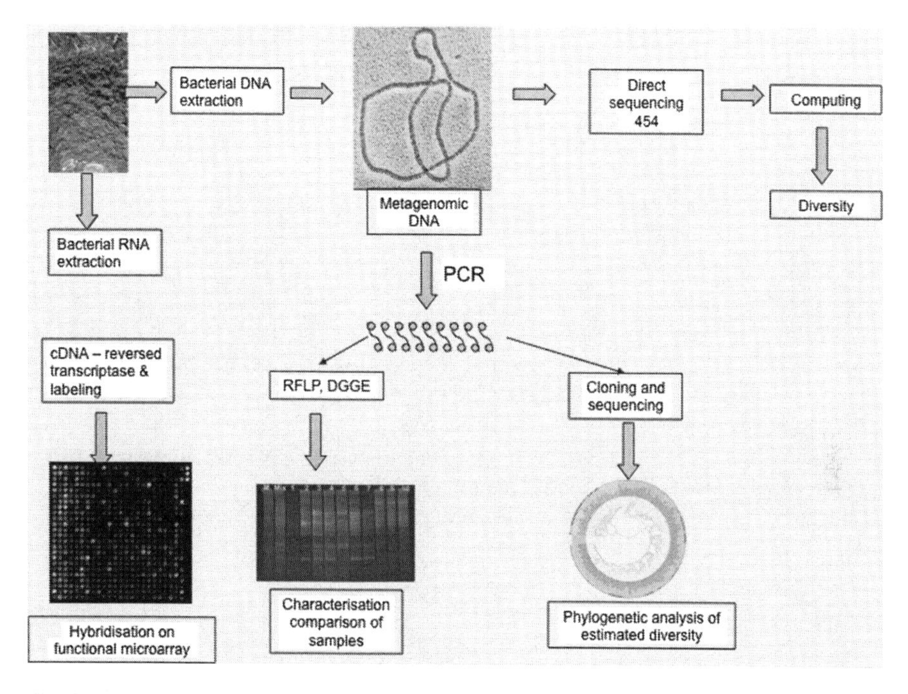

Fig. 3.2 Proposed workflow for molecular analysis of bacterial diversity of soil

directly isolated, its quality is often diminished by problems such as mechanical shearing owing to bead-beating, for example, or due to the effects of humic acid contamination. When the direct DNA extraction is not applied, the Nycodenz extraction method may be recommended. In brief (Guazzaroni et al. 2010), 1–5 g of soil or sediment is suspended in 5–40 ml of TE (Tris/EDTA) buffer at pH 8 and inverted a number of times, then the samples are mixed to separate the cells from the solid matrix and centrifuged at low speed to eliminate the bulky soil particles. The supernatant is separated and again centrifuged for 15–30 min at 6,000×*g* at 4°C and the pellet obtained used for DNA preparation according to the Nycodenz extraction technique (Lindahl 1996). A whitish band of microbial biomass is then obtained at the interface of the Nycodenz layer and the aqueous layer in the ultracentrifugation step. DNA has been successfully isolated from compost, rhizosphere soils, and pristine and contaminated sediments. For more detailed information refer Guazzaroni et al. (2010), Bertrand et al. (2005) and Poté et al. (2010).

3.3.2 DNA/RNA Extractions from Bulk Soil

A general problem in direct DNA or RNA extraction procedures concerns cell lysis. Three different approaches of cell disruption are possible: physical, chemical and enzymatic. Physical disruption may include methods such as freeze-thawing, freeze-boiling, the use of bead mills, bead-beating, mortar-mill grinding, grinding under liquid nitrogen and ultrasonication. These methods often result in the DNA shearing and yield poor quality DNA. Therefore chemical disruptions are most commonly used, involving treatment with sodium dodecyl sulphate (SDS) to dissolve the hydrophobic material of the cell membranes; this is often used in combination with heat treatment and with chelating agents as EDTA, and a variety of phosphate and Tris buffers. To remove (or at least partially remove) the humic acids that may interfere with a PCR reaction at a later stage of the procedure, cytyltrimethyl-ammonium bromide (CTAB) and polyvinylpolypyrrolidone (PVPP) are added. Enzymes such as lysozyme, achromopeptidase and proteinase K are also commonly added in order to remove contaminating material from the nucleic acid preparation.

Later in the procedure, in principle, two alternative major DNA extraction approaches may be followed – the phenol chloroform method, followed by purification and cleaning up, and commercially available DNA extraction kits. Many variants of the phenol chloroform method have been developed in order to match the specific characteristics of the sample. If the Nycodenz approach has been followed, a relatively straightforward DNA preparation procedure is suggested by Guazzaroni et al. (2010). When commercially available DNA extraction kits are used, various options are available and information can easily be obtained from the manufacturer. The efficiency of DNA extraction may be evaluated so that improvements can be made to the above-mentioned Nycodenz approach (Poté et al. 2010). Extraction of DNA of high quality is probably the most crucial step in metagenomic approaches, and the selection of a suitable method may often be informed by in-house experience obtained through trial and error.

In the case of RNA extraction, precautions must be taken both to neutralize the activity of RNAs that are ubiquitous in environmental samples and to circumvent the short life of mRNA. These are covered in Sect. 3.3.4.1.

3.3.3 *Methods that Explore the Overall Bacterial Community*

The "who is out there?" question is in most cases only looking for information on the most abundant or conspicuous traits of micro-organisms in a given soil sample, and even then the answer is biased by the efficiencies of sample treatment, DNA extraction and the PCR-based methods applied to it.

3.3.3.1 Denaturing Gradient Gel Electrophoresis and Terminal Restriction Fragment Length Polymorphism

Denaturing gradient gel electrophoresis (DGGE) is based on the different melting behaviours of double-stranded DNAs of different G + C contents during denaturing gradient electrophoresis (Muyze et al. 1993). In this approach, a small part (200–400 bp) of the 16S rRNA gene is amplified with general primers, and the amplicons obtained are separated on the basis of their denaturation capacities in a salt gradient according to the %G + C of each fragment. (When fragments are separated in a temperature gradient, it is referred to as Temperature Gradient Gel Electrophoresis, or TGGE.) In order to avoid complete denaturation, a GC clamp is attached to the amplicons prior to their separation. The banding pattern obtained can be used in a direct comparative analysis in order to follow changes of the dominant bacterial flora – for example, as a result of alterations to external parameters such as temperature or other stressors. Bacterial guilds that are present in less than 1–2% of the total population will most probably not be picked up by this kind of approach. Further analysis may address the identities of the individual bands; they can be sequenced after being cut from the gel, taking care to avoid contamination by adjacent bands. Nonetheless, the identification is biased because only a small part of the 16S rRNA gene has been amplified; also, because different alleles of a single genome may result in different bands, it means that the assumption that one band corresponds to one species is misleading in many cases. The DDGE approach can be refined by selective amplification, using specific primers that match well-defined phylogenetic groups in given ecosystems, such as lactic acid bacteria in beer or dairy products.

In the case of terminal restriction fragment length polymorphism (T-RFLP), separation is based on terminal fragments of various lengths that depend upon the different cutting sites of the endonuclease employed (e.g., RFLP and ARDRA) (Liu et al. 1997).

3.3.3.2 Cloning of 16S rRNA Gene Fragments into *Escherichia coli*

DNA extracts can be used as sources for cloning of 16S rRNA fragments into *E. coli*. Commercial kits for this purpose are easy to use and the 16S rRNA gene-containing clones obtained must then be sequenced using the established Sanger method. Although this approach is theoretically possible at the bench, it is time-consuming if one wants to reach a statistically supported view on global diversity in the sample, and this approach has therefore been overtaken by faster and more cost-effective sequencing methods.

3.3.3.3 Direct Sequencing

Recently, DNA sequencing has been redesigned completely by the so-called pyrosequencing (e.g., 454 sequencing). In this approach a hundred thousand clones or more are created and sequenced in such a way that partial (400 bp) 16S rRNA gene sequences are obtained after an automated assembling and annotating process. Analysis and interpretation of the data is the most time-consuming part. Despite this enormous quantitative improvement, the above-mentioned demands of steps such as cell lysis, DNA extraction and PCR at least partly remain. Furthermore, knowledge of the diversity of the microbiota in an ecosystem does not necessarily provide information on the functionality expressed by its members.

3.3.4 Methods that Reveal the Functionality of the Bacterial Community

3.3.4.1 Micro-array Approach

Once we know "Who is out there" in an ecosystem we can proceed to the question: "Who is doing what?" This of course reflects the functionalities that are expressed by the constituent organisms. While the genome can be regarded as "the hardware", the expression of it can be regarded as "the software". The first step in expression is the transcription of the genome into mRNA that varies partly with changing of the external conditions in the ecosystem. The first part of the procedure is the extraction of mRNA, and this needs special precautions because of its short half-life time and the presence of RNases that will destroy it. Immediate and rapid freezing of the samples in liquid nitrogen is a prerequisite and must be accompanied by pre-treatment of the solutions to be used with, for example, DEPC (diethylpyrocarbonate) as an RNase inhibitor, while the baking of all glassware and keeping samples on ice during all of the preparation steps are essential. For more details on mRNA extraction methods, refer Borneman and Triplett (1997) and Bürgmann et al. (2003). RNA is then amplified with reversed transcriptase into cDNA which can then be used on the micro-array.

Ideally, a micro-array should be constructed on the basis of the functional genes of (all) the organisms that were discovered in the original soil sample by the earlier diversity study. The information on the functional genes for the organisms detected by the diversity study (or of very close relatives) should be retrieved from the whole-genome sequences available in public databases. Because of the present insufficiency of genomic information, an overall examination of the overall functionality of a bacterial community is generally not yet possible, but some more narrowly focused examinations have been made, for example, the diversity of the methanotrophs (Bodrossy et al. 2003; Stralis-Pavese et al. 2004).

References

Altemüller HJ, Vliet-Lanoe B (1990) Soil thin section fluorescence microscopy. In: Douglas LA (ed) Soil micromorphology: a basic and applied science. Elsevier, Amsterdam, pp 565–579

Bertrand H, Poly F, Van VT, Lombard N, Nalin R, Vogel TM, Simonet P (2005) High molecular weight DNA recovery from soils prerequisite for biotechnological metagenomic library construction. J Microb Methods 62:1–11

Bodrossy L, Stralis-Pavese N, Murrell JC, Radajewski S, Weilharter A, Sessitsch A (2003) Development and validation of a diagnostic microbial microarray for methanotrophs. Environ Microbiol 5:566–582

Borneman J, Triplett EW (1997) Rapid and direct method for extraction of RNA from soil. Soil Biochem 29:1621–1624

Bürgmann H, Widmer F, Sigler WV, Zeyer J (2003) mRNA extraction and reverse transcription-PCR protocol for detection of *nifH* gene expression by *Azotobacter vinelandii* in soil. Appl Environ Microbiol 69:1928–1935

Danilatos GD (1997) Environmental scanning electron microscopy. In: Gai PL (ed) In-situ microscopy in materials research. Kluwer Academic, Dordrecht, pp 14–44

DeLeo PC, Baveye P (1996) Enumeration and biomass estimation of bacteria in aquifer microcosm studies by flow cytometry. Appl Environ Microbiol 62:4580–4586

Guazzaroni M-E, Golyshin PN, Ferrer M (2010) Analysis of complex microbial communities through metagenomic survey. In: Marco D (ed) Metagenomics: theory, methods and applications. Caister Academic, Norfolk, pp 55–77

Heylen K (2007) Study of the genetic basis of denitrification in pure culture denitrifiers isolated from activated sludge and soil. PhD Thesis, Laboratory of Microbiology (LM-UGent), Ghent University, Belgium

Jaeger CH, Lindow SE, Miller W, Clark W, Firestone MK (1999) Mapping of sugar and amino acid availabililty in soil around roots with bacterial sensors of sucrose and tryptophan. Appl Environ Microbiol 65:2685–2690

Janssen PH, Yates PS, Grinton BE, Taylor PM, Sait M (2002) Improved culturablility of soil bacteria and isolation in pure culture of novel members of the divisions Acidobacteria, Actinobacteria, Proteobacteria and Verrucomicrobia. Appl Environ Microbiol 68:2391–2396

Leadbetter JR (2003) Cultivation of recalcitrant microbes: cells are alive, well and revealing their secrets in the 21st century laboratory. Curr Opin Microbiol 6:274–281

Li Y, Warren AD, Tuovinen OH (2004) Fluorescence microscopy for visualization of soil microorganisms – a review. Biol Fertil Soils 39:301–311

Lindahl V (1996) Improved soil dispersion procedures for total bacteria counts, extraction of indigenous bacteria and cell survival. J Microbiol Methods 25:279–286

Liu WT, Marsh TL, Cheng H, Formey LJ (1997) Characterization of microbial diversity by determining thermal restriction fragment length polymorphisms of genes encoding 16S rRNA. Appl Environ Microbiol 63:4516–4522

Mandic-Mulec I, Prosser JI (2011) Diversity of endospore-forming bacteria in soil: characterization and driving mechanisms. In: Logan NA (ed) Endospore forming soil bacteria (Soil Biology 27). Springer, Heidelberg, doi: 10.1007/978-3-642-19577-2

Meikle A, Amin-Hanjani S, Glover LA, Killham K, Prosser JI (1995) Matric potential and the survival and acitivity of a *Pseudomonas fluorescens* inoculum in soil. Soil Biol Biochem 27:881–892

Muyze G, De Waal EC, Uitterlinden AG (1993) Profiling of complex microbial populations by denaturation gradient gel electrophoresis analysis of polymerase chain reaction-amplified genes coding for 16S rRNA. Appl Environ Microbiol 59:695–700

Nichols D (2007) Cultivation gives context to the microbial ecologist. FEMS Microb Ecol 60:351–357

Poté J, Bravo AG, Mavingui P, Ariztegui D, Wildi W (2010) Evaluation of quantitative recovery of bacterial cells and DNA from different lake sediments by Nycodenz density gradient centrifugation. Ecol Indic 10:234–240

Pratscher J, Stichtemot C, Fichtl K, Schleifer K-H, Braker G (2009) Application of recognition of individual genes fluorescence in situ hybridization (RING-FISH) to detect nitrite reductase genes (nirK) of denitrifiers in pure cultures and environmental samples. Appl Environ Microbiol 75:802–810

Santaella C, Schue M, Berge O, Heulin T, Achouak W (2008) The exopolysacharide of *Rhizobium* sp. YAS34 is not necessary for biofilm formation of *Arabidopsis thaliana* and *Brassica napus* roots but contributes to colonization. Environ Microbiol 10:2150–2163

Setlow P (2003) Spore germination. Curr Opin Microbiol 6:550–556

Stralis-Pavese N, Sessitsch A, Weilharter A, Reichenauer T, Riesing J, Csontos J, Murrell JC, Bodrossy L (2004) Optimisation of diagnostic microarray for application in analysing landfill methanotroph communities under different plant covers. Environ Microbiol 6:347–363

Timmusk S, Grantcharov N, Wagner EGH (2005) *Paenibacillus polymyxa* invades plant roots and forms biofilms. Appl Environ Microbiol 71:7292–7300

Tringe SG, Rubin EM (2005) Metagenomics: DNA sequencing of environmental samples. Nat Rev Genet 6:805–814

Watt M, Hugenholz P, White R, Vinall K (2006) Numbers and locations of native bacteria on field-grown wheat roots quantified by fluorescence in situ hybridization (FISH). Environ Microbiol 8:871–884

Wolgelius RA, Morris PM, Kertesz MA, Chardon E, Stark AIR, Warren M, Brydie JR (2007) Mineral surface reactivity and mass transfer in environmental mineralogy. Eur J Mineral 19:297–307

Wu CH, Hwang Y-C, Lee W, Mulchandani TK, Yates MV, Chen W (2008) Detection of recombinant *Pseudomonas putida* in the wheat rhizosphere by fluorescence in situ hybridization targeting mRNA and rRNA. Appl Microbiol Biotechnol 79:511–518

Chapter 4
Exploring Diversity of Cultivable Aerobic Endospore-forming Bacteria: From Pasteurization to Procedures Without Heat-Shock Selection

O. Berge, P. Mavingui, and T. Heulin

4.1 Introduction

The genus *Bacillus sensu lato* represents one of the most diverse groups of bacteria, and includes aerobic and facultatively anaerobic, rod-shaped, Gram-positive spore-forming bacteria. Recently, 16S rRNA gene sequence analysis has allowed the

O. Berge (✉)
CEA, IBEB, Lab Ecol Microb Rhizosphere & Environ Extrem (LEMiRE), Saint-Paul-lez-Durance 13108, France
and
CNRS, UMR 6191, Saint-Paul-lez-Durance 13108, France
and
Aix-Marseille Université, Saint-Paul-lez-Durance 13108, France
and
Centre de Recherche d'Avignon, INRA, UR407 Pathologie Végétale, Domaine Saint Maurice BP 94, 84143 Montfavet cedex, France
e-mail: odile.berge@avignon.inra.fr

P. Mavingui
CNRS, UMR 5557, Laboratoire d'Ecologie Microbienne, Université de Lyon, 69000 Lyon, France
and
Université Lyon 1, 69622 Villeurbanne, France
e-mail: patrick.mavingui@univ-lyon1.fr

T. Heulin
CEA, IBEB, Lab Ecol Microb Rhizosphere & Environ Extrem (LEMiRE), Saint-Paul-lez-Durance 13108, France
and
CNRS, UMR 6191, Saint-Paul-lez-Durance 13108, France
and
Aix-Marseille Université, Saint-Paul-lez-Durance 13108, France
e-mail: thierry.heulin@cea.fr

N.A. Logan and P. De Vos (eds.), *Endospore-forming Soil Bacteria*, Soil Biology 27, DOI 10.1007/978-3-642-19577-8_4, recognition of many distinct genera of aerobic endospore-formers, including: *Bacillus*, *Alicyclobacillus*, *Paenibacillus*, *Brevibacillus*, *Aneurinibacillus*, *Virgibacillus* and *Gracilibacillus*. In this chapter, the term "aerobic endospore-forming bacteria" (AEFB) is used to indicate all the members of these genera. Searching for "*Bacillus* OR spore-forming bacteria" in the NCBI database yielded 62,000 hits at the time of writing.

AEFB are ubiquitous, and they inhabit diverse and contrasted niches in agro-ecosystems, including extreme environments. The wide distribution of AEFB is linked to their wide-ranging nutritional requirements, growth conditions and metabolic diversities. For instance, the strains belonging to the *Bacillus cereus* group are among the most abundant cultivable soil bacteria worldwide and they display a wide diversity (von Stetten et al. 1999). In cultivated soils, AEFB play important roles in fertility and plant nutrition (Francis et al. 2010). This is well illustrated by the *Paenibacillus* species that were reported to possess nitrogenase activity (Achouak et al. 1999) and that have been shown to be involved in the soil nitrogen cycle (Seldin et al. 1983; Gouzou et al. 1995). AEFB species often produce antibiotics, or secrete proteins and enzymes, that make them candidates for biotechnological applications (Lal and Tabacchioni 2009). They have also been studied and commercialized as probiotics for animal and human nutrition and health (Fuller 1992; Sanders et al. 2003; Patel et al. 2009). In cultivated soils, they are also able to rapidly and completely biotransform herbicides (Batisson et al. 2009).

To understand better the roles, dynamics and community structures of AEFB in natural environments such as soil, high-quality methods for their detection and analysis are needed. Many species from soils have been already described, but the taxonomic and functional diversity of this group is largely unknown and has to be explored further. The study of AEFB diversity has been a classical Pasteurian or conventional approach until recently. It is based on bacterial cultivation giving access to isolates. The development of analyses of diversity that are independent of cultivation, as presented in De Vos (2011), have allowed access to non-cultivable microorganisms. These techniques are promising: for example, new species of *Paenibacillus* were revealed by a PCR-DGGE approach using genus-specific primers (da Silva et al. 2003; da Mota et al. 2005; Coelho et al. 2007) and are complementary to isolation methods. However, the isolation of strains allows us to characterize them extensively, and eventually to exploit them for industrial or agronomical purposes, making this approach very useful still.

In the first part of this chapter, we report on the culture-dependent approaches that are mainly based on spore selection and provide data on their helpfulness in discovering the large diversity of AEFB. As with any cultivation method, the spore selection strategy has some limits that could skew the study of AEFB population diversities in soils. Alternative methods that allow isolation of both spores and vegetative cells are therefore presented in the second part of the chapter.

4.2 Isolation of Aerobic Endospore-forming Bacteria from Soils by Spore Selection

In almost all environments, and especially in soils, most AEFB species are not among the dominant microorganisms. In temperate soils, *Paenibacillus polymyxa* represents about 0.1% of the total cultivable bacteria (Mavingui et al. 1990, 1992). As for most bacterial species in nature, their isolation and study is severely hampered by the lack of efficient methods to select for them specifically from environmental samples. Strategies such as enrichment techniques have been developed to overcome this problem. Enrichment can be a very powerful tool available for isolation, but it is dependent on the ratio of bacteria to be selected to bacteria to be counter-selected. It does not allow direct bacterial enumeration. Moreover, enrichment steps could introduce biases in the diversity of the population by favouring some bacterial genotypes; this is a general problem in microbiology.

The most widely used procedure today for isolation of AEFB populations still relies on spore selection (Fig. 4.1). The rationale of the strategy is based on the ability of AEFB to produce spores that are resistant to many physical and chemical agents whereas their vegetative counterparts are not. For *Bacillus subtilis* it has been shown that this tolerance is due to the presence of a resistant spore coat (Shapiro and Setlow 2006) and also to spore-specific DNA-binding proteins (Setlow et al. 2000). Applying treatments such as heat, irradiation and various chemicals facilitates selection of spores from among a mixture of cells. After recovery, germination of the bacterial spores can be obtained by incubation in appropriate growth conditions. An overview and discussion of key steps in the protocols for spore isolation follow.

4.2.1 Soil Sample Preparation

Usually, spores are recovered from a soil suspension. Treatments are then necessary to ensure good extraction and dispersion of spores from soil aggregates, in order to allow subsequent growth as separate colonies on a solid medium, as is done for all bacteria being isolated from a complex matrix such as soil. Nicholson and Law (1999) extracted spores from soil samples using the chelating resin Chelex 100 followed by a NaBr density gradient centrifugation. They showed that spores preferentially adhere to large soil aggregates. Extraction of spores from soil requires steps for dissociation from other soil components. To that end, soil can be suspended in water or saline buffer, with an agitation in the presence of beads and detergents or with sterile sand. The procedures described in the literature for diluting and dispersing the soil vary; for instance, in one case 0.1 g of soil was suspended in 950 μl of phosphate buffered saline and submitted to vertical shaking for 60 min at room temperature (Tzeneva et al. 2004); in another case 10 g soil was suspended in 90 ml of dispersion and stabilizing buffer containing 0.2% sodium

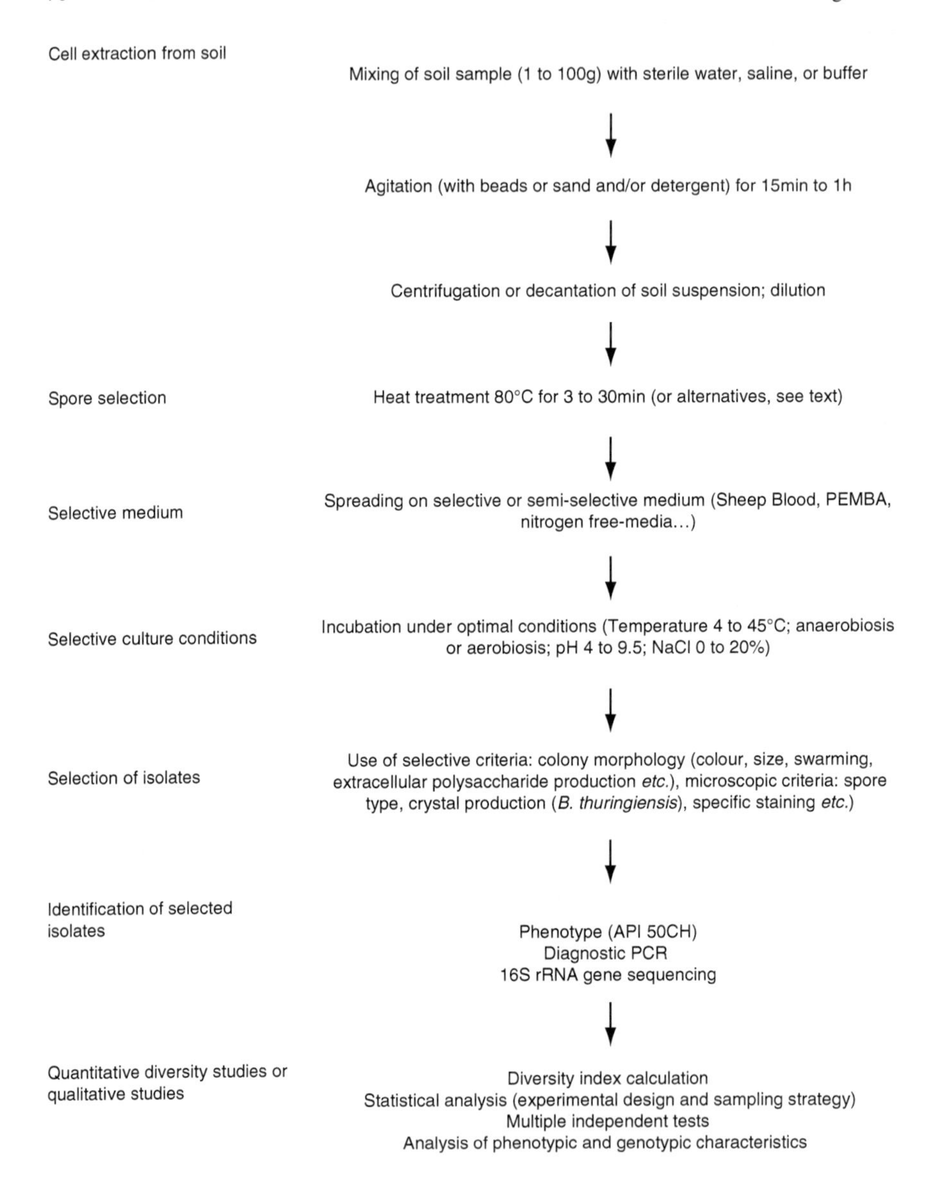

Fig. 4.1 General protocol for isolation of aerobic endospore-forming populations by spore selection

polyphosphate and 0.05% Tween 80, then the suspension was shaken at 200 rpm for 10 min with glass beads (von Stetten et al. 1999); finally, Collier et al. (2005) proposed placing the soil sample in a 30 ml universal tube with 5 ml of 0.85% sterile saline and 0.5 g sterile sand and then vortex mixing for 30 s.

4.2.2 Heat Treatment

The most frequently used technique for spore selection is heat treatment or pasteurization. Suspensions of soil samples are heated for given times at temperatures that kill vegetative cells. Spores from the heated sample are then incubated in optimal conditions to encourage germination (Fig. 4.1). This technique is powerful because it is highly selective. It eliminates all non-spore-forming microorganisms from any sample and is very efficient for obtaining bacterial populations from spores. Emberger (1970) extensively studied the physico-chemical and physical conditions needed to isolate *Bacillus* species efficiently. The most suitable treatment was heating, and the recommended temperatures ranged from 65 to 70°C for 15 min. Claus and Berkeley (1986) proposed heating samples at 80°C for 10 min, and this is a procedure widely reported in the literature. However, heat treatment may have to be adapted for some types of spore formers, because the endospores of some bacterial strains are not as heat tolerant as others. The incubation time used may vary from 3 (Jara et al. 2006) to 30 min. It is recommended to start heating at a relatively low temperature (e.g. 70 or 75°C) and progressively increase it until an optimum temperature is achieved.

To isolate AEFB, some authors have taken advantage of the tolerance of spores to other stresses. Koransky et al. (1978) concluded that treatment with 50% ethanol for 1 h is an effective technique for selectively isolating spore-forming bacteria from a mixed culture, as effective as heat treatment of 80°C for 15 min. Patel et al. (2009) confirmed this finding by isolating *Bacillus* strains from food wastes both by heating at 65°C for 45 min and using incubation with ethanol. Soil drying can be also used as a selective pressure, as endospores produced by AEFB have a remarkable tolerance of desiccation and may survive for long periods in such conditions. Drying treatment is probably gentler than heating or ethanol incubation. To study the diversity of *Bacillus* genotypes in soil samples from a reserve in Egypt, Eman et al. (2006) killed vegetative cells by adding chloroform (1% v/v); however, this technique has not been validated.

An interesting selection process, differing from classical heat treatment, was developed by Travers et al. (1987) for isolating *Bacillus thuringiensis*. In this process, named “acetate selection”, germination of *B. thuringiensis* spores was selectively inhibited by sodium acetate (0.25 M), while most of the undesired spore formers were allowed to germinate. Next, all of the non-sporulated bacteria were eliminated by heat treatment at 80°C for 3 min. The surviving spores were then plated and allowed to germinate on a rich agar medium. Even if some other *Bacillus* species are also selected by this method, such as *Bacillus sphaericus* and *B. cereus*, this technique is frequently used to study the diversity of *B. thuringiensis* worldwide (Martin and Travers 1989; Helgason et al. 1998; Hongyu et al. 2000; Uribe et al. 2003). A modification of the method promotes greater spore yields by stimulating sporulation before applying the stress shock. For instance, some authors suggested suspending one gram of soil in 50 ml of a sporulation medium, then incubating at 37°C under shaking for 48 h before killing vegetative cells and then heat treating (Eman et al. 2006), whereas others proposed incubating the soil suspensions in nutrient broth at different temperatures for 5 days to allow a better maturation of spores (Walker et al. 1998).

As opposed to using heat treatment to select for spores, Bizzarri and Bishop (2007) were interested in specifically recovering vegetative cells of *B. thuringiensis*. Both spores and vegetative cells of AEFB will produce colonies if environmental samples are plated onto conventional growth media; so, they selected vegetative forms on a medium containing the germination inhibitor TAME (Nα-*p*-tosyl-L-arginine methyl ester) and showed that these bacteria exist in active forms on the phylloplane.

4.2.3 Selective Growth Media

After selection of spores by heat and complementary treatments, soil samples are plated onto growth media to obtain isolates and to study the diversity of the corresponding populations. Depending on the targeted bacterial species, different selective strategies will be chosen.

To monitor all AEFB, rich media such as nutrient agar are often used after heat treatment. Many different AEFB are able to grow on such media and they allow the study of their diversity. In this way, Walker et al. (1998) obtained diverse AEFB isolates to screen for their antifungal activity against *Botrytis cinerea* and *Pythium* species. Garabito et al. (1998) used such a medium containing yeast extract, protease-peptone, glucose and soil extract to isolate AEFB from different saline soils. Their isolates were extremely halotolerant and their diversity indicated that they may have had important ecological roles in these hypersaline environments.

When targeting a particular species, or a group of closely related species, more specific media and/or selective techniques are necessary. This is especially true when the species sought is in relatively low abundance; this was the case with *B. thuringiensis* (Travers et al. 1987), which represented 0.5–0.005% among all *Bacillus* strains isolated from the soil samples tested. Some growth media have been developed to be selective or semi-selective for one species. This selectivity is based on known characteristics of the given species such as the use of a special substrate or the resistance to an antibiotic. These media, combined with specific physico-chemical conditions such as temperature, pH, NaCl concentration or aeration, may lead to efficient isolation of AEFB strains. The most specific factors are the growth conditions and the most efficient outcome will be selection for the species. For example, Stefanic and Mandic-Mulec (2009) studying distribution of *B. subtilis* at micro-scale in soil, used the properties of strains belonging to this species to be catalase positive, to convert pyruvate to acetoin, to hydrolyse starch and to be unable to grow anaerobically on agar. Thus, after heat treatment (80°C, 15 min), soil suspensions were spread on tryptose blood agar and the emerging colonies were screened for these metabolic characters.

The strategy is different for AEFB groups that are abundant in a particular environment. For instance, *B. cereus*-like organisms are known to be abundant in cultivable soils. This group of bacteria comprises, besides *B. cereus sensu stricto* that is frequently involved in food spoilage and in food poisoning, the insect pathogen *B. thuringiensis*, the human pathogen *B. anthracis*, *B. mycoides* that produces rhizoid colonies, *B. pseudomycoides* and the psychrotolerant *B. weihenstephanensis*. *B. cereus* and *B. thuringiensis* are closely related and genomic studies have concluded that they should be merged into a single species. However, the name *B. thuringiensis* is retained for those strains that produce crystalline parasporal inclusions. Several selective media for the detection and enumeration of *B. cereus* in food have been developed (reviewed by van Netten and Kramer 1992). Among them, the PEMBA medium (Polymyxin pyruvate Egg yolk Mannitol Bromothymol blue Agar) developed by Holbrook and Andersson (1980) is commonly used to study the diversity of *B. cereus* group members in soil samples. von Stetten et al. (1999) employed PEMBA after heat treatment (80°C for 10 min) of soil suspensions and showed that the geographic distribution of psychrotolerant and mesophilic isolates of *B. cereus sensu lato* correlated significantly with the annual average temperature. This protocol was also used to study the diversity of *B. cereus* and *B. megaterium* in honey (Lopez and Alippi 2007, 2009). In another study (Guinebretière et al. 2003), spores of *B. cereus* were grown on MYPA (Mannitol egg Yolk Polymyxin Agar; Lancette and Harmon 1980) after heat treatment (80°C for 15 min) of soil and plant samples. These authors found high concentrations of *B. cereus* spores in the soil in which zucchini were grown and showed that they contributed to the contamination of industrially cooked food.

In the case of extremophilic AEFB, the growth conditions to which they are adapted, such as at high temperature and/or high pH, are selective enough to allow screening for isolates of these species (Ward and Cockson 1972; Smith et al. 2009). Logan et al. (2000) used a combination of temperature (15–65°C) and pH (4.5–6.5) conditions as unique selective pressures during growth on a semi-selective medium after heat treatment (80°C, 10 min) of soil suspensions to isolate a new moderately thermophilic species, *Bacillus fumarioli*, from Antarctic geothermal sites. Using the same method, Rodríguez-Díaz et al. (2005) examined 12 samples of soils from Alexander Island in Antarctica, for the presence of AEFB. A novel species, harbouring the *nifH* gene and able to grow at 4°C, was isolated and named *Paenibacillus wynnii*. The species *Alicyclobacillus acidoterrestris* and *Alicyclobacillus acidocaldarius* are both thermo-acidophilic AEFB that cause spoilage of acid food products. To study the role of *Alicyclobacillus* species in the spoilage of fruit juice, analysis of their diversity in soil was performed (Groenewald et al. 2009); after heat treatment (80°C for 10 min), soil samples were incubated in yeast-starch broth, adjusted to a final pH of 4, at 45°C for 24 h as an enrichment step, then dilutions were spread on plates of the same medium solidified with agar, followed by incubation at 45°C for 72–120 h. Spore-forming rods were then selected and tested by PCR amplification of a DNA fragment lying between coordinates 1254 and 1388 of the 16S rRNA gene. The majority of isolates belonged to the species *A. acidoterrestris*, but *A. acidocaldarius* was also isolated. Using a similar approach, facultative alkalitolerant and halotolerant AEFB were isolated from soils surrounding Borax Lake, Oregon (Smith et al. 2009); in this case, the medium was supplemented with 20% NaCl and the pH adjusted to 9.5. In addition, heat treatment (80°C for 15 min) was used instead of aerobic enrichment culture. The collection of strains obtained represented a rich source of alkalitolerant and halotolerant AEFB that could be of potential ecological and/or commercial interest.

The nitrogen-fixing ability of some AEFB could be exploited as a selection criterion. Some nitrogen-fixing *Paenibacillus* have been isolated from soil, rhizospheres and roots of crops, mainly cereals such as wheat (Heulin et al. 1994; Beneduzi et al. 2008) and maize (Berge et al. 1991, 2002; Seldin et al. 1998). Their various properties in relation to plant growth such as nitrogen fixation, production of auxin-like molecules (Lebuhn et al. 1997) as well as production of chitinase and antifungal molecules (Mavingui and Heulin 1994) made the analysis of their intra-specific diversity of great interest (Mavingui et al. 1992; von der Weid et al. 2000; da Mota et al. 2002; Beneduzi et al. 2008). Rennie (1981) and Seldin et al. (1983) developed an isolation procedure based on pasteurization and growth on N-free media in anaerobic jars. After heat treatment (80°C for 10 min), appropriate soil dilutions were plated on modified Line's thiamine–biotin medium and incubated for 7 days at 28°C anaerobically. Individual colonies were tested for aerobic growth and nitrogen-fixing capacity by measuring acetylene reduction.

4.2.4 *Other Selective Criteria*

Since the semi-selective media used for the isolation of a specific AEFB do not eliminate all other co-occurring species found in soil, it is necessary to have additional techniques to obtain isolates belonging to the targeted species. To that purpose, standard bacteriological criteria can be used. These include colony morphology (size, shape, margin, colour, opacity and consistency), characteristics of the spore within the cell, the vegetative and sporangial morphologies or the production of crystals observed under a phase-contrast microscope (Logan et al. 2000). da Mota et al. (2002) selected bright yellow, convex, shiny and mucoid colonies growing anaerobically on thiamine–biotin agar as presumptive *P. polymyxa* colonies, whereas Mavingui et al. (1990) used sucrose agar medium to discriminate isolates of *P. polymyxa* strains.

To screen for members of the *B. cereus* group, suspected colonies grown on PEMBA medium were identified by their shapes, colour, rhizoidal growth and the presence of lecithinase haloes, followed by a microscopic examination of bacterial smears by the lipid globule staining procedure, the presence of crystalline inclusions and spores within cells as well as for the size and shape of the vegetative cells (Holbrook and Andersson 1980; Lopez and Alippi 2007). Interestingly, the selected isolates were as confirmed as members of the *B. cereus* group by PCR amplification of 16S rDNA and cold shock protein genes (Francis et al. 1998). *B. thuringiensis* is able to produce an insecticidal crystal protein and this property is largely used to screen colonies of this species after heat treatment (Travers et al. 1987). Many authors have investigated the diversity of *B. thuringiensis* in different ecosystems in order to obtain biological insecticides against local insects, with studies in China (Su et al. 2007; Gao et al. 2008; Hongyu et al. 2000), Colombia (Jara et al. 2006), Iran (Jouzani et al. 2008), New Zealand (Chilcott and Wigley 1993) and Spain (Quesada-Moraga et al. 2004). This very simple criterion is also used to study the genetic structures of *B. cereus*/*B. thuringiensis* populations, since these bacteria are common in nature and known to be potentially pathogenic to both humans and insects (Helgason et al. 1998). More recently, Guinebretière et al. (2008) established seven major phylogenetic groups in the *B. cereus* complex. They recommended that these phylogenetic groups be used instead of current phenotypic species, especially when studying the risk of pathogenicity.

Finally, all isolates must be fully identified prior to their use for diversity analysis. Biochemical tests such as API 50 CHB (bioMérieux) are routinely performed (von Stetten et al. 1999; Seldin et al. 1998; von der Weid et al. 2000; Guemouri-Athmani et al. 2000) as well as rapid diagnostic PCR on isolated colonies. When appropriate markers are available, the latter technique is useful because of the short time required to carry it out and the reliability of the results. PCR amplification was successfully used for *Paenibacillus larvae* (Govan et al. 1999), *Paenibacillus durum* (formerly *Paenibacillus azotofixans*) (Seldin et al. 1998), and *P. polymyxa* (von der Weid et al. 2000). Alternatively, a region of the 16S rRNA gene may be amplified and sequenced (Groenewald et al. 2009; Smith et al. 2009).

4.3 Isolation of Aerobic Endospore-Formers Without Spore Selection

Studies of population diversities of AEFB originating from air-dried, ethanol- or heat-treated samples, hence from endospores that are metabolically inactive, have their pitfalls (Claus and Berkeley 1986). These studies that rely upon spore selection may adversely affect the size and diversities of the populations recovered, owing to the long-term survival of spores under adverse conditions. On the other hand, it may be assumed that high numbers of endospores found in a given sample indicate that earlier activity of vegetative AEFB cells has occurred in the environment from which the sample was taken. However, in a particular habitat and at a given time, the spore diversity does not necessarily represent the diversity of active bacteria. This is particularly true in soil compartments where these AEFB are active, as in rhizospheres or litters. These soil environments are favourable to vegetative growth because of the presence of carbon and energy sources such as plant exudates and fresh organic matter. Alternative isolation techniques to pasteurization were then developed in order to isolate representatives of both spores and vegetative cells.

The simplest situation is the one where AEFB are among the most abundant species. Their isolation is relatively easier and could be achieved by simple spread-plating on a selective medium (Tzeneva et al. 2004). For example, a selective medium containing polymyxin B sulphate (5 μg/ml) and penicillin G (4 μg/ml) has been employed as the basis for AEFB selection (Saleh et al. 1969). Collier et al. (2005) used a commercially available selective agar (based on PEMBA) to remove the need for heat treatment. This efficient selective medium for the isolation and tentative identification of *B. cereus sensu lato* allowed the authors to determine the prevalence of toxin gene diversity among *B. thuringiensis*/*B. cereus* in the phyllospheres of plants and in soil. This medium has also been used for the detection of both spores and vegetative cells of the *B. cereus* group in contaminated foods (Rosenquist et al. 2005).

However, most soil AEFB are not among the most dominant soil bacterial species, and selective or semi-selective media are not available for their isolation. Due to the lack of a selective medium for isolating *Bacillus* (now *Paenibacillus*) *polymyxa*, an alternative approach using serological properties was used. Van Vuurde (1987) showed that antibodies coated onto a suitable solid phase enable selective trapping of homologous phytopathogenic bacteria as well as immunologically related bacteria by immunoaffinity. In 1990, the application of the immunoisolation to soil *Bacillus* species was performed for the first time (Mavingui et al. 1990); the authors used polyclonal antibodies (PcAbs) raised in rabbit against whole-cell antigens of one *P. polymyxa* strain. All *P. polymyxa* reference strains that were tested cross-reacted equally well with these PcAbs. In contrast, *Bacillus* (now *Paenibacillus*) *macerans* ATCC 8244^T and *Bacillus circulans* NCIB 9374^T did not cross-react. This specificity of PcAbs was then exploited to trap *P. polymyxa* strains from soil. Briefly, microtitre plates were coated with these partially purified PcAbs and appropriate dilutions of soil in phosphate buffer were added into the wells. After incubation at 4°C, the unbound bacteria were washed out twice with phosphate buffer (pH 7.2), then the bound bacteria were desorbed in 0.1 M KCl (pH 5.5) with the help of manual mechanical action of the pipette tip to scrape the bottom of the well. This immunocapture was found to be semi-selective, as other bacterial species could be detected and removed during the next step (selection on sucrose medium). Finally, the desorbed bacteria were plated on sucrose (4%) agar medium. The typical production of exopolysaccharide (levan) by *P. polymyxa* colonies made their recognition and isolation easy (Mavingui et al. 1990). A general protocol for this technique is presented in Fig. 4.2. In a study of co-adaptation of *P. polymyxa* populations and wheat roots using this immunotrapping technique, isolated cells within this species represented 0.1–1% of the cultivable bacteria (Mavingui et al. 1992). The authors isolated 130 strains of *P. polymyxa* from non-rhizosphere soil, rhizosphere soil and the rhizoplane of wheat, covering a large diversity within the species. Genetic and phenotypic analyses of these strains showed that roots of wheat selected a subpopulation of *P. polymyxa*, probably from the natural soil populations (Mavingui et al. 1992). This finding suggested that, over the time frame of this study, the diversity of the soil bacterial population was lowered owing to selection by the plant roots.

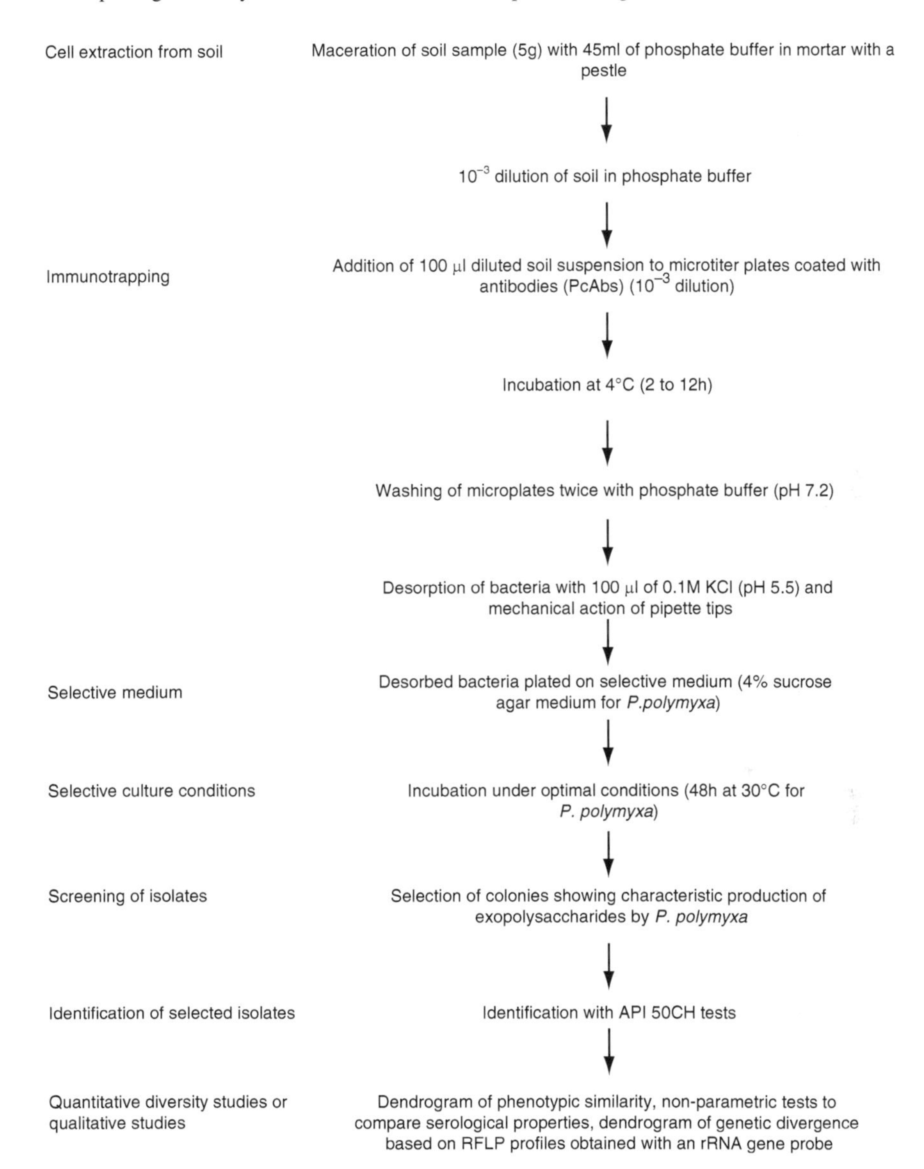

Fig. 4.2 Example of a protocol of immunotrapping for isolation of *Paenibacillus polymyxa* populations

Actually, the study of AEFB diversity is mainly a descriptive endeavour, but the motivations behind these studies arose from specific hypotheses about the nature of biodiversity and its impact on ecological processes. Testing of such hypotheses involves careful general experimental design, sampling procedures, statistical analyses and multiple independent assays, as recommended by Morris et al. (2002).

For such studies of diversity, it is necessary to compare a large number of strains. Immunotrapping appears to be a specific enough and rapid enough method to overcome this problem. It is relatively simple and easy to perform, and microplate numbers could be increased to replicate samples and enhance the number of isolates. Using the immunotrapping strategy, Guemouri-Athmani et al. (2000) studied the diversity of *P. polymyxa* in wheat rhizosphere grown in different Algerian soils that had distinct wheat-cultivation histories. One hundred and eleven strains were immunotrapped in total, and statistical comparison of their diversity was thus possible. Even in a sandy soil that had recently been cultivated with wheat and which contained low numbers of *P. polymyxa,* some strains were successfully isolated.

A variant of this technique is the use of immunomagnetic separation to isolate bacteria from the environment. Antibody-coated magnetizable beads are used to trap specific cells from heterogeneous suspensions, and subsequent separation of the target cells from the suspension is carried out with a high-strength magnetic field. Güven and Mutlu (2000) developed a high-specificity immunomagnetic technique for the detection and selective isolation of *Pseudomonas syringae* pv. phaseolicola, a pathogen of beans. In another study, over 300 strains of *Azospirillum* were isolated from the rhizosphere of wheat by immunomagnetic separation (Han and New 1998). The authors estimated between 30 and 50% recovery of *Azospirillum*, which is a 160-fold increase in comparison with classical methods. Immunomagnetic separation was little affected by the presence of plant extracts or soil particles and therefore could be used to test large numbers of samples in combination with PCR. So it could represent a very useful technique to trap AEFB in soil and rhizosphere samples.

More recently, novel techniques that combine molecular techniques and cultivation to isolate AEFB without spore selection have been developed (see De Vos 2011). For example, new *Bacillus* species found to be abundant in a Dutch grassland soil by a culture-independent approach have been further isolated from several of the soils, after plating soil suspensions on different media and incubating at different temperatures. High numbers of colonies (>1,000) were screened by multiplex PCR with specific primers. Sequencing of 16S rRNA genes ascertained the identity of isolated colonies (Felske et al. 2003, 2004; Tzeneva et al. 2004). This approach is very interesting because it makes possible the isolation of new lineages of AEFB that were previously detected as "non-cultivable", and gives ideas for the development of new media to cultivate them. However, the technique has to be improved to enhance the number of isolates per species, as this is necessary to measure the diversity.

4.4 Conclusion

Most AEFB form populations with low abundances in soil, and the effective and accurate description of their diversity needs adapted techniques. Most AEFB diversity studies still rely on traditional isolation following spore selection. This procedure is simple and easy to use and yields most of the sporulated cells from different environments, but it must be complemented by other selective techniques to study the less abundant populations of spore formers. These approaches may give views of diversity, but it is through the filter of sporulation. To avoid this bias, and have an access to AEFB strains arising from both spores and vegetative cells, there are some techniques like immunocapture and/or selective media combined with molecular techniques, such as colony PCR and/or miniaturization of screening on microplates. They may give a better view of aerobic endospore-former diversity and allow the study of strains tailored to the aim of the research. It is important to evaluate the real benefits of these alternative techniques, regarding the intended objectives of the research, and to compare them with spore selection. All together, these approaches will definitely help the clarification and quantification of AEFB diversity in soils and other environments. The developments of new technologies, such as high-throughput sequencing, give new perspectives and will improve AEFB studies in the near future.

References

Achouak W, Normand P, Heulin T (1999) Comparative phylogeny of *rrs* and *nifH* genes in the *Bacillaceae*. Int J Syst Evol Microbiol 49:961–967

Batisson I, Crouzet O, Besse-Hoggan P, Sancelme M, Mangot JF, Mallet C, Bohatier J (2009) Isolation and characterization of mesotrione-degrading *Bacillus* sp. from soil. Environ Pollut 157:1195–1201

Beneduzi A, Peres D, da Costa PB, Bodanese Zanettini MH, Passaglia LM (2008) Genetic and phenotypic diversity of plant-growth-promoting bacilli isolated from wheat fields in southern Brazil. Res Microbiol 159:244–250

Berge O, Heulin T, Balandreau J (1991) Diversity of diazotroph populations in the rhizosphere of maize (*zea mays* L.) growing on different French soils. Biol Fertil Soils 11:210–215

Berge O, Guinebretiere MH, Achouak W, Normand P, Heulin T (2002) *Paenibacillus graminis* sp. nov. and *Paenibacillus odorifer* sp. nov., isolated from plant roots, soil and food. Int J Syst Evol Microbiol 52:607–616

Bizzarri MF, Bishop AH (2007) Recovery of *Bacillus thuringiensis* in vegetative form from the phylloplane of clover (*Trifolium hybridum*) during a growing season. J Invertebr Pathol 94:38–47

Chilcott CN, Wigley PJ (1993) Isolation and toxicity of *Bacillus thuringiensis* from soil and insect habitats in New Zealand. J Invertebr Pathol 61:244–247

Claus D, Berkeley RCW (1986) Genus *Bacillus* Cohn 1872. In: Sneath PHA, Mair NS, Sharpe ME, Holt JG (eds) Bergey's manual of systematic bacteriology, vol 2. William & Wilkins, Baltimore, pp 1114–1120

Coelho MR, Da Mota FF, Carneiro NP, Marriel IE, Paiva E, Rosado AS, Seldin L (2007) Diversity of *Paenibacillus* spp. in the rhizosphere of four sorghum (*Sorghum bicolor*) cultivars sown with two contrasting levels of nitrogen fertilizer assessed by *rpoB*-based PCR-DGGE and sequencing analysis. J Microbiol Biotechnol 17:753–760

Collier FA, Elliot SL, Ellis RJ (2005) Spatial variation in *Bacillus thuringiensis/cereus* populations within the phyllosphere of broad-leaved dock (*Rumex obtusifolius*) and surrounding habitats. FEMS Microbiol Ecol 54:417–425

da Mota FF, Nobrega A, Marriel IE, Paiva E, Seldin L (2002) Genetic diversity of *Paenibacillus polymyxa* populations isolated from the rhizosphere of four cultivars of maize (*Zea mays*) planted in Cerrado soil. Appl Soil Ecol 20:119–132 da Mota FF, Gomes EA, Paiva E, Seldin L (2005) Assessment of the diversity of *Paenibacillus* species in environmental samples by a novel *rpoB*-based PCR-DGGE method. FEMS Microbiol Ecol 53:317–328

da Silva KR, Salles JF, Seldin L, van Elsas JD (2003) Application of a novel *Paenibacillus*-specific PCR-DGGE method and sequence analysis to assess the diversity of *Paenibacillus* spp. in the maize rhizosphere. J Microbiol Methods 54:213–231

De Vos P (2011) Studying the bacterial diversity of the soil by culture-independent approaches. In: Logan NA (ed) Endospore forming soil bacteria (Soil Biology 27). Springer, Heidelberg, doi:10.1007/978-3-642-19577-3

Eman AHM, Mikiko A, Ghanem KM, Abdel-Fattah YR, Nakagawa Y, El-Helow ER (2006) Diversity of *Bacillus* genotypes in soil samples from El-Omayed biosphere reserve in Egypt. J Cult Collect 5:78–84

Emberger O (1970) Cultivation methods for the detection of aerobic spore-forming bacteria. Zentralbl Bakteriol Parasitenkd Infektionskr Hyg 125:555–565

Felske AD, Heyrman J, Balcaen A, De Vos P (2003) Multiplex PCR screening of soil isolates for novel *Bacillus*-related lineages. J Microbiol Methods 55:447–458

Felske AD, Tzeneva V, Heyrman J, Langeveld MA, Akkermans AD, De Vos P (2004) Isolation and biodiversity of hitherto undescribed soil bacteria related to *Bacillus niacini*. Microb Ecol 48:111–119

Francis KP, Mayer R, von Stetten F, Stewart GSAB, Scherer S (1998) Discrimination of psychrotrophic and mesophilic strains of the *Bacillus cereus* group by PCR targeting of major cold shock protein genes. Appl Environ Microbiol 64:3525–3529

Francis I, Holsters M, Vereecke D (2010) The Gram-positive side of plant-microbe interactions. Environ Microbiol 12:1–12

Fuller R (1992) Probiotics: the scientific basis. Chapman & Hall, London, 392 pp

Gao M, Li R, Dai S, Wu Y, Yi D (2008) Diversity of *Bacillus thuringiensis* strains from soil in China and their pesticidal activities. Biol Control 44:380–388

Garabito MJ, Márquez MC, Ventosa A (1998) Halotolerant *Bacillus* diversity in hypersaline environments. Can J Microbiol 44:95–102

Gouzou L, Cheneby D, Nicolardot B, Heulin T (1995) Dynamics of the diazotroph *Bacillus polymyxa* in the rhizosphere of wheat (*Triticum aestivum* L.) after inoculation and its effect on uptake of ^{15}N-labelled fertilizer. Eur J Agron 4:47–54

Govan VA, Allsopp MH, Davison S (1999) A PCR detection method for rapid identification of *Paenibacillus larvae*. Appl Environ Microbiol 65:2243–2245

Groenewald WH, Gouws PA, Witthuhn RC (2009) Isolation, identification and typification of *Alicyclobacillus acidoterrestris* and *Alicyclobacillus acidocaldarius* strains from orchard soil and the fruit processing environment in South Africa. Food Microbiol 26:71–76

Guemouri-Athmani S, Berge O, Bourrain M, Mavingui P, Thiéry JM, Bhatnagar T, Heulin T (2000) Diversity of *Paenibacillus polymyxa* populations in the rhizosphere of wheat (*Triticum durum*) in Algerian soils. Eur J Soil Biol 36:149–159

Guinebretière MH, Girardin H, Dargaignaratz C, Carlin F, Nguyen-The C (2003) Contamination flows of *Bacillus cereus* and spore-forming aerobic bacteria in a cooked, pasteurized and chilled zucchini puree processing line. Int J Food Microbiol 82:223–232

Guinebretière MH, Thompson FL, Sorokin A, Normand P, Dawyndt P, Ehling-Schulz M, Svensson B, Sanchis V, Nguyen-The C, Heyndrickx M, De Vos P (2008) Ecological diversification in the *Bacillus cereus* Group. Environ Microbiol 10:851–865

Güven K, Mutlu MB (2000) Development of immunomagnetic separation technique for isolation of *Pseudomonas syringae* pv. *phaseolicola*. Folia Microbiol (Praha) 45(4):321–324

Han SO, New PB (1998) Isolation of *Azospirillum* spp from natural soils by immunomagnetic separation. Soil Biol Biochem 30:975–981

Helgason E, Caugant DA, Lecadet MM, Chen Y, Mahillon J, Lovgren A, Hegna I, Kvaloy K, Kolstø AB (1998) Genetic diversity of *Bacillus cereus*/*B. thuringiensis* isolates from natural sources. Curr Microbiol 37:80–87

Heulin T, Berge O, Mavingui P, Gouzou L, Hebbar KP, Balandreau J (1994) *Bacillus polymyxa* and *Rahnella aquatilis*, the dominant N_2-fixing bacteria associated with wheat rhizosphere in French soils. Eur J Soil Biol 30:35–42

Holbrook R, Andersson JM (1980) An improved selective and diagnostic medium for the isolation and enumeration of *Bacillus cereus* in foods. Can J Microbiol 26:753–759

Hongyu Z, Ziniu Y, Wangxi D (2000) Composition and ecological distribution of cry proteins and their genotypes of *Bacillus thuringiensis* isolates from warehouses in China. J Invertebr Pathol 76:191–197

Jara S, Maduell P, Orduz S (2006) Diversity of *Bacillus thuringiensis* strains in the maize and bean phylloplane and their respective soils in Colombia. J Appl Microbiol 101:117–124

Jouzani GS, Abad AP, Seifinejad A, Marzban R, Kariman K, Maleki B (2008) Distribution and diversity of Dipteran-specific *cry* and *cyt* genes in native *Bacillus thuringiensis* strains obtained from different ecosystems of Iran. J Ind Microbiol Biotechnol 35:83–94

Koransky JR, Allen SD, Dowell VR (1978) Use of ethanol for selective isolation of spore forming microorganisms. Appl Environ Microbiol 35:762–765

Lal S, Tabacchioni S (2009) Ecology and biotechnological potential of *Paenibacillus polymyxa*: a minireview. Indian J Microbiol 49:2–10

Lancette GA, Harmon SM (1980) Enumeration and confirmation of *Bacillus cereus* in foods: collaborative study. J Assoc Off Anal Chem 63:581–586

Lebuhn M, Heulin T, Hartmann A (1997) Production of auxin and other indolic and phenolic compounds by *Paenibacillus polymyxa* strains isolated from different proximity to plant roots. FEMS Microbiol Ecol 22:325–334

Logan NA, Lebbe L, Hoste B, Goris J, Forsyth G, Heyndrickx M, Murray BL, Syme N, Wynn-Williams DD, De Vos P (2000) Aerobic endospore-forming bacteria from geothermal environments in northern Victoria Land, Antarctica, and Candlemas Island, South Sandwich archipelago, with the proposal of *Bacillus fumarioli* sp. nov. Int J Syst Evol Microbiol 50:1741–1753

Lopez AC, Alippi AM (2007) Phenotypic and genotypic diversity of *Bacillus cereus* isolates recovered from honey. Int J Food Microbiol 117:175–184

Lopez AC, Alippi AM (2009) Diversity of *Bacillus megaterium* isolates cultured from honeys. LWT Food Sci Technol 42:212–219

Martin PA, Travers RS (1989) Worldwide abundance and distribution of *Bacillus thuringiensis* isolates. Appl Environ Microbiol 55:2437–2442

Mavingui P, Heulin T (1994) In vitro chitinase and antifungal activity of a soil, rhizosphere and rhizoplane population of *Bacillus polymyxa*. Soil Biol Biochem 26:801–803

Mavingui P, Berge O, Heulin T (1990) Immunotrapping of *Bacillus polymyxa* in soil and in the rhizosphere of wheat. Symbiosis 9:215–221

Mavingui P, Laguerre G, Berge O, Heulin T (1992) Genetic and phenotypic diversity of *Bacillus polymyxa* in soil and in the wheat rhizosphere. Appl Environ Microbiol 58:1894–1903

Morris CE, Bardin M, Berge O, Frey-Klett P, Fromin N, Girardin H, Guinebretière MH, Lebaron P, Thiery JM, Troussellier M (2002) Microbial biodiversity: approaches to experimental design and hypothesis testing in primary scientific literature from 1975 to 1999. Microbiol Mol Biol Rev 66:592–616

Nicholson WL, Law JF (1999) Method for purification of bacterial endospores from soils: UV resistance of natural Sonoran desert soil populations of *Bacillus* spp. with reference to *B. subtilis* strain 168. J Microbiol Methods 35:13–21

Patel AK, Ahire JJ, Pawar SP, Chaudhari BL, Chincholkar SB (2009) Comparative accounts of probiotic characteristics of *Bacillus* spp. isolated from food wastes. Food Res Int 42:505–510

Quesada-Moraga E, Garcia-Tovar E, Valverde-Garcia P, Santiago-Alvarez C (2004) Isolation, geographical diversity and insecticidal activity of *Bacillus thuringiensis* from soils in Spain. Microbiol Res 159:59–71

Rennie RJ (1981) A single medium for the isolation of acetylene-reducing (dinitrogen-fixing) bacteria from soils. Can J Microbiol 27:8–14

Rodríguez-Díaz M, Lebbe L, Rodelas B, Heyrman J, De Vos P, Logan NA (2005) *Paenibacillus wynnii* sp. nov., a novel species harbouring the *nifH* gene, isolated from Alexander Island, Antarctica. Int J Syst Evol Microbiol 55:2093–2099

Rosenquist H, Smidt L, Andersen SR, Jensen GB, Wilcks A (2005) Occurrence and significance of *Bacillus cereus* and *Bacillus thuringiensis* in ready-to-eat food. FEMS Microbiol Lett 250:129–136

Saleh SM, Harris RF, Allen N (1969) Method for determining *Bacillus thuringiensis* var. *thuringiensis* Berliner in soil. Can J Microbiol 15:1101–1104

Sanders ME, Morelli L, Tompkins TA (2003) Spore-formers as human probiotics: *Bacillus*, *Sporolactobacillus*, and *Brevibacillus*. Comp Rev Food Sci Food Saf 2:102–110

Seldin L, Van Elsas J, Penido E (1983) *Bacillus* nitrogen fixers from Brazilian soils. Plant Soil 70:243–255

Seldin L, Rosado AS, da Cruz DW, Nobrega A, van Elsas JD, Paiva E (1998) Comparison of *Paenibacillus azotofixans* strains isolated from rhizoplane, rhizosphere, and non-root-associated soil from maize planted in two different Brazilian soils. Appl Environ Microbiol 64:3860–3868

Setlow B, McGinnis KA, Ragkousi K, Setlow P (2000) Effects of major spore-specific DNA binding proteins on *Bacillus subtilis* sporulation and spore properties. J Bacteriol 182:6906–6912

Shapiro MP, Setlow P (2006) Mechanisms of *Bacillus subtilis* spore killing by and resistance to an acidic Fe-EDTA-iodide-ethanol formulation. J Appl Microbiol 100:746–753

Smith SA, Benardini JA, Strap JL, Crawford RL (2009) Diversity of aerobic and facultative alkalitolerant and halotolerant endospore formers in soil from the Alvord Basin, Oregon. Syst Appl Microbiol 32:233–244

Stefanic P, Mandic-Mulec I (2009) Social interactions and distribution of *Bacillus subtilis* pherotypes at microscale. J Bacteriol 191:1756–1764

Su X, Shu C, Zhang J, Huang D, Tan J, Song F (2007) Identification and distribution of *Bacillus thuringiensis* isolates from primeval forests in Yunnan and Hainan provinces and Northeast Region of China. Agric Sci China 6:1343–1351

Travers RS, Martin PA, Reichelderfer CF (1987) Selective process for efficient isolation of soil *Bacillus* spp. Appl Environ Microbiol 53:1263–1266

Tzeneva VA, Li Y, Felske AD, de Vos WM, Akkermans AD, Vaughan EE, Smidt H (2004) Development and application of a selective PCR-denaturing gradient gel electrophoresis approach to detect a recently cultivated *Bacillus* group predominant in soil. Appl Environ Microbiol 70:5801–5809

Uribe D, Martinez W, Ceron J (2003) Distribution and diversity of *cry* genes in native strains of *Bacillus thuringiensis* obtained from different ecosystems from Colombia. J Invertebr Pathol 82:119–127

van Netten P, Kramer JM (1992) Media for the detection and enumeration of *Bacillus cereus* in foods: a review. Int J Food Microbiol 17:85–99

van Vuurde JWL (1987) New approach in detecting phytopathogenic bacteria by combined immunoisolation and immunoidentification assays. EPPO Bull 17:139–148

von der Weid I, Paiva E, Nobrega A, van Elsas JD, Seldin L (2000) Diversity of *Paenibacillus polymyxa* strains isolated from the rhizosphere of maize planted in Cerrado soil. Res Microbiol 151:369–381

von Stetten F, Mayr R, Scherer S (1999) Climatic influence on mesophilic *Bacillus cereus* and psychrotolerant *Bacillus weihenstephanensis* populations in tropical, temperate and alpine soil. Environ Microbiol 1:503–515

Walker R, Powell AA, Seddon B (1998) *Bacillus* isolates from the spermosphere of peas and dwarf French beans with antifungal activity against *Botrytis cinerea* and *Pythium* species. J Appl Microbiol 84:791–801

Ward J, Cockson A (1972) Studies on a thermophilic *Bacillus*: its isolation, properties, and temperature coefficient of growth. J Bacteriol 112:1040–1042

Chapter 5
Life Cycle and Gene Exchange

Xiaomin Hu and Jacques Mahillon

5.1 Introduction

Many aerobic endospore-forming soil bacteria belong to the family *Bacillaceae*, which contains important species, either for their potential industrial applications or their pathogenic/opportunistic behaviours. For instance, *Bacillus subtilis* and its close relatives represent a significant source of industrial enzymes (e.g., amylases or proteases), and much of the commercial interest in these bacteria arises from their capacity to secrete these enzymes at gram per litre concentrations. Similarly, because of its entomocidal properties, *Bacillus thuringiensis* has been used commercially as biopesticide in the control of insect pests for more than 40 years. At the other end of the spectrum, *Bacillus anthracis* is notorious as the etiological agent of anthrax and *Bacillus cereus sensu stricto* is an opportunistic pathogen that has been implicated in cases of food poisoning, periodontitis and endophthalmitis. *Bacillus* species are also ubiquitous, covering most habitats, from soil, fresh or deteriorating plant material to human and animal faeces. Some species, obligatory halophilic, thermophilic or alkaliphilic, have also been isolated from diverse extreme environments. How do all these microbes adjust to these complex and unusual environmental conditions? In this chapter, the specialized life cycles, the optimized physiological behaviours and the horizontal gene exchange among representative *Bacillus* species are used to illustrate the adaptive ecology and lifestyles of aerobic endospore-forming soil bacteria.

X. Hu
Key Laboratory of Agricultural and Environmental Microbiology, Wuhan Institute of Virology, Chinese Academy of Sciences, Wuhan 430071, China

J. Mahillon (✉)
Laboratory of Food and Environmental Microbiology – Earth and Life Institute, Université catholique de Louvain, Croix du Sud, 2/12, 1348 Louvain-la-Neuve, Belgium
e-mail: Jacques.mahillon@uclouvain.be

N.A. Logan and P. De Vos (eds.), *Endospore-forming Soil Bacteria*, Soil Biology 27,
DOI 10.1007/978-3-642-19577-8_5, © Springer-Verlag Berlin Heidelberg 2011

5.2 Life Cycle and Sporulation

For most bacteria, under laboratory conditions, the standard life cycle consists of four phases: lag (adaptation), log (exponential), stationary and decline phases. In natural environments though, where nutrients are less accessible, it is thought that bacteria are mostly in a state of low metabolism, either at rest or in "dormancy". For aerobic endospore-forming soil bacteria (mostly *Bacillus* species), the formation of spores is an additional developmental strategy designed for survival in harsh environments where these bacteria predominate. In this cryptobiotic state of dormancy, the endospore is the most durable type of cell found in nature and can remain viable for long periods of time, perhaps thousands of years (Cano and Borucki 1995; Vreeland et al. 2000). When the soil environmental conditions (e.g., temperature, relative humidity, pH value and nutrients) become favourable, spores germinate by a successive process: firstly, spores are partially hydrated, cations (e.g., Ca^{2+}) and the dipicolinic acid (DPA) that are the important spore components are released from the spore core, the cortex is degraded and the core is hydrated (Paidhungat et al. 2001; Cowan et al. 2004). If growth conditions permit, cells multiply and produce daughter cells in the vegetative cell cycle. Interestingly, besides nutrients, the oxygen concentration is also a sporulation factor (Sarrafzadeh and Navarro 2006; Park et al. 2009). For instance, *B. thuringiensis* and *B. anthracis* have been reported to require oxygen to trigger sporulation (Roth et al. 1955; Sarrafzadeh and Navarro 2006).

Endospores of aerobic soil-dwelling bacteria have higher resistances to withstand heat, dehydration and other physical stresses than vegetative cells. However, if unacceptable stress or too much damage has occurred during spore dormancy, the spore may lose its ability to return to life. As an example, although spores of the entomopathogen *B. thuringiensis* can persist in the soil for several years (Smith and Barry 1998), there is a rapid decline in spore viability and toxicity during the first few weeks after the spray application of this organism as a biopesticide (Petras and Casida 1985; Tamez-Guerra et al. 2000). The decline may be due to UV damage when exposed to sunlight in nature, because spores of many *Bacillus* species have been found to be sensitive to UV light (Luna et al. 2008; Zhao et al. 2009); in *B. anthracis*, photo-induced repair of UV damage is absent in spores of the Sterne vaccine strain (Knudson 1986). After exposure to sunlight or UV radiation for several hours, a significant negative effect on the survival of *B. anthracis* spores has been observed (Lindeque and Turnbull 1994; Nicholson and Galeano 2003; Coohill and Sagripanti 2008).

In addition to a full life cycle in soil, many endospore-forming organisms are also adapted to a lifestyle in insect (e.g., *B. thuringiensis* and *Bacillus* – now *Lysinibacillus* – *sphaericus*) or mammalian hosts (e.g., *B. anthracis*). Once the spores, or vegetative cells, have been taken up by the host, they multiply and may either provoke (in a pathogenic mode), or not provoke (in a symbiotic mode), disease. When bacteria produce insecticidal proteins, the infected insects become sick, the bacteria grow in the dead carcases until nutrients become limited, then sporulation occurs and eventually spores are released into the environment (Jensen et al. 2003). That is the case for *B. thuringiensis* and *L. sphaericus*. *B. thuringiensis* harbours *cry* genes expressing insecticidal crystal delta-endotoxins (Cry toxins) during sporulation. This feature distinguishes *B. thuringiensis* from *B. cereus sensu stricto*. Furthermore, several *B. thuringiensis* isolates also produce vegetative insecticidal proteins (VIPs) (Selvapandiyan et al. 2001; Bhalla et al. 2005). Due to these insecticidal toxins, *B. thuringiensis* is used as biopesticide for the control of certain insect larvae of lepidopteran, dipteran and coleopteran species (van Frankenhuyzen 2009). *B. thuringiensis* isolates are also active against other microbes such as nematodes and protozoa (Feitelson et al. 1992; Liébano et al. 2006). Similarly, certain *L. sphaericus* strains exhibit toxicity against mosquito larvae and are used in insect control programmes to reduce the populations of vector species that transmit tropical diseases, such as malaria, filariasis, yellow fever or dengue fever. The mosquitocidal properties are due to the action of binary toxin (Bin proteins), which forms crystal inclusions during sporulation, and mosquitocidal toxins (Mtx proteins), produced during vegetative growth (Priest et al. 1997). Several *L. sphaericus* strains also produce another two-component toxin (Cry48 and Cry49 proteins) upon sporulation (Jones et al. 2007).

B. anthracis is a mammalian pathogen. It has become notorious as potential bioweapon, causing lethal inhalation anthrax, owing to its ability to synthesize the anthrax toxin proteins and the poly-D-glutamic acid capsule (Mock and Fouet 2001; Okinaka et al. 1999). Anthrax has been linked with endemic soil environmentals long before *B. anthracis* was identified as the causative agent (Rayer 1850; Davaine 1863). A current model of *B. anthracis* ecology relies on its pathogenicity: when the spores, which are normally present in the soil, are taken up by an animal host through oral, cutaneous or pulmonary routes, the spores germinate to become vegetative cells that can proliferate to high numbers in different body parts (e.g., lymph glands) and produce toxins, ultimately leading to the death of the animal (Jensen et al. 2003; Stenfors Arnesen et al. 2008; Koehler 2002, 2009). Death of the host and contact of infected tissues with air results in a return to the spore form of the bacterium into the environment. Thus, the spore–vegetative cell–spore cycle is essential for the pathogenic lifestyle of this extremely virulent bacterium. Spores are considered to be the predominant form of *B. anthracis* outside the host. Like those of all other *Bacillus* species, *B. anthracis* spores are highly resistant to adverse environmental conditions, but compared to other *Bacillus* species, *B. anthracis* is more dependent on sporulation for survival (Turnbull 2002). Vegetative cells appear to survive poorly in simple environments such as water and soil (Saile and Koehler 2006; Turnbull 2002). However, although the soil or other natural environment is generally adverse, a few reports stated that spores could actually germinate when conditions are favourable. Van ness (1971) reported that soil pH above 6.0 and temperature above 15.5 °C in which decaying grass and other organic matter constitute the nutrients necessary for the germination of *B. anthracis* spores will favour outbreaks of anthrax. It is also noteworthy that growth of *B. anthracis* outside a host often leads to loss of virulence caused by loss of virulent plasmids pXO1 or pXO2 (Dragon and Rennie 1995; Jensen et al. 2003).

Generally, members of the Gram-positive, aerobic endospore-forming genus *Bacillus* occur in a range of environments, from soil, water, insect guts and food and dairy processing surfaces. The natural survival environment is complicated, ever changing and often hostile. To adapt to various survival pressures, they have evolved a series of versatile physiological functions and strategies for survival, utilizing all the benefits and avoiding all the disadvantages. In addition to spores, cells can become mobile, genetically competent, produce extracellular matrices or secrete toxins, or can even "communicate" with each other and "make decisions" for certain social behaviours, which allow them to cannibalize their neighbours.

5.3 Motility and Chemotaxis

It has been proposed that the majority of bacterial species are motile during at least a part of their life cycle (Fenchel 2002). This mobility trait helps the bacteria to access resources and escape from adverse conditions. Chemotaxis is a universal phenomenon whereby bacteria monitor the environmental stressors and adjust themselves to allow optimal growth and survival in soil ecosystems or other natural habitats. By sensing changes in the concentrations of various organic compounds required for survival or by detecting different physical signals, motile bacteria move towards more favourable environments (or escape from inappropriate conditions). Many aerobic spore-formers have flagella, but their motility may be flagellum dependent or independent (Henrichsen 1972; Fraser and Hughes 1999; Kinsinger et al. 2003).

The flagellum-dependent mode of motility has been well studied. When flagella rotate in the counterclockwise (CCW) direction, multiple rotating flagella assemble together to form a bundle, promoting the bacteria forward in a motion of "smooth swimming"; when the flagella rotate in the clockwise (CW) direction, the component flagella of the bundle become separated and a "tumbling" motion may take place. A variety of signal transduction proteins and flagellar structure components are associated with such flagellum-dependent mobility. It has been suggested that the flagellum-independent mode, in a sliding or spreading manner, is not an active form of movement and mainly functions as a surface colonization activity (Harshey 2003).

Flagellum-dependent motility and chemotaxis have been extensively studied in *B. subtilis*. As in other bacteria, there are three basic processes in the *B. subtilis* chemotactic signal transduction pathway: (1) signal reception by bacterial membrane receptors, (2) signal relay from the membrane receptors to the flagellar motor and (3) signal adaptation (or desensitization) of the original stimulus (Lux and Shi 2004; Rao et al. 2008). When *B. subtilis* moves towards higher concentrations of attractant (e.g., asparagine) by chance, the attractant binds to chemoreceptor proteins (which are normally transmembrane proteins) and activates the methylated receptor. The activated receptor can bind and stimulate the main chemotaxis protein CheA that will transfer the phosphoryl group from phosphoryl-CheA to its cognate response regulator CheY. The phosphoryl-chemotaxis protein activates the flagella gene operon and cause CCW rotation of the flagella, which produces "smooth swimming" and, hence, continued migration towards higher attractant concentrations. If the bacterium moves towards lower attractant concentrations by chance, then attractant molecules will tend to leave the binding sensor protein. The ultimate consequence is that the bacteria will be more likely to tumble so that the cells are reorientated, and therefore move in a more favourable direction subsequently. Simultaneously, a time-delayed signal adaptation (or desensitization) mechanism will function to "calm down" the activated chemoreceptors (Rao et al. 2008). The major adaptation system requires two enzymes, CheR and CheB, to maintain equilibrium of methylation/demethylation by adding or removing methyl groups. Two other adaptation systems, involving CheC, CheD and CheV, which are not found in all bacteria (e.g., absent from *Escherichia coli*), are found in *B. subtilis* (Karatan et al. 2001; Muff and Ordal 2007).

5.4 Cell–Cell Communication

To optimize population survival, bacterial cells need to have communication and decision-making capabilities that enable the single-celled individuals to coordinate growth, movement and biochemical activities so as to allow them to function efficiently and utilize the most optimal growth conditions, and to adjust their life cycles in a collective mode. Quorum sensing (QS) is a signal-mediated, cell–cell communication system that couples bacterial cell density to a synchronized gene expression, therefore controlling vital processes in the life cycle: these include competence, sporulation, motility and biofilm formation (Boyer and Wisniewski-Dyé 2009). It also regulates other biological phenomena such as bioluminescence, symbiosis, conjugation or virulence (Falcão et al. 2004; Nealson and Hastings 2006; White and Winans 2007).

The first and most intensely studied QS system is the LuxI/LuxR bioluminescent system of the Gram-negative marine bacterium *Vibrio fischeri* (Nealson and Hastings 1979; Engebrecht and Silverman 1984). The bacterial cross-talk within the population of *V. fischeri* relies on the secretion of a signal molecule by the bacterial cell, i.e. an autoinducer (HSL, *N*-(3-oxohexanoyl)-Homoserine Lactone) (Kaplan and Greenberg 1985). The concentration of autoinducer is correlated with bacterial cell density, constituting an indicator of the growth status of the population. When the accumulation of the HSL autoinducer reaches a specific threshold along with the high cell-population density, the signal will be sensed by autoinducer sensors (i.e., LuxR-like proteins), leading to the binding of autoinducer to the sensors which, in turn, trigger the light emission of the population (Dunlap 1999; Callahan and Dunlap 2000; Antunes et al. 2007).

For endospore-forming organisms, QS systems impact on their life cycles by regulating a variety of processes in response to increasing cell-population densities in stationary phase, so enabling populations to adapt rapidly to environmental changes. The basic mechanism is to regulate the phosphorelay (e.g., sporulation phosphorelay signal transduction), which is an important process that activates a variety of kinases involved in vital biological activities (Burbulys et al. 1991; Bischofs et al. 2009). Two types of QS pheromones are found in endospore-forming organisms, autoinducers and secreted signal peptides. In most cases, small, secreted peptides (e.g., ComX and PapR) are used as pheromones in QS systems. The autoinducer-2 (AI-2) has been found in *B. cereus* ATCC10987, where it has an inhibitory effect on biofilm formation (Auger *et al.* 2006), and in *B. anthracis*, where it displays functional LuxS-like activity and induces luminescence in a *Vibrio harveyi* reporter strain (Jones and Blaser 2003). Many peptide-based quorum-signalling systems indirectly regulate transcription by controlling phosphoryl signalling (Reading and Sperandio 2006). In general, the concentration of the secreted peptide pheromones increases with cell-population density, and the interaction of the peptide pheromones with the two-component sensor kinases initiates the phosphorylation of a cognate-response regulator protein, allowing it to bind DNA and alter the transcription of the QS-controlled target genes (Miller and Bassler 2001; Reading and Sperandio 2006). This may be illustrated by the following QS systems found in *Bacillus*.

In *B. subtilis,* the peptide ComX is secreted and accumulated as cell density increases. The signal is detected and transduced by the sensor kinase ComP to the response regulator ComA, so initiating the phosphorylation of ComA. Phospho-ComA triggers the expression of the master regulator ComK, promoting the commitment of the cells to the competence pathway (Comella and Grossman 2005). The effect of ComA is also regulated by a QS system mediated by peptides that act as phosphatase regulators (Phr) of receptor aspartyl phosphatases (Rap), which is probably a common form in endospore-forming bacteria as it is also found in other *Bacillus* species (Bongiorni et al. 2006). The Phr precursor peptides are first secreted, then cleaved into active pentapeptides, which correspond to the C-terminal end of the precursor. These released pheromonal oligopeptides diffuse into and potentially accumulate in the medium (Lanigan-Gerdes et al. 2008). The extracellular pentapeptides are reimported into the cytoplasm by an ABC-type oligopeptide transporter (Lazazzera 2001). The intracellular Phr signals then interact with a subset of Rap and subsequently lead to a regulatory shift of phosphorelays (Bischofs et al. 2009). Since phosphorelay is required to activate a variety of environmentally controlled kinases, it is easy to understand why spore formation is affected not only by a high cell density, but also by harsh environmental signals (e.g., starvation). There are at least 11 *rap-phr* genes in *B. subtilis*, which code for a family of signals and cognate phosphatases, constituting a complex quorum signalling network and promoting certain collective behaviours within a population, e.g. motility, sporulation and biofilm formation (Pottathil and Lazazzera 2003; Bischofs et al. 2009).

In the *B. cereus* group, the PlcR-PapR QS system controls the expression of a large range of genes, including most of the extracellular virulence factors produced by *B. cereus* and *B. anthracis*. PlcR is a transcriptional regulator, which is activated by a small QS effector signalling peptide PapR, the product of a gene located immediately downstream of *plcR* (Slamti and Lereclus 2002). PapR is secreted by the bacterial cell, is processed and a C-terminal heptapeptide active form of PapR (not a pentapeptide as previously suggested) is reimported through the oligopeptide permease (Opp) system. The now-intracellular PapR-derived signal binds to and activates the pleiotropic regulator PlcR (Bouillaut et al. 2008), which in turn binds to a specific DNA sequence called the "PlcR box", located upstream from its controlled genes and at various distances ahead of the −35 box of the sigma A (σ^A) promoter (Agaisse et al. 1999). A recent study identified a whole PlcR consensus sequence as "wTATGnAwwwwTnCATAw" by directed mutagenesis in *B. cereus* ATCC14579. The binding of PlcR to its box is necessary but not in itself sufficient to activate the transcription of genes located downstream, which also involves some sigma factors, such as σ^A, σ^B or σ^H. Forty-five genes are found to be controlled by 28 plcR "boxes", including a wide array of genes related to virulence (e.g., phospholipases, proteases or toxins), cell protection (e.g., bacteriocins, transporters or cell wall biogenesis) and environment sensing (e.g., two-component sensors, chemotaxis proteins or GGDEF family regulators) (Gohar et al. 2008). Thus, PlcR is a master gene regulator. By the specific interaction between PlcR and PapR, *B. cereus sensu lato* strains can monitor a large number of environmental signals, such as food deprivation and self cell-density, and regulate the transcription of genes designed to adaptation of cell growth and life cycle of the population.

5.5 Multicellular Phenotypes and Biofilms

Multicellular behaviour, another mechanism associated with optimizing growth and survival, is considered to be a preferred lifestyle option for a variety of prokaryotes. By multicellular solidarity, the bacterial population can achieve functions that are impossible for single cell, assisting the population in nutrition absorption and defence strengthening (Shapiro 1998; Lee et al. 2007b). Biofilm formation is one of the recognized multicellular behaviours in bacteria. It is a matrix-aggregation form of many bacteria on a surface, or interface, which is related to QS systems to assess population size or cell density (Irie and Parsek 2008). In liquid culture, bacteria form two growing populations, the majority being in free-living mode (termed planktonic), and some cells occur as parts of biofilms, in the presence of an "attachment" surface. Planktonic cells usually express flagella, whereas biofilm-forming cells produce an extracellular matrix. Exopolysaccharides (EPS), surface proteins and sometimes nucleic acids are involved into the assembling of the extracellular biofilm matrix (Branda et al. 2005; Lasa 2006; Vu et al. 2009).

The mechanisms for building biofilms vary among different species and under different environmental conditions. As the best-studied soil microorganism, *B. subtilis* is well known for its ability to become competent and undergo sporulation in response to starvation and high population densities. Biofilm formation by *B. subtilis* has also been well studied and has become a model for studying biofilm formation by endospore-forming soil bacteria. Biofilm formation by *B. subtilis* can be initiated by specific environmental signals, such as nutrient and oxygen availability, or even by a variety of structurally unrelated natural products produced by bacteria, including the *B. subtilis* non-ribosomal peptide surfactin (López et al. 2009a). During biofilm development, certain responses take place, such as the repression of flagellar genes and the hyperexpression of genes for adhesion and ribosomal protein formation (Kobayashi 2008). A variety of genes, proteins and complex molecular mechanisms are involved in *B. subtilis* biofilm formation, of which two operons *eps* and *yqxM* are required for matrix synthesis. The *eps* operon consists of 15 genes (*epsA-O*), responsible for the production of the EPS component, and *yqxM-sipW-tasA* operon is responsible for the production and secretion of the major protein component of the matrix, TasA (Kearns et al. 2005; López et al. 2009b). The regulator SinI is thought to induce the *eps* and *yqxM* operons by antagonizing their repressor SinR (Kearns et al. 2005; Chai et al. 2008). Recently, the transcriptional repressor SlrR/SlrA was also found to be a key regulator for biofilm formation. It represses transcription of flagellar genes, and, like SinI, activates transcription of the *eps* and *yqxM* operons by antagonizing SinR (Kobayashi 2008).

The *B. subtilis* biofilm comprises heterogeneous populations and displays cellular differentiation with distinct cell types: planktonic cell, matrix-producing cells and spores. Interestingly, each cell type preferentially localizes in distinct regions within the biofilm, but the localization and percentage of each cell type is dynamic throughout the development of the community (Vlamakis et al. 2008). It is remarkable that spore formation is coupled to the formation of an architecturally complex community of cells. When in a mutant the extracellular matrix is not produced, the sporulation of the population is observed to be deficient (Vlamakis et al. 2008).

Many pathogenic *B. anthracis* and *B. cereus* have been observed forming biofilms on various surfaces such as glass wool, stainless steel or at solid–liquid interfaces (Ryu and Beuchat 2005; Auger et al. 2009; Lee et al. 2007b). A biofilm has a higher resistance to antibiotics and a more active host immune system than does the single-celled planktonic counterpart, and this probably enhances the effects of the pathogenic bacteria on human health. For instance, the biofilm organisms of *B. anthracis* display an obvious stronger antibiotic resistance than the planktonic populations (Lee et al. 2007b).

5.6 Horizontal Gene Transfer

5.6.1 Genome Plasticity and Evolution

Bacterial diversity can be seen at two levels. At the "macro-level", the explosion of genome sequencing data and the use of metagenomic approaches have revealed an enormous diversity among bacteria and archaea, including the existence of microorganisms never cultivated under laboratory conditions. At the "micro-level" recent analyses of multiple bacterial genomes from single or close-related species have brought a new genomic concept: the existence of the core genome, the dispensable genome and pan-genome (Medini et al. 2005). The *core* genome is the pool of genes shared by all the strains of the same taxon, including all genes required for the basic biologic functions and the major phenotypic traits used for taxonomy. The *dispensable* genome is the pool of genes present in some but not all strains of the same taxon, including genes which might not be essential for bacterial growth but that contribute to the species' diversity, and may confer selective advantages, such as antibiotic resistance, pathogenicity or adaptation to different niches. The *pan* genome represents the total gene repertoire of a bacterial taxon, consisting of both the core and dispensable genomes. Although new genes may originate via gene duplication followed by diversification, horizontal gene transfer (HGT) is the simplest way to increase the size of the population's pan-genome by providing an influx of novel genetic material from unrelated organisms. HGT is defined as the movement of genetic material between bacteria other than to their own offspring.

5.6.2 *Mobile Genetic Elements*

Mobile genetic elements (MGEs), the "mobile" genetic material, are segments of DNA encoding enzymes and other proteins that mediate their DNA movement via HGT. MGEs include phages, plasmids, integrative and conjugative elements (ICE) and their associated hitchhiking elements: Insertion sequences (IS), Transposons (Tn) and Introns.

The transfer of MGEs can happen between bacterial cells (intercellular mobility) or within the host genome (intracellular mobility). Intercellular movement of DNA takes three forms in prokaryotes: transformation, conjugation and transduction. The intracellular mechanisms transfer DNA between the microbial genome and a plasmid or phage genome in a genetic recombinant mode, usually by recombination and transposition, allowing DNA fragments to be copied or excised from the original genome location and inserted into target loci elsewhere of the same host.

5.6.2.1 Transformation

Transformation is the uptake and incorporation of naked DNA (circular or linear) from the environment. Many bacteria are constitutively or inductively competent, allowing the free DNA from the environment to bind to and pass across the cell surface. The incoming double-stranded DNA (dsDNA) is transformed to single-stranded DNA (ssDNA) and transported across the cytoplasmic membrane in an energy-dependent manner. If the imported DNA is to be stably maintained in the cell, it should be compatible and able to escape elimination by the DNA repair mechanisms. When a plasmid taken from the environment re-circularizes, or a non-replicating DNA segment integrates into the host genome by homologous recombination, new genetic material becomes a heritable part of the genome (Chen et al. 2005; Zaneveld et al. 2008). To date, the characterization of the dsDNA binding and translocation structures is incomplete. For most naturally transformable bacteria studied, similar proteins involved in DNA uptake are used, and these are also needed for the assembly of type IV pili and the type II secretion system (TIISS) (Zaneveld et al. 2008). The transformation mechanism in which the type IV pili and TIISS components are involved will be further illustrated by the model of *B. subtilis* transformation.

Both *Bacillus licheniformis* and *B. subtilis* are natural transformable spore-forming soil bacteria, but only the latter has been extensively studied so far. The competence of *B. subtilis* can be induced in response to environmental stress (e.g., nutrient limitation, presence of antibiotics or DNA damage). A specialized membrane-associated machinery is required for uptake of exogenous dsDNA (Claverys et al. 2009). Naked dsDNA is first bound to the competent cell surface without the requirement of sequence specificity. Fragmentation of dsDNA, which generates an average length of 13.5–18 kb breaks in *B. subtilis*, occurs during the binding process (Dubnau 1999). One non-transported strand of the dsDNA fragment is degraded while the other strand is imported across the membrane (Chen and Dubnau 2003) and becomes nuclease resistant once intracellular. The involvement of pilus-related components in the transformation process is observed, including some protein components needed for assembly of type IV pilus formation and the TIISS. A pseudopilus structure, recently identified (Chen et al. 2006; Craig and Li 2008), consists of the major pilin-like protein (ComGC) and three minor pre-pilin proteins (ComGD, ComGE and ComGG). The pre-pilin proteins are processed by a prepilin peptidase ComC and assembled (or polymerized) to the outside of the membrane with the aid of the ATPase ComGA and membrane protein ComGB. By a hypothetical DNA-binding protein, the polymerized pseudopilus attaches exogenous DNA. The depolymerization is driven by a putative proton-motive force, which allows the exogenous DNA to traverse the peptidoglycan (Claverys et al. 2009). A membrane binding protein, ComEA, functions as DNA receptor to interact with the exogenous dsDNA (Inamine and Dubnau 1995; Provvedi and Dubnau 1999). While one strand of ssDNA is hydrolysed by an endonuclease, the other one is delivered through the membrane channel (ComEC) with the aid of the DNA translocase (ComFA) and ssDNA binding protein (SSB) or another DNA transport/recombinant protein DprA (also called Smf) (Tadesse and Graumann 2007; Claverys et al. 2009). *ComK* encodes the master transcriptional regulatory protein ComK, which activates the expression of many genes, including the genes needed for competence (Berka et al. 2002; Ogura and Tanaka 2009). Recently, two newly identified genes *yutB* and *yrzD* (or *comN*) were found to regulate the *comE* operon (*comEA-EB-EC-ER*), independently of ComK, at the both transcriptional and post-transcriptional levels (Ogura and Tanaka 2009).

The transforming DNA will face two possible fates, either integration into the host genome by homologous recombination, which will ensure stability during the following generations, or elimination during cell division (Chen et al. 2005). The recombination protein (RecA) forms a filament around the incoming ssDNA and mediates a homologous search with host chromosome DNA. The integration of homologous DNA into the recipient plasmid is independent of RecA but requires at least RecO and RecU, which accumulate at the polar DNA uptake machinery (Kidane et al. 2009).

For other *Bacillus* species under laboratory conditions, such as *B. cereus sensu lato*, various optimal electroporation protocols have been used to obtain the *Bacillus* transformants using plasmid DNA (Belliveau and Trevors 1989; Mahillon et al 1989; Turgeon et al. 2006). Nevertheless, there is not much information available on their natural competent capabilities or transformation mechanisms. With the present blooming of genome projects, a large number of *Bacillus* genomes have been sequenced and these reveal the presence of many genes potentially involved in competence and its regulation in transformation.

5.6.2.2 Transduction

Bacteriophages are abundant in the natural environment and are more stable than naked DNA because the protein coat provides protection. Phage transduction is a mechanism by which bacterial genes are incorporated into bacteriophage particles and transferred to other bacteria. Transduction happens through either the lytic cycle or the lysogenic cycle. If the lysogenic cycle is adopted, the phage genome is integrated into the bacterial chromosome, where it can remain stable, as prophage, for generations. If the prophage is induced (e.g., by UV light), the phage genome is excised from the bacterial chromosome and initiates the lytic cycle, which ultimately results in lysis of the cell and the release of phage particles. Phage can transfer bacterial DNA via either a specialized or a generalized mode. Temperate phages integrate their DNA at specific attachment sites in the host genome by integrase-mediated recombination. When the integrated phage is incorrectly excised from the bacterial chromosome, small neighbouring sequences on either side of the prophage genome can remain attached and be packaged along with the phage genome. The resulting "recombinant" phage can be transferred to a new host where the bacterial DNA will be incorporated into the genome. Generalized transducing phages integrate random bacterial chromosomal DNA into their capsids. This "mistake" will lead to generalized transduction, whereby any bacterial gene may be transferred among bacteria (Zaneveld et al. 2008).

Transduction has been demonstrated in endospore-forming soil bacteria. For instance, previous studies demonstrated the utility of the generalized transducing bacteriophage CP-51 and CP-54 (both isolated from soil), which can mediate the transfer of plasmids and chromosomal markers among *B. anthracis*, *B. cereus* and *B. thuringiensis* (Thorne 1968, 1978; Yelton and Thorne 1971). Both phages SP-10 and SP-15, also isolated from soil samples, have the ability to propagate on *B. subtilis* and *B. licheniformis* and to mediate general transduction in either species when homologous integration takes place (Taylor and Thorne 1963). Transduction is a common genetic tool used to stably introduce a foreign gene into the genome of a host cell. This can be exemplified by the utilization of phi29 and pBS related phages in *B. subtilis* as genetic tools for gene recombination technique.

5.6.2.3 Conjugation

Unlike transformation, in which any segment of DNA, regardless of the sequence, can be theoretically transferred horizontally, conjugation is a process associated with particular plasmids or transposons. Self-transmissible plasmids, conjugative transposons and ICEs all transfer using this mechanism (Burrus et al. 2002). MGEs are structured as a combination of *cis*-acting sequences, required for recognition specificity, and of genes coding for the *trans*-acting components that constitute the DNA translocation machinery (Merlin et al. 2000). An origin of transfer (*oriT*) element and an entire set of transfer apparatus are required for conjugal transfer of self-transmissible plasmids. In addition to these features, site-specific recombinases and their cognate repeated sequences are present in ICEs, which allow their integration and excision from the host genome (Scott and Churchward 1995; Burrus et al. 2002).

The simplest conjugative transfer is when a plasmid is self-transmissible and does not require the presence of other plasmids. These plasmids are said to be "conjugative". For plasmids that are not self-transmissible, the transfer can be conducted by mobilization with the help of a conjugative plasmid, if they contain an *oriT*. However, in some cases (mainly in Gram-positive bacteria), a "mobilization" (*mob*) gene is also involved to promote the use of the conjugation machinery encoded by another, co-resident element. Such plasmids are called "mobilizable". Interestingly, "non-mobilizable" plasmids may also be transferred by an aggregation-mediated conjugation system, independent of the presence of *oriT* and *mob* (Andrup et al. 1996). This was illustrated by the conjugative transfer of *B. thuringiensis* subsp. *israelensis* plasmid pXO16 (see below). In some cases, another mechanism allows non-conjugative plasmids to be transferred. In fact, whenever a fusion takes place between the conjugative and non-conjugative plasmids, the resulting "fused" molecules will be transferred into the recipient cell. The recombination events could result from homologous recombination or be promoted by the presence of ISs or transposons, leading to the formation of a transient cointegrate molecules. This mechanism has been referred to as "conduction", as opposed to the "donation" observed in the case of the Mob-*oriT* plasmids. Finally, it is worth mentioning that when the conduction takes place with chromosome, transfer of any chromosomal gene can be achieved. This phenomenon has been described as High frequency recombination (Hfr) conjugation by the *E. coli* F-plasmid (Boyer 1966) but can, in principle, occur in any bacterium.

DNA transfer by conjugation happens in a cell contact-dependent fashion. A cell-envelope-spanning translocation channel is necessary for building up this contact, which is implemented either by sex pili for Gram-negative bacteria or by complex (and not well identified) surface adhesins for Gram-positive bacteria (Harrington and Rogerson 1990; Grohmann et al. 2003). Two protein complexes are believed to be essential for the initiation of conjugation and the formation of a translocation channel. One is the relaxosome and the other is the Mating pair formation (Mpf). As conjugation only mediates the transport of ssDNA into the cytoplasm, in order to separate the two strands, the DNA is nicked by a relaxase. The relaxase binds to the *cis*-element *oriT*, remains covalently bound to the ssDNA and forms a relaxosome–DNA complex. With the aid of a coupling protein, the processed ssDNA from the donor is brought to the Mpf complex, which spans the inner membrane, periplasm, peptidoglycan and outer membrane of the donor cell. Both ATP energy and proton motive force are needed for this DNA translocation process (Schröder and Lanka 2005).

Six secretion systems (Types I, II, III, IV, V and VI) have been identified in bacteria. The Type IV Secretion Systems (T4SSs) are special due to their ability to transfer both proteins and nucleoprotein complexes. T4SSs are usually encoded by multiple genes and are made up of multisubunit cell-envelope-spanning structures comprising a secretion channel (Christie and Cascales 2005). The T4SS was originally characterized on the Ti (tumour-inducing) plasmid of the Gram-negative *Agrobacterium tumefaciens* (Stachel and Nester 1986). It is formed by 11 VirB proteins (VirB1–VirB11) encoded by a single operon and requires two additional proteins, VirD2 and VirD4. VirB1 to VirB11 represent the core of the Mpf complex, providing a likely channel for the DNA to pass through (Christie 2004). VirD2 is the relaxase that recruits the transferred DNA and VirD4 is the coupling protein, mediating interaction between the relaxosome and Mpf systems (Atmakuri et al. 2004; Yeo and Waksman 2004). Many conjugative plasmids bear the VirB/D4-like subunits, such as R388 (from *E. coli*) in Gram-negative bacteria or pIP501 (from *Streptococcus agalactiae*) in Gram-positive bacteria (Middleton et al. 2005; Abajy et al. 2007; Van der Auwera 2007).

B. thuringiensis is one of the spore-forming soil bacteria best studied for conjugation. The first report of conjugation in *B. thuringiensis* dates back to 1982 (González et al. 1982). Since then, a variety of conjugative plasmids originating from *B. thuringiensis* have been discovered. These plasmids are able to transfer from *B. thuringiensis* donors to other *B. cereus* group members, to other *Bacillus* spp. [such as *B. subtilis*, *B. licheniformis* or *L. sphaericus* (Jensen et al. 1996; Gammon et al. 2006)] or even to *Listeria* species (J. Godziewski, P. Modrie, J. Mahillon unpublished data). Some of these large plasmids not only have the capability of horizontal transfer between different hosts, but are also responsible for the production of different Cry toxins forming insecticidal crystal protein (ICP) inclusions. For instance, pHT73 from *B. thuringiensis* subsp. *kurstaki* carries *cry* toxin genes and is conjugative (González et al. 1982; Hu et al. 2004). pBtoxis from *B. thuringiensis* subsp. *israelensis* is a non-conjugative *cry*-plasmid, but it can be mobilized by other plasmids from *B. thuringiensis* (Hu et al. 2005; Gammon et al. 2006). As already mentioned, another rather interesting conjugative plasmid is the 350 kb, aggregation-mediated element pXO16 (Jensen et al. 1995) from *B. thuringiensis* subsp. *israelensis*, a bacterium highly toxic to dipteran larvae. The aggregation phenotype (Agr^+) involves a proteinaceous molecule on the cell surface, but apparently no aggregation-inducing pheromone (Jensen et al. 1995).

The transfer of pXO16 is very efficient (reaching 100%) and fast (needing only 3.5–4 min) but is limited to a narrow host range. Furthermore, it is able to mobilize not only mobilizable plasmids, but also non-mobilizable plasmids (Andrup et al. 1996; Timmery et al. 2009). The molecular basis for the pXO16 exceptional and unique transfer system remains to be elucidated.

Two self-transmissible plasmids, pAW63 and pBT9727, display a very high level of sequence similarity and synteny with *B. anthracis* plasmid pXO2, with the exception of a ~30-kb PAI that contains the pXO2-specific anthrax capsule genes. Besides sharing similar replication genes, all three plasmids (pXO2, pAW63 and pBT9727) possess a 40-kb transfer region containing homologues of key components of the T4SS system (Van der Auwera 2007; Van der Auwera et al. 2005; 2008). Nevertheless, unlike pAW63 and pBT9727, pXO2 is not self-transmissible (Reddy et al. 1987). In a recent study, a large number of *B. cereus* group strains from soil were screened for plasmids with pXO2-like replicons and *virB4*-, *virB11*- and *virD4*-like T4SS genes. It is interesting that most of the plasmids that possessed the pXO2-like replicon and transfer region were capable of promoting their own transfer as well as that of small mobilizable plasmids (pUB110 and pBC16) (Hu et al. 2009b). It was also suggested that the common ancestral form of pXO2-like plasmids was conjugative, and that some descending lineages underwent genetic drift, leading to loss of transfer capability. Until now, neither pXO1 from *B. anthracis* nor any pXO1-like plasmid found in other *B. cereus* group members has been found to be self-transmissible (Hu et al. 2009b). However, as for pXO2, pXO1 can be mobilized to plasmid-cured *B. anthracis* or to *B. cereus* recipients by the conjugative element pXO14, suggesting the potential mobility of both *B. anthracis* virulence plasmids (Reddy et al. 1987).

In a recent study, the comparison based on plasmidic versus chromosomal genetic backgrounds of a set of sympatric soil-borne *B. cereus* group isolates (i.e., organisms whose ranges overlap or are identical), using multilocus sequence typing (MLST) analysis, revealed, for some of the isolates, the same chromosomal sequence type but different pXO1- and pXO2-like replicon contents. These isolates were collected from soil samples from two neighbouring sites, suggesting probable instances of horizontal transmission within the *B. cereus* group in genuine soil ecosystem (Hu et al. 2009a).

B. thuringiensis is widely distributed in nature, often infecting insect larvae and commonly found in soil. Although most conjugational studies on *B. thuringiensis* are performed in laboratory broth conditions, plasmid transfer has been observed in natural environmental ecosystems such as insect larvae, soil and river water (Vilas-Bôas et al. 2000; Thomas et al. 2001; Yuan et al. 2007). *B. thuringiensis* spores can germinate and multiply in insect larvae or in soil. Studies have found that plasmids can move between donor and recipient strains of *B. thuringiensis* in soil or in larvae of lepidopteran (e.g., *Galleria mellonella*, *Spodoptera littoralis* and *Lacanobia oleracea*) and dipteran insects (e.g., *Aedes aegypti*) as they do in broth culture (Thomas et al. 2000, 2001; Vilas-Bôas et al. 2000). Nevertheless, in comparison with laboratory broth culture, and because of unfavourable conditions for survival, such as the highly alkaline conditions and competition from the background bacterial population present in larval guts, or the poor nutrient levels in soil or water, some reports have suggested that most vegetative cells die and the remaining cells sporulate quickly, and therefore these natural environment ecosystems are not favourable niches for *B. thuringiensis* multiplication and conjugation (Thomas et al. 2001; Ferreira et al. 2003).

Many strains of *B. thuringiensis* contain a complex array of plasmids, including numerous elements coding for delta-endotoxins. The toxin genes and plasmids pool diversity and the plasticity of *B. thuringiensis* strains suggest that conjugation may be an important means for dissemination of these genes and plasmids among *Bacillus* populations in nature. One of the major reasons for the complex activity spectrum of *B. thuringiensis* may be the growth of more than one strain of *B. thuringiensis* in susceptible larvae (or in other suitable environments) allowing plasmid transfer, hence creating new combinations of delta-endotoxins. In addition, plasmid transfer could occur among related bacteria during growth within an insect. Indeed, *B. thuringiensis* plasmids could also be transferred to other spore-forming bacteria from soil samples (Jarrett and Stephenson 1990). This indicates that the potential exists for new combinations of crystal toxins in nature. Thus, it may due to gene exchange phenomenon between two very different species that *cry* or *cry*-like genes were found in some non-*B. thuringiensis* species, e.g. the anaerobic bacterium *Clostridium bifermentans* subsp. *malaysia* and *L. sphaericus* (Barloy et al. 1996; Jones et al. 2007).

The conjugation phenomenon is also observed in other aerobic spore-forming soil bacteria, e.g. *B. subtilis*. The 95-kb plasmid p19 and 65-kb plasmid pLS20 of *B. subtilis* are self-transmissible (Poluektova et al. 2004; Itaya et al. 2006). pLS30 of *B. subtilis* contains *mob* gene and its recognition sequence, *oriT*, that features itself as a mobile plasmid between *Bacillus* species (Sakaya et al. 2006). Another interesting element is ICE*Bs1* from *B. subtilis*. When DNA damage or high concentrations of potential mating partners that lack the element are monitored, the integrated conjugative element ICE*Bs1* may excise from the chromosome of *B. subtilis* and transfer to recipients (Lee and Grossman 2007). ICE*Bs1* is integrated into the chromosomal *trnS-leu2* genes of some *B. subtilis* strains (Auchtung et al. 2005). The excision/integration of ICE*Bs1* is regulated by a Rap-phr system. When DNA damage (triggering SOS response) or high densities of potential mating partners not carrying the element are monitored, ICE*Bs1* can excise from the chromosome by site-specific recombination between its terminal 60 bp direct repeats (DR) to form a circular intermediate. At least two genes are involved in this process, the *int* and *xis* genes which encode the integrase and excisionase required for excision through site-specific recombination. A single strand of the circular intermediate may then transfer to a recipient cell in a conjugation-like mode, which also requires the identification of a *cis*-acting *oriT* and a DNA relaxase (Lee and Grossman 2007; Lee et al. 2007a). An ICE*Bs1*-encoded repressor ImmR regulates the expression of genes involved into excision and transfer (Auchtung et al. 2007). It is worth noting that the transfer of ICE*Bs1* is efficient and its insertion site specific, which could certainly bring new perspectives in genetic engineering Gram-positive bacteria.

5.6.2.4 Transposition and Recombination

In contrast to transformation, transduction and conjugation, which transfer DNA from one cell to another across the cell envelope via complex protein structures, intracellular genomic rearrangement can take place by intra- or intermolecular homologous recombination, site-specific recombination or transposition. Whereas homologous recombination refers to genetic material reciprocally exchanged between two similar or identical DNA molecules, transposition and site-specific recombination are processes in which a transposable element (TE) is copied or excised from its original location and inserted into a new target position. TEs comprise IS, transposons and conjugative transposons, carrying enzymes (integrase or transposase) that catalyse the DNA strand cleavages and transfers necessary for their movement. They are mostly delineated by short inverted repeated (IR) sequences and flanked by DR. Generally, the transposition requires the transposase to recognize the IRs of the element and to catalyse the cleavage and rejoining of the appropriate DNA strands by a series of transesterification reactions. By far the major class of TEs within transposons known at present encodes so-called DDE (Glutamate–Glutamate–Aspartate) transposases (TPase). The DRs are created at each end of the transposon after the transposon is inserted into the host DNA and the gaps are filled in (Chandler and Mahillon 2002; Mahillon and Chandler 1998).

The first *Bacillus* TE to be characterized was IS*231* from the entomopathogenic bacterium *B. thuringiensis,* where it is associated with delta-endotoxin genes (Mahillon et al. 1985). Since then, the iso-IS*231* elements have been observed among all the members of the *B. cereus sensu lato* group (Mahillon 2002). At present, large numbers of TEs are found in *Bacillus* species, mostly IS elements and some transposons. The largest IS family in *Bacillus* is IS*4*. Members of this group are mainly distributed among *B. cereus sensu lato* members but are also found in other endospore-forming species such as *B. subtilis* (IS*4Bsu1*), *Bacillus halodurans* (IS*641*), *Geobacillus stearothermophilus* (IS*4712*) and *Geobacillus kaustophilus* (S*Gka3*). IS*6* is also a large family observed in *Bacillus*, as illustrated by the IS*240*-like elements from *B. thuringiensis* and IS*Bwe2* from *B. weihenstephanensis.*

Another important *Bacillus* IS family is IS*3*. Members of this group generally have two consecutive and partially overlapping open reading frames. IS*655* from *B. halodurans*, IS*Bce13* from "*B. cereus* subsp. *cytotoxis*" NVH 391–98 and IS*Bt1* from *B. thuringiensis* all belong to this family (see http://www-is.biotoul.fr/). Two copies of IS*1627*, also belonging to the IS*3* family, were observed flanking a 45-kb segment of the pXO1 virulence plasmid of *B. anthracis*. Sixteen types of ISs, IS*641* to IS*643*, IS*650* to IS*658*, IS*660*, IS*662* and IS*663* and IS*Bha1* were identified in the genome of alkaliphilic *B. halodurans* C-125, which can be classified into 12 families (Takami et al. 2001). Besides these three major IS families, *Bacillus* species also contains various numbers of IS belonging to other families, e.g. IS*21*, IS*30*, IS*66*, IS*110*, IS*200*, IS*256*, IS*481*, IS*630*, IS*982* or IS*1182*. (For an updated IS database, see ISFinder at http://www-is.biotoul.fr/.)

In addition to IS elements, transposons are also found in *Bacillus,* represented by Tn*4430* (Mahillon and Lereclus 1988) and Tn*5401* (Baum 1995) from

B. thuringiensis, and Tn*XO1* from *B. anthracis* (Van der Auwera and Mahillon 2005). Compared to IS elements, transposons contain not only transposition-related cassettes but also "foreign" genes unrelated with transposition in many cases, such as antibiotic resistance genes or metabolic-related genes. In addition, the conjugative transposons and the ICEs combine the dual properties of transposition and conjugation, which makes it possible to mediate both the intracellular translocation and intercellular transfer of DNA. These are exemplified by the Tn*916*-like gene cluster and ICEBs1 in *B. subtilis*, respectively (Kunst et al. 1997; Lee et al. 2007a).

Besides IS and transposons, the mobile introns also mediate the transfer of genetic material in a transposition-related mechanism (when inserting into ectopic genomic locations) or in a so-called homing process (when inserting into cognate intron-less DNA sites) (Lambowitz and Belfort 1993). There are two main intron groups, I and II, which have been characterized in prokaryotic and eukaryotic genomes. They are catalytic RNAs (ribozymes) that are mobile and capable of self-splicing, and some of them have been shown experimentally to be able to invade new DNA sites and transfer between species. Homing of an intron into an intronless allele occurs by a process that is catalysed by the intron-encoded protein (IEP) (Guhan and Muniyappa 2003). Recently, the occurrence of introns in 29 sequenced genomes of the *B. cereus sensu lato* group has been analysed, and 73 group I introns and 77 group II introns (Tourasse and Kolstø 2008) were identified. It is interesting that group II introns tend to be located within mobile DNA elements such as plasmids, IS elements, transposons or pathogenicity islands. For instance, the group II introns B.th.I.1 and B.th.I.2 have been found in the T4SS transfer genetic region of the conjugative plasmid pAW63 from *B. thuringiensis* (Van der Auwera et al. 2005, 2008).

5.7 Concluding Remarks

The aerobic endospore-forming soil bacteria live in complex and changing environments and ecosystems. The formation of endospores during a certain period of the life cycle is certainly a key feature of these bacteria, which makes them dormant and resistant to drought, heat or other harsh conditions. Yet, their ability to become genetically competent for DNA uptake, to swim or swarm, to permanently "communicate" among themselves or with other community members, to act collectively and to perform many other complex and versatile physiological behaviours have made it possible for a large number of *Bacillus* species to enjoy the most optimal growth and survival in inhospitable environments. Moreover, gene exchange events between these organisms play important roles for their ecological adaptation and evolution. Through HGT, these bacteria can obtain chromosomal genes and gene clusters, as well as plasmids, transposons and prophage bearing adaptive genetic determinants. This pool of "non-essential", more "exotic", genes in fact drives the bacterial communities to new environments and is therefore essential for the expansion of their ecological niches.

References

Abajy MY, Kopec J, Schiwon K, Burzynski M, Döring M, Bohn C, Grohmann E (2007) A type IV-secretion-like system is required for conjugative DNA transport of broad-host-range plasmid pIP501 in Gram-positive bacteria. J Bacteriol 189:2487–2496

Agaisse H, Gominet M, Økstad OA, Kolstø AB, Lereclus D (1999) PlcR is a pleiotropic regulator of extracellular virulence factor gene expression in *Bacillus thuringiensis*. Mol Microbiol 32:1043–1053

Andrup L, Jørgensen O, Wilcks A, Smidt L, Jensen GB (1996) Mobilization of "nonmobilizable" plasmids by the aggregation-mediated conjugation system of *Bacillus thuringiensis*. Plasmid 36:75–85

Antunes LC, Schaefer AL, Ferreira RB, Qin N, Stevens AM, Ruby EG, Greenberg EP (2007) Transcriptome analysis of the *Vibrio fischeri* LuxR-LuxI regulon. J Bacteriol 189:8387–8391

Atmakuri K, Cascales E, Christie PJ (2004) Energetic components VirD4, VirB11 and VirB4 mediate early DNA transfer reactions required for bacterial type IV secretion. Mol Microbiol 54:1199–11211

Auchtung JM, Lee CA, Monson RE, Lehman AP, Grossman AD (2005) Regulation of a *Bacillus subtilis* mobile genetic element by intercellular signaling and the global DNA damage response. Proc Natl Acad Sci USA 102:12554–12559

Auchtung JM, Lee CA, Garrison KL, Grossman AD (2007) Identification and characterization of the immunity repressor (ImmR) that controls the mobile genetic element ICEBs1 of *Bacillus subtilis*. Mol Microbiol 64:1515–1528

Auger S, Krin E, Aymerich S, Gohar M (2006) Autoinducer 2 affects biofilm formation by *Bacillus cereus*. Appl Environ Microbiol 72:937–941

Auger S, Ramarao N, Faille C, Fouet A, Aymerich S, Gohar M (2009) Biofilm formation and cell-surface properties among pathogenic and non-pathogenic strains of the *Bacillus cereus* group. Appl Environ Microbiol 75:6616–6618

Barloy F, Delécluse A, Nicolas L, Lecadet MM (1996) Cloning and expression of the first anaerobic toxin gene from *Clostridium bifermentans* subsp. *malaysia*, encoding a new mosquitocidal protein with homologies to *Bacillus thuringiensis* delta endotoxins. J Bacteriol 178:3099–3105

Baum JA (1995) TnpI recombinase: identification of sites within Tn*5401* required for TnpI binding and site-specific recombination. J Bacteriol 177:4036–4042

Belliveau BH, Trevors JT (1989) Transformation of *B. cereus* vegetative cells by electroporation. Appl Environ Microbiol 55:1649–1652

Berka RM, Hahn J, Albano M, Draskovic I, Persuh M, Cui X, Sloma A, Widner W, Dubnau D (2002) Microarray analysis of the *Bacillus subtilis* K-state: genome-wide expression changes dependent on ComK. Mol Microbiol 43:1331–1345

Bhalla R, Dalal M, Panguluri SK, Jagadish B, Mandaokar AD, Singh AK, Kumar PA (2005) Isolation, characterization and expression of a novel vegetative insecticidal protein gene of *Bacillus thuringiensis*. FEMS Microbiol Lett 243:467–472

Bischofs IB, Hug JA, Liu AW, Wolf DM, Arkin AP (2009) Complexity in bacterial cell-cell communication: quorum signal integration and subpopulation signaling in the *Bacillus subtilis* phosphorelay. Proc Natl Acad Sci USA 106:6459–6464

Bongiorni C, Stoessel R, Shoemaker D, Perego M (2006) Rap phosphatase of virulence plasmid pXO1 inhibits *Bacillus anthracis* sporulation. J Bacteriol 188:487–498

Bouillaut L, Perchat S, Arold S, Zorrilla S, Slamti L, Henry C, Gohar M, Declerck N, Lereclus D (2008) Molecular basis for group-specific activation of the virulence regulator PlcR by PapR heptapeptides. Nucleic Acid Res 36:3791–3801

Boyer H (1966) Conjugation in *Escherichia coli*. J Bacteriol 91:1767–1772

Boyer M, Wisniewski-Dyé F (2009) Cell-cell signalling in bacteria: not simply a matter of quorum. FEMS Microbiol Ecol 70:1–19

Branda SS, Vik S, Friedman L, Kolter R (2005) Biofilms: the matrix revisited. Trends Microbiol 13:20–26
Burbulys D, Trach KA, Hoch JA (1991) Initiation of sporulation in *Bacillus subtilis* is controlled by a multicomponent phosphorelay. Cell 64:545–552
Burrus V, Pavlovic G, Decaris B, Guédon G (2002) Conjugative transposons: the tip of the iceberg. Mol Microbiol 46:601–610
Callahan SM, Dunlap PV (2000) LuxR- and acyl-homoserine-lactone-controlled non-lux genes define a quorum-sensing regulon in *Vibrio fischeri*. J Bacteriol 182:2811–2822
Cano RJ, Borucki MK (1995) Revival and identification of bacterial spores in 20- to 40-million-year-old Dominican amber. Science 268:1060–1064
Chai Y, Chu F, Kolter R, Losick R (2008) Biostability and biofilm formation in *Bacillus subtilis*. Mol Microbiol 67:254–263
Chandler M, Mahillon J (2002) Insertion sequences revisited. In: Craig NL, Craigie R, Gellert M, Lambowitz AM (eds) Mobile DNA II. American Society for Microbiology, Washington DC, pp 305–366
Chen I, Dubnau D (2003) DNA transport during transformation. Front Biosci 8:s544–s556
Chen I, Christie PJ, Dubnau D (2005) The ins and outs of DNA transfer in bacteria. Science 310:1456–1460
Chen I, Provvedi R, Dubnau D (2006) A macromolecular complex formed by a pilin-like protein in competent *Bacillus subtilis*. J Biol Chem 281:21720–21727
Christie PJ (2004) Type IV secretion: the *Agrobacterium* VirB/D4 and related conjugation systems. Biochim Biophys Acta 1694:219–234
Christie PJ, Cascales E (2005) Structural and dynamic properties of bacterial type IV secretion systems. Mol Membr Biol 22:51–61
Claverys JP, Martin B, Polard P (2009) The genetic transformation machinery: composition, localization, and mechanism. FEMS Microbiol Rev 33:643–656
Comella N, Grossman AD (2005) Conservation of genes and processes controlled by the quorum response in bacteria: characterization of genes controlled by the quorum-sensing transcription factor ComA in *Bacillus subtilis*. Mol Microbiol 57:1159–1174
Coohill TP, Sagripanti JL (2008) Overview of the inactivation by 254 nm ultraviolet radiation of bacteria with particular relevance to biodefense. Photochem Photobiol 84:1084–1090
Cowan AE, Olivastro EM, Koppel DE, Loshon CA, Setlow B, Setlow P (2004) Lipids in the inner membrane of dormant spores of *Bacillus* species are largely immobile. Proc Natl Acad Sci USA 101:7733–7738
Craig L, Li J (2008) Type IV pili: paradoxes in form and function. Curr Opin Struct Biol 18:267–277
Davaine C (1863) Recherches sur les infusoires du sang dans la maladie de la pustule maligne. C R Acad Sci 60:1296–1299
Dragon DC, Rennie RP (1995) The ecology of anthrax spores: tough but not invincible. Can Vet J 36:295–301
Dubnau D (1999) DNA uptake in bacteria. Annu Rev Microbiol 53:217–244
Dunlap PV (1999) Quorum regulation of luminescence in *Vibrio fischeri*. J Mol Microbiol Biotechnol 1:5–12
Engebrecht J, Silverman M (1984) Identification of genes and gene products necessary for bacterial bioluminescence. Proc Natl Acad Sci USA 81:4154–4158
Falcão JP, Sharp F, Sperandio V (2004) Cell-to-cell signaling in intestinal pathogens. Curr Issues Intest Microbiol 5:9–17
Feitelson JS, Payne J, Kim L (1992) *Bacillus thuringiensis*: insects and beyond. Bio/Technology 10:271–275
Fenchel T (2002) Microbial behavior in a heterogeneous world. Science 296:1068–1071
Ferreira LHPL, Suzuki MT, Itano EN, Ono MA, Arantes OMN (2003) Ecological aspects of *Bacillus thuringiensis* in an Oxisol. Scientia Agricola 60:199–222
Fraser GM, Hughes C (1999) Swarming motility. Curr Opin Microbiol 2:630–635

Gammon K, Jones GW, Hope SJ, de Oliveira CM, Regis L, Silva Filha MH, Dancer BN, Berry C (2006) Conjugal transfer of a toxin-coding megaplasmid from *Bacillus thuringiensis* subsp. *israelensis* to mosquitocidal strains of *Bacillus sphaericus*. Appl Environ Microbiol 72:1766–1770

Gohar M, Faegri K, Perchat S, Ravnum S, Økstad OA, Gominet M, Kolstø AB, Lereclus D (2008) The PlcR virulence regulon of *Bacillus cereus*. PLoS One 3:e2793

González JM Jr, Brown BJ, Carlton BC (1982) Transfer of *Bacillus thuringiensis* plasmids among strains of *B. thuringiensis* and *B. cereus*. Proc Natl Acad Sci USA 79:6951–6955

Grohmann E, Muth G, Espinosa M (2003) Conjugative plasmid transfer in Gram-positive bacteria. Microbiol Mol Biol Rev 67:277–301

Guhan N, Muniyappa K (2003) Structural and functional characteristics of homing endonucleases. Crit Rev Biochem Mol Biol 38:199–248

Harrington LC, Rogerson AC (1990) The F pilus of *Escherichia coli* appears to support stable DNA transfer in the absence of wall-to-wall contact between cells. J Bacteriol 172:7263–7264

Harshey RM (2003) Bacterial motility on a surface: many ways to a common goal. Annu Rev Microbiol 57:249–273

Henrichsen J (1972) Bacterial surface translocation: a survey and a classification. Bacteriol Rev 36:478–503

Hu X, Hansen BM, Eilenberg J, Hendriksen NB, Smidt L, Yuan Z, Jensen GB (2004) Conjugative transfer, stability and expression of a plasmid encoding a *cry*1Ac gene in *Bacillus cereus* group strains. FEMS Microbiol Lett 231:45–52

Hu X, Hansen BM, Johansen JE, Hendriksen NB, Smidt L, Jensen GB, Yuan Z (2005) Transfer and expression of the mosquitocidal plasmid pBtoxis in *Bacillus cereus* group strains. FEMS Microbiol Lett 245:239–247

Hu X, Swiecicka I, Timmery S, Mahillon J (2009a) Sympatric soil communities of *Bacillus cereus sensu lato*: population structure and potential plasmid dynamics of pXO1- and pXO2-like elements. FEMS Microbiol Ecol 70:344–355

Hu X, Van der Auwera G, Timmery S, Zhu L, Mahillon J (2009b) Distribution, diversity, and potential mobility of extrachromosomal elements related to the *Bacillus anthracis* pXO1 and pXO2 virulence plasmids. Appl Environ Microbiol 75:3016–3028

Inamine GS, Dubnau D (1995) ComEA, a *Bacillus subtilis* integral membrane protein required for genetic transformation, is needed for both DNA binding and transport. J Bacteriol 177:3045–3051

Irie Y, Parsek MR (2008) Quorum sensing and microbial biofilms. Curr Top Microbiol Immunol 322:67–84

Itaya M, Sakaya N, Matsunaga S, Fujita K, Kaneko S (2006) Conjugational transfer kinetics of pLS20 between *Bacillus subtilis* in liquid medium. Biosci Biotechnol Biochem 70:740–742

Jarrett P, Stephenson M (1990) Plasmid transfer between strains of *Bacillus thuringiensis* infecting *Galleria mellonella* and *Spodoptera littoralis*. Appl Environ Microbiol 56:1608–1614

Jensen GB, Wilcks A, Petersen SS, Damgaard J, Baum JA, Andrup L (1995) The genetic basis of the aggregation system in *Bacillus thuringiensis* subsp. *israelensis* is located on the large conjugative plasmid pXO16. J Bacteriol 177:2914–2917

Jensen GB, Andrup L, Wilcks A, Smidt L, Poulsen OM (1996) The aggregation-mediated conjugation system of *Bacillus thuringiensis* subsp. *israelensis*: host range and kinetics of transfer. Curr Microbiol 33:228–236

Jensen GB, Hansen BM, Eilenberg J, Mahillon J (2003) The hidden lifestyles of *Bacillus cereus* and relatives. Environ Microbiol 5:631–640

Jones MB, Blaser MJ (2003) Detection of a *luxS*-signaling molecule in *Bacillus anthracis*. Infect Immun 71:3914–3919

Jones GW, Nielsen-Leroux C, Yang Y, Yuan Z, Dumas VF, Monnerat RG, Berry C (2007) A new Cry toxin with a unique two-component dependency from *Bacillus sphaericus*. FASEB J 21:4112–4120

Kaplan HB, Greenberg EP (1985) Diffusion of autoinducer is involved in regulation of the *Vibrio fischeri* luminescence system. J Bacteriol 163:1210–1214

Karatan E, Saulmon MM, Bunn MW, Ordal GW (2001) Phosphorylation of the response regulator CheV is required for adaptation to attractants during *Bacillus subtilis* chemotaxis. J Biol Chem 276:43618–43626

Kearns DB, Chu F, Branda SS, Kolter R, Losick R (2005) A master regulator for biofilm formation by *Bacillus subtilis*. Mol Microbiol 55:739–749

Kidane D, Carrasco B, Manfredi C, Rothmaier K, Ayora S, Tadesse S, Alonso JC, Graumann PL (2009) Evidence for different pathways during horizontal gene transfer in competent *Bacillus subtilis* cells. PLoS Genet 5:e1000630

Kinsinger RF, Shirk MC, Fall R (2003) Rapid surface motility in *Bacillus subtilis* is dependent on extracellular surfactin and potassium ion. J Bacteriol 185:5627–5631

Knudson GB (1986) Photoreactivation of ultraviolet-irradiated, plasmid-bearing, and plasmid-free strains of *Bacillus anthracis*. Appl Environ Microbiol 52:444–449

Kobayashi K (2008) SlrR/SlrA controls the initiation of biofilm formation in *Bacillus subtilis*. Mol Microbiol 69:1399–1410

Koehler TM (2002) *Bacillus anthracis* genetics and virulence gene regulation. Curr Top Microbiol Immunol 271:143–164

Koehler TM (2009) *Bacillus anthracis* physiology and genetics. Mol Aspects Med 30:386–396

Kunst F, Ogasawara N, Moszer I, Albertini AM, Alloni G, Azevedo V, Bertero MG, Bessières P, Bolotin A, Borchert S, Borriss R, Boursier L, Brans A, Braun M, Brignell SC, Bron S, Brouillet S, Bruschi CV, Caldwell B, Capuano V, Carter NM, Choi S-K, Codani J-J, Connerton IF, Cummings NJ, Daniel RA, Denizot F, Devine KM, Düsterhöft A, Ehrlich SD, Emmerson RT, Entian KD, Errington J, Fabret C, Ferrari E, Foulger D, Fritz C, Fujita M, Fujita Y, Fuma S, Galizzi A, Galleron N, Ghim S-Y, Glaser P, Goffeau A, Golightly EJ, Grandi G, Guiseppi G, Guy BJ, Haga K, Haiech J, Harwood CR, Hénaut A, Hilbert H, Holsappel S, Hosono S, Hullo M-F, Itaya M, Jones L, Joris B, Karamata D, Kasahara Y, Klaer-Blanchard M, Klein C, Kobayashi Y, Koetter P, Koningstein G, Krogh S, Kumano M, Kurita K, Lapidus A, Lardinois L, Lauber J, Lazarevic V, Lee S-M, Levine A, Liu H, Masuda S, Mauël C, Médigue C, Medina N, Mellado RP, Mizuno M, Moestl D, Nakai S, Noback M, Noone D, O'Reilly M, Ogawa K, Ogiwara A, Oudega B, Park S-H, Parro V, Pohl TM, Portetelle D, Porwollik S, Prescott AM, Presecan E, Pujic P, Purnelle B, Rapoport G, Rey M, Reynolds S, Rieger M, Rivolta C, Rocha E, Roche B, Rose M, Sadaie Y, Sato T, Scanlan E, Schleich S, Schroeter R, Scoffone F, Sekiguchi J, Sekowska A, Seror SJ, Serror P, Shin B-S, Soldo B, Sorokin A, Tacconi E, Takagi T, Takahashi H, Takemaru K, Tacheuchi M, Tamakoshi A, Tanaka R, Terpstra P, Tognoni A, Tosato V, Uchiyama S, Vandenbol M, Vannier F, Vassarotti A, Viari A, Wambutt R, Wedler E, Wedler H, Weitzenegger T, Winters P, Wipat A, Yamamoto H, Yamane K, Yasumoto K, Yata K, Yoshida K, Yoshikawa H-F, Zumstein E, Yoshikawa H, Danchin A (1997) The complete genome sequence of the Gram-positive bacterium *Bacillus subtilis*. Nature 390:249–256

Lambowitz AM, Belfort M (1993) Introns as mobile genetic elements. Annu Rev Biochem 62:587–622

Lanigan-Gerdes S, Briceno G, Dooley AN, Faull KF, Lazazzera BA (2008) Identification of residues important for cleavage of the extracellular signaling peptide CSF of *Bacillus subtilis* from its precursor protein. J Bacteriol 190:6668–6675

Lasa I (2006) Towards the identification of the common features of bacterial biofilm development. Int Microbiol 9:21–28

Lazazzera BA (2001) The intracellular function of extracellular signaling peptides. Peptides 22:1519–1527

Lee CA, Grossman AD (2007) Identification of the origin of transfer (*oriT*) and DNA relaxase required for conjugation of the integrative and conjugative element ICEBs1 of *Bacillus subtilis*. J Bacteriol 189:7254–7261

Lee CA, Auchtung JM, Monson RE, Grossman AD (2007a) Identification and characterization of int (integrase), xis (excisionase) and chromosomal attachment sites of the integrative and conjugative element ICEBs1 of *Bacillus subtilis*. Mol Microbiol 66:1356–1369

Lee K, Costerton JW, Ravel J, Auerbach RK, Wagner DM, Keim P, Leid JG (2007b) Phenotypic and functional characterization of *Bacillus anthracis* biofilms. Microbiology 153:1693–1701

Liébano E, Bravo A, Herrera D, Godínes E, Vargas P, Zamudio F (2006) Use of *Bacillus thuringiensis* toxin as an alternative method of control against *Haemonchus contortus*. Ann NY Acad Sci 1081:347–354

Lindeque PM, Turnbull PC (1994) Ecology and epidemiology of anthrax in the Etosha National Park, Namibia. Onderstepoort J Vet Res 61:71–83

López D, Fischbach MA, Chu F, Losick R, Kolter R (2009a) Structurally diverse natural products that cause potassium leakage trigger multicellularity in *Bacillus subtilis*. Proc Natl Acad Sci USA 106:280–285

López D, Vlamakis H, Kolter R (2009b) Generation of multiple cell types in *Bacillus subtilis*. FEMS Microbiol Rev 33:152–163

Luna VA, Cannons AC, Amuso PT, Cattani J (2008) The inactivation and removal of airborne *Bacillus atrophaeus* endospores from air circulation systems using UVC and HEPA filters. J Appl Microbiol 104:489–498

Lux R, Shi W (2004) Chemotaxis-guided movements in bacteria. Crit Rev Oral Biol Med 15:207–220

Mahillon J (2002) Insertion sequence elements and transposons in *Bacillus*. In: Berkeley R, Heyndrickx M, Logan N, de Vos P (eds) *Applications and systematics of* Bacillus *and relatives*. Blackwell Science, London, pp 236–253

Mahillon J, Chandler M (1998) Insertion sequences. Microbiol Mol Biol Rev 62:725–774

Mahillon J, Lereclus D (1988) Structural and functional analysis of Tn*4430*: identification of an integrase-like protein involved in the co-integrate-resolution process. EMBO J 7:1515–1526

Mahillon J, Seurinck J, van Rompuy L, Delcour J, Zabeau M (1985) Nucleotide sequence and structural organization of an insertion sequence element (IS*231*) from *Bacillus thuringiensis* strain Berliner 1715. EMBO J 4:3895–3899

Mahillon J, Chungjatupornchai W, Decock J, Dierickx S, Michiels F, Peferoen M, Joos H (1989) Transformation of *Bacillus thuringiensis* by electroporation. FEMS Microbiol Lett 60:205–210

Medini D, Donati C, Tettelin H, Masignani V, Rappuoli R (2005) The microbial pan-genome. Curr Opin Genet Dev 15:589–594

Merlin C, Mahillon J, Nesvera J, Toussaint A (2000) Gene recruiters and transporters: the molecular structure of bacterial mobile elements. In: Thomas C (ed) The horizontal gene pool. Harwood Academic, London, pp 363–409

Middleton R, Sjölander K, Krishnamurthy N, Foley J, Zambryski P (2005) Predicted hexameric structure of the *Agrobacterium* VirB4 C terminus suggests VirB4 acts as a docking site during type IV secretion. Proc Natl Acad Sci USA 102:1685–1690

Miller MB, Bassler BL (2001) Quorum sensing in bacteria. Annu Rev Microbiol 55:165–199

Mock M, Fouet A (2001) Anthrax. Annu Rev Microbiol 55:647–671

Muff TJ, Ordal GW (2007) The CheC phosphatase regulates chemotactic adaptation through CheD. J Biol Chem 282:34120–34128

Nealson KH, Hastings JW (1979) Bacterial bioluminescence: its control and ecological significance. Microbiol Rev 43:496–518

Nealson KH, Hastings JW (2006) Quorum sensing on a global scale: massive numbers of bioluminescent bacteria make milky seas. Appl Environ Microbiol 72:2295–2297

Nicholson WL, Galeano B (2003) UV resistance of *Bacillus anthracis* spores revisited: validation of *Bacillus subtilis* spores as UV surrogates for spores of *B. anthracis* Sterne. Appl Environ Microbiol 69:1327–1330

Ogura M, Tanaka T (2009) The *Bacillus subtilis* late competence operon *comE* is transcriptionally regulated by *yutB* and under post-transcription initiation control by *comN* (*yrzD*). J Bacteriol 191:949–958

Okinaka RT, Cloud K, Hampton O, Hoffmaster A, Hill K, Keim P, Koehler T, Lamke G, Kumano S, Manter D, Martinez Y, Ricke D, Svensson R, Jackson P (1999) Sequence, assembly, and analysis of pXO1 and pXO2. J Appl Microbiol 87:261–262
Paidhungat M, Ragkousi K, Setlow P (2001) Genetic requirements for induction of germination of spores of *Bacillus subtilis* by Ca(2+)-dipicolinate. J Bacteriol 183:4886–4893
Park S, Rittmann BE, Bae W (2009) Life-cycle kinetic model for endospore-forming bacteria, including germination and sporulation. Biotechnol Bioeng 104:1012–1024
Petras SF, Casida LE Jr (1985) Survival of *Bacillus thuringiensis* spores in soil. Appl Environ Microbiol 50:1496–1501
Poluektova EU, Fedorina EA, Lotareva OV, Prozorov AA (2004) Plasmid transfer in bacilli by a self-transmissible plasmid p19 from a *Bacillus subtilis* soil strain. Plasmid 52:212–217
Pottathil M, Lazazzera BA (2003) The extracellular Phr peptide-rap phosphatase signaling circuit of *Bacillus subtilis*. Front Biosci 8:D32–D45
Priest FG, Ebdrup L, Zahner V, Carter P (1997) Distribution and characterization of mosquitocidal toxin genes in some strains of *Bacillus sphaericus*. Appl Environ Microbiol 63:1195–1198
Provvedi R, Dubnau D (1999) ComEA is a DNA receptor for transformation of competent *Bacillus subtilis*. Mol Microbiol 31:271–280
Rao CV, Glekas GD, Ordal GW (2008) The three adaptation systems of *Bacillus subtilis* chemotaxis. Trends Microbiol 16:480–487
Rayer P (1850) Inoculation du sang de rate. C R Soc Biol Paris 2:141–144
Reading NC, Sperandio V (2006) Quorum sensing: the many languages of bacteria. FEMS Microbiol Lett 254:1–11
Reddy A, Battisti L, Thorne CB (1987) Identification of self-transmissible plasmids in four *Bacillus thuringiensis* subspecies. J Bacteriol 169:5263–5270
Roth NG, Livery DH, Hodge HM (1955) Influence of oxygen uptake and age of culture on sporulation of *Bacillus anthracis* and *Bacillus globigii*. J Bacteriol 69:455–459
Ryu JH, Beuchat LR (2005) Biofilm formation and sporulation by *Bacillus cereus* on a stainless steel surface and subsequent resistance of vegetative cells and spores to chlorine, chlorine dioxide, and a peroxyacetic acid-based sanitizer. J Food Prot 68:2614–2622
Saile E, Koehler TM (2006) *Bacillus anthracis* multiplication, persistence, and genetic exchange in the rhizosphere of grass plants. Appl Environ Microbiol 72:3168–3174
Sakaya N, Kaneko S, Matsunaga S, Itaya M (2006) Experimental basis for a stable plasmid, pLS30, to shuttle between *Bacillus subtilis* species by conjugational transfer. J Biochem 139:557–561
Sarrafzadeh MH, Navarro JM (2006) The effect of oxygen on the sporulation, delta-endotoxin synthesis and toxicity of *Bacillus thuringiensis* H14. World J Microbiol Biotechnol 22:305–310
Schröder G, Lanka E (2005) The mating pair formation system of conjugative plasmids: a versatile secretion machinery for transfer of proteins and DNA. Plasmid 54:1–25
Scott JR, Churchward GG (1995) Conjugative transposition. Annu Rev Microbiol 49:367–397
Selvapandiyan A, Arora N, Rajagopal R, Jalali SK, Venkatesan T, Singh SP, Bhatnagar RK (2001) Toxicity analysis of N- and C-terminus-deleted vegetative insecticidal protein from *Bacillus thuringiensis*. Appl Environ Microbiol 67:5855–5858
Shapiro JA (1998) Thinking about bacterial populations as multicellular organisms. Annu Rev Microbiol 52:81–104
Slamti L, Lereclus D (2002) A cell-cell signaling peptide activates the PlcR virulence regulon in bacteria of the *Bacillus cereus* group. EMBO J 21:4550–4559
Smith RA, Barry JW (1998) Environmental persistence of *Bacillus thuringiensis* spores following aerial application. J Invert Pathol 71:263–267
Stachel SE, Nester EW (1986) The genetic and transcriptional organization of the *vir* region of the A6 Ti plasmid of *Agrobacterium tumefaciens*. EMBO J 5:1445–1454
Stenfors Arnesen LP, Fagerlund A, Granum PE (2008) From soil to gut: *Bacillus cereus* and its food poisoning toxins. FEMS Microbiol Rev 32:579–606

Tadesse S, Graumann PL (2007) DprA/Smf protein localizes at the DNA uptake machinery in competent *Bacillus subtilis* cells. BMC Microbiol 7:105

Takami H, Han CG, Takaki Y, Ohtsubo E (2001) Identification and distribution of new insertion sequences in the genome of alkaliphilic *Bacillus halodurans* C-125. J Bacteriol 183:4345–4356

Tamez-Guerra P, McGuire MR, Behle RW, Shasha BS, Wong LJ (2000) Assessment of microencapsulated formulations for improved residual activity of *Bacillus thuringiensis*. J Econ Entomol 93:219–225

Taylor MJ, Thorne CB (1963) Transduction of *Bacillus licheniformis* and *Bacillus subtilis* by each of two phages. J Bacteriol 86:452–461

Thomas DJ, Morgan JA, Whipps JM, Saunders JR (2000) Plasmid transfer between the *Bacillus thuringiensis* subspecies *kurstaki* and *tenebrionis* in laboratory culture and soil and in lepidopteran and coleopteran larvae. Appl Environ Microbiol 66:118–124

Thomas DJ, Morgan JA, Whipps JM, Saunders JR (2001) Plasmid transfer between *Bacillus thuringiensis* subsp. *israelensis* strains in laboratory culture, river water, and dipteran larvae. Appl Environ Microbiol 67:330–338

Thorne CB (1968) Transduction in *Bacillus cereus* and *Bacillus anthracis*. Bacteriol Rev 32:358–361

Thorne CB (1978) Transduction in *Bacillus thuringiensis*. Appl Environ Microbiol 35:1109–1115

Timmery S, Modrie P, Minet O, Mahillon J (2009) Plasmid capture by the *Bacillus thuringiensis* conjugative plasmid pXO16. J Bacteriol 191:2197–2205

Tourasse NJ, Kolstø AB (2008) Survey of group I and group II introns in 29 sequenced genomes of the *Bacillus cereus* group: insights into their spread and evolution. Nucleic Acids Res 36:4529–4548

Turgeon N, Laflamme C, Ho J, Duchaine C (2006) Elaboration of an electroporation protocol for *Bacillus cereus* ATCC 14579. J Microbiol Methods 67:543–548

Turnbull PCB (2002) Introduction: anthrax history, disease and ecology. Curr Top Microbiol Immunol 271:1–19

Van der Auwera GA (2007) pAW63: a molecular wanderer in the *Bacillus cereus* gene pool. Ph.D. thesis. UCL, 296 pp

Van der Auwera G, Mahillon J (2005) TnXO1, a germination-associated class II transposon from *Bacillus anthracis*. Plasmid 53:251–257

Van der Auwera GA, Andrup L, Mahillon J (2005) Conjugative plasmid pAW63 brings new insights into the genesis of the *Bacillus anthracis* virulence plasmid pXO2 and of the *Bacillus thuringiensis* plasmid pBT9727. BMC Genomics 6:103

Van der Auwera GA, Timmery S, Mahillon J (2008) Self-transfer and mobilisation capabilities of the pXO2-like plasmid pBT9727 from *Bacillus thuringiensis* subsp. *konkukian* 97–27. Plasmid 59:134–138

van Frankenhuyzen K (2009) Insecticidal activity of *Bacillus thuringiensis* crystal proteins. J Invert Pathol 101:1–16

Van Ness GB (1971) Ecology of anthrax. Science 172:1303–1307

Vilas-Bôas LA, Vilas-Bôas GF, Saridakis HO, Lemos MV, Lereclus D, Arantes OM (2000) Survival and conjugation of *Bacillus thuringiensis* in a soil microcosm. FEMS Microbiol Ecol 31:255–259

Vlamakis H, Aguilar C, Losick R, Kolter R (2008) Control of cell fate by the formation of an architecturally complex bacterial community. Genes Dev 22:945–953

Vreeland RH, Rosenzweig WD, Powers DW (2000) Isolation of a 250 million-year-old halotolerant bacterium from a primary salt crystal. Nature 407:897–900

Vu B, Chen M, Crawford RJ, Ivanova EP (2009) Bacterial extracellular polysaccharides involved in biofilm formation. Molecules 14:2535–2554

White CE, Winans SC (2007) Cell-cell communication in the plant pathogen *Agrobacterium tumefaciens*. Philos Trans R Soc Lond B Biol Sci 362:1135–1148

Yelton DB, Thorne CB (1971) Comparison of *Bacillus cereus* bacteriophages CP-51 and CP-53. J Virol 8:242–253

Yeo HJ, Waksman G (2004) Unveiling molecular scaffolds of the type IV secretion system. J Bacteriol 186:1919–1926

Yuan YM, Hu XM, Liu HZ, Hansen BM, Yan JP, Yuan ZM (2007) Kinetics of plasmid transfer among *Bacillus cereus* group strains within lepidopteran larvae. Arch Microbiol 187:425–431

Zaneveld JR, Nemergut DR, Knight R (2008) Are all horizontal gene transfers created equal? Prospects for mechanism-based studies of HGT patterns. Microbiology 154:1–15

Zhao J, Krishna V, Hua B, Moudgil B, Koopman B (2009) Effect of UVA irradiance on photocatalytic and UVA inactivation of *Bacillus cereus* spores. J Photochem Photobiol B 94:96–100

Chapter 6
Studying the Life Cycle of Aerobic Endospore-forming Bacteria in Soil

Volker S. Brözel, Yun Luo, and Sebastien Vilain

6.1 Introduction

Aerobic spore-forming bacteria of the genus *Bacillus* are commonly isolated from soils, at a range of depths, altitudes, and under various climatic conditions (Mishustin 1972; von Stetten et al. 1999; Garbeva et al. 2003). The most commonly observed soil-isolated *Bacillus* are members of the *B. cereus* group (von Stetten et al. 1999). *B. cereus sensu lato* comprises the species *B. cereus*, *B. anthracis*, *B. thuringiensis*, *B. mycoides*, *B. pseudomycoides* and *B. weihenstephanensis* (Jensen et al. 2003; Priest et al. 2004). Members of this group are genetically very closely related, but their precise phylogenetic and taxonomic relationships are still debated (Helgason et al. 2004). *B. cereus* is widely reported as a soil bacterium and also as a food poisoning bacterium that can occasionally be an opportunistic human pathogen (Schoeni and Wong 2005). Some isolates of *B. cereus* are found in the rhizosphere and some produce antibiotics active against certain plant pathogenic fungi (Stabb et al. 1994; Handelsman and Stabb 1996). *B. thuringiensis* is pathogenic to various insects by producing plasmid-encoded insecticidal proteins (Chattopadhyay et al. 2004), and *B. anthracis* is a pathogen of mammals (Ivanova et al. 2003). All may reportedly be isolated from soils, but our knowledge of their ecology is far from complete.

V.S. Brözel (✉)
Department of Biology and Microbiology, South Dakota State University, SNP252B, Rotunda Lane, Box 2140D, Brookings, SD 57007, USA
e-mail: volker.brozel@sdstate.edu

Y. Luo
Department of Microbiology, Cornell University, Ithaca, NY, USA
e-mail: yl423@cornell.edu

S. Vilain
Laboratoire de Biotechnologie des Protéines Recombinantes à Visée Santé (EA4135), ESTBB, Université Victor Segalen Bordeaux 2, 33076 Bordeaux cedex, France
e-mail: Sebastien.vilain@u-bordeaux2

N.A. Logan and P. De Vos (eds.), *Endospore-forming Soil Bacteria*, Soil Biology 27,
DOI 10.1007/978-3-642-19577-8_6, © Springer-Verlag Berlin Heidelberg 2011

6.1.1 *Does* Bacillus *Grow in Soil?*

Early microbiologists viewed *Bacillus* as "saprophytic organisms whose natural habitat is the soil" (Henrici 1934). *Bacillus* species were reported to occur in soil as spores, germinating and becoming active only when readily decomposable organic matter was available (Waksman 1932). Some later reports implied growth of *B. thuringiensis* in soil by suggesting that it germinated in soil at a pH above 6.0 and temperatures above 15.5°C (Saleh et al. 1970). Some have drawn correlations between anthrax outbreaks and soil pH, arguing that *B. anthracis* is able to grow in specific soil types (Van Ness 1971). *B. megaterium* is able to grow in soil, as supported by proof for in situ transcription of the exponentially expressed protease gene *nprM* (Honerlage et al. 1995). Yet all recent literature implies that *B. cereus sensu lato* does not grow in soil (Jensen et al. 2003; Dragon et al. 2005). The paradigm would hold that *B. cereus sensu lato* germinate and grow in an animal host, resulting in either symbiotic or pathogenic interactions. So *B. anthracis* is able to grow in various mammalian species, *B. thuringiensis* proliferates in the guts of various insects before killing its host, and *B. cereus* is a gut commensal of various insects (Margulis et al. 1998; Jensen et al. 2003). Defecation or death of the host leads to release of cells and spores into the soil where the vegetative cells may sporulate and survive until their uptake by another host (Dragon and Rennie 1995; Jensen et al. 2003; Dragon et al. 2005).

6.1.2 *A Microbiologist's Perspective on Soil*

Soil is one of two sources and sinks of nutrients and wastes required to make Earth supportive of life. It is Earth's most vital organ next to the oceans. While we know a considerable amount about the physical structure, chemistry and microbial diversity in soil, there is a dearth of knowledge regarding the dynamic processes occurring between the bacterial community and soil organic chemistry. Litterfall constitutes a considerable import of organic matter to forest soil, composed of 78% leaves, 20% flowers, seeds and related organs, and 2% twigs and bark (Rihani et al. 1995). While litter decomposition is driven largely by fungal activity (Tiunov and Scheu 2004), the high density of bacteria in soils is only sustainable if this group has access to a significant proportion of organic matter. Soils support up to 1.5×10^{10} bacteria per gram (Torsvik et al. 1990; Torsvik and Øvreås 2002), suggesting that bacteria form a significant part of the organic-matter-consuming community. The diversity of soil bacterial communities is also high (Rappe and Giovannoni 2003), but the majority of this diversity remains uncharacterized beyond its 16S rDNA gene pool (Pace 1997; Fierer et al. 2007). The role played by bacteria in soils, and specifically in the conversion of organic compounds in soil, is undisputed and yet poorly understood. While taxa involved in select processes such as the nitrogen cycle and methane production and consumption have been studied extensively in situ, the ecophysiological characteristics of the majority of soil bacteria are largely unknown (Fierer et al. 2007). Information on the available gene pool through metagenomic approaches has progressed in recent years (Daniel 2004; Schloss and Handelsman 2006), but presence of a gene cannot be equated to function. While metagenomic approaches have shed light on the diversity and distribution of biocatalytic potential in soils, little is known regarding the seasonal flux of organic matter as it is mineralized through bacterial activity during the course of a year (Fierer and Jackson 2006).

Soil is a complex assemblage composed of a range of mineral components with varying concentrations of organic and inorganic matter. Historically, the organic matter is divided into solid soil organic matter (SOM) and dissolved organic matter (DOM). The DOM in soils is a cocktail of sugars derived from plant polymers, aromatic organic compounds derived from lignin, some oligomeric sugar derivatives derived from cellulose and hemicellulose, and fatty acids between C_{14} and C_{54}, believed to derive from both plant wall material and dead bacteria (Huang et al. 1998; Kalbitz et al. 2000). A proportion of the DOM is collectively termed humic acid. The concentrations of solutes such as amino acids range from 0.1 to 5 μM. Monoprotic acids (e.g. formate, acetate and lactate) range from 1 μM to 1,000 μM, and di- and trivalent low molecular organic acids (e.g. oxalate, malate and citrate) from 0.1 to 50 μM (Strobel 2001; Pizzeghello et al. 2006). Monomeric intermediates such as carboxylic acids and amino acids have residence times in the order of hours in soils (Jones et al. 2005; Van Hees et al. 2005). Carbohydrates such as mono-, di- and oligosaccharides vary in their presence and concentration (Lynch 1982; Guggenberger and Zech 1993a, b; Kaiser et al. 2001; Kalbitz et al. 2003). While many soil-associated species of bacteria are able to metabolize glucose very rapidly and to almost zero basal levels under laboratory conditions, glucose is present in soils up to 100 μM concentrations (Olsen and Bakken 1987; Schneckenberger et al. 2008; Liebeke et al. 2009), and has a very long residence time in soils, in the order of 100 days (Saggar et al. 1999).

6.1.3 Growth of Bacillus *in Soil*

6.1.3.1 Growth in Liquid Soil Extract

Aqueous extracts of soil have been used as basis in agar culture media for the isolation of bacteria from soil for decades (Olsen and Bakken 1987). Recently, we reported an approach to preparing both liquid and solid medium prepared from soil without exposure to the high temperature of autoclaving – soil-extracted soluble organic matter (SESOM). A protocol is given below (see Sect. 6.3.1.1). Agar can be prepared by mixing filter-sterilized SESOM with highly concentrated autoclaved agar (see Sect. 6.3.1.2, below). SESOM prepared from a variety of soils is able to support growth of *B. cereus* ATCC 14579 (Fig. 6.1), and a number of other Gram-positive and Gram-negative species isolated from soil (Davis et al. 2005; Liebeke et al. 2009).

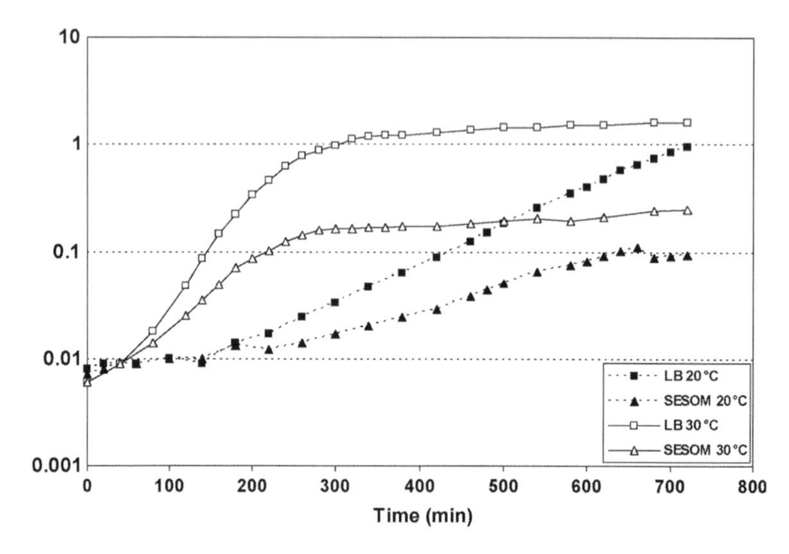

Fig. 6.1 Growth curves of *B. cereus* in SESOM and LB broth, incubated at 20 and 30°C while shaking at 150 rpm

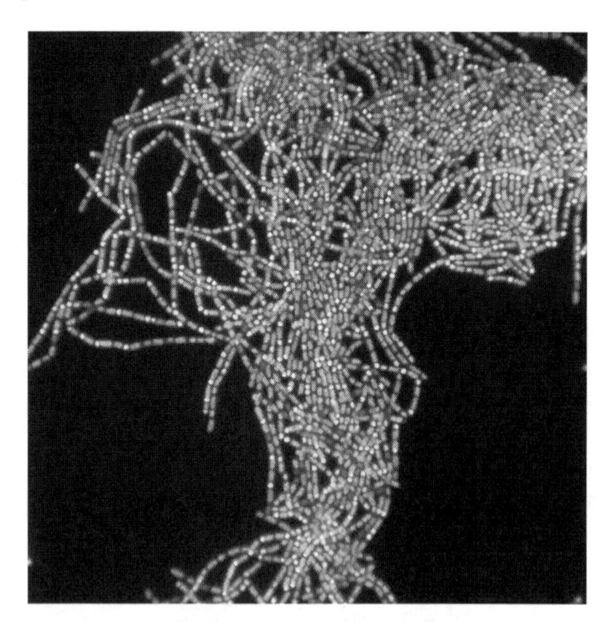

Fig. 6.2 Morphology of self-assembling biofilms of *B. cereus* growing aerobically in SESOM, while shaking at 150 rpm to transition phase at 30°C. Culture samples were stained with SYTO 9 and propidium iodide, and visualized by fluorescence microscopy

Furthermore, SESOM supports the germination of spores, albeit with a rather extended lag phase (Vilain et al. 2006). During transition phase, *B. cereus* begins to form long chains that coalesce to form coherent bundles of filaments or clumps

(Fig. 6.2). While *B. mycoides* displays this phenotype in various culture media, presumably in a constitutive manner (Di Franco et al. 2002), *B. cereus* does so in response to growth in SESOM. This self-assembly of a multicellular structure is reminiscent of the pellicle formed at the liquid–air interface by *B. subtilis* (Branda et al. 2001). While a pellicle forms at the surface of stagnant liquid, and is easily disturbed, clumps of *B. cereus* form in the bulk liquid despite shaking. These clumps are robust, withstanding repeated pipetting, and are indicative of strong terminal and lateral cell–cell interactions.

Biofilms, that is to say interface-associated multicellular assemblages, have been described as the second bacterial mode of growth after the planktonic free-floating or phase (Costerton et al. 1995). Biofilms are structured communities of microbial cells occurring at interfaces, and are mostly encased in a self-produced polymeric matrix (O'Toole et al. 2000; Sutherland 2001). Such biofilm-supporting interfaces include solid–liquid, liquid–air and oil–water interfaces (Costerton et al. 1995; Davey and O'Toole 2000; Branda et al. 2001). The clumps formed by *B. cereus* in SESOM could be viewed as representing a unique class of biofilm that forms in the absence of an interface. In addition, *B. cereus* is also able to form classical biofilms at solid–liquid interfaces (Oosthuizen et al. 2002; Vilain and Brozel 2006).

Biofilm formation is thought to begin when bacteria sense environmental conditions that trigger the transition from planktonic existence to life associated with interface. Nutritional content of the medium, temperature, osmolarity, pH, iron availability, and oxygen tension have been reported as environmental cues that can influence biofilm formation (Davey and O'Toole 2000). The adherence of cells to interfaces is described as critical for biofilm formation, and a variety of surface appendages, surface proteins and adhesins have been reported to assist in adherence (O'Toole et al. 2000; Rickard et al. 2003). Adherent cells accumulate to form clusters or microcolonies by growth and/or surface translocation. With time, microcolonies develop into a mature biofilm, characterized by multicellular architecture and extracellular polymeric substance (EPS) or extracellular matrix (Davey and O'Toole 2000). Biofilm architecture is influenced by culture conditions, such as nutrient availability and flow condition (Wolfaardt et al. 1994; Stoodley et al. 1999, 2002; Davey and O'Toole 2000). *B. subtilis* forms fruiting body-like aerial projections that extend from the surfaces of the pellicle biofilm, and the tips of these fruiting bodies serve as preferential sites for spore formation (Branda et al. 2001). Cells in clumps of *B. cereus* produce multiple granules of polyhydroxyalkanoate granules (Luo et al. 2007), and later progress to sporulate. Spores remain constrained in the extracellular matrix; this matrix contains both DNA and proteins. Presence of DNA was indicated both by fluoresecence when stained with either propidium iodide or DAPI (4′,6-diamidino-2-phenylindole), and by removal of the corresponding extracellular fluorescence following treatment with DNase. We have recently also shown that eDNA (extracellular DNA) is required for biofilm formation by *B. cereus* (Vilain et al. 2009). Similarly, proteins were shown by staining with Sypro Red, Proteinase K treatment, and SDS-PAGE of the extracellular matrix. The matrix contained six dominant proteins, three homologues of OppA, enolase, glutamate dehydrogenase (not reported in the genome sequence but present in the raw sequence data of *B. cereus* ATCC 14579), and the ATP synthase beta chain.

6.1.3.2 Growth in Artificial Soil Microcosms

The study of bacterial growth in soil poses numerous challenges, including sterilization and optical constraints. We therefore opted to produce artificial soil microcosms (ASMs) using a 4:1 ratio of sand and Ca-montmorillonite saturated with SESOM (Sect. 6.3.1.3). As SESOM contains a diluted subset of the SOM, ASMs do not represent true soil, but do offer several advantages for studying growth of bacteria in soil. Spores of *B. cereus* inoculated at the centre of the microcosm germinate, grow to a larger population, and subsequently sporulate (Vilain et al. 2006). This indicates that *B. cereus* is able to grow in free soil without the proximity of rhizosphere, and without the requirement for insects that was held previously (Margulis et al. 1998; Jensen et al. 2003). Maintenance of *B. cereus* populations, including germination, growth and sporulation can therefore occur in bulk soil.

When inoculated at a specific point in the ASM matrix, *B. cereus* appears several mm further on the agar surface after several days, indicating translocation through the soil matrix (Fig. 6.3). Microscopic investigation of BacLight Live/Dead stained soil taken from inside the ASM revealed bundles of chains (Fig. 6.3c). A similar phenomenon was reported for a *Bacillus* spreading across an agar surface, and termed sliding (Henrichsen 1972). Sliding appears to be a mechanism driven by elongation and division of cells in chains while constrained within the particulate soil environment, leading to movement of the filaments between gaps in the soil matrix. The observed switch to form bundles of chains, observed in liquid in SESOM, may be ecologically relevant as it provides a mechanism for the species to translocate through the soil matrix. We are currently screening for genes that contribute to this phenotype. Initial screens have yielded a mutant that forms single chains, apparently deficient in lateral chain interactions (Fig. 6.4a). The transposon (Tn917) insertion was mapped to the *galE* gene, encoding galactose epimerase. While no extracellular polysaccharide has been described for *B. cereus* ATCC 14579, analysis of the genome sequence predicts a galactose-containing polysaccharide. The *galE* gene forms part of the 25-member putative operon (Ivanova et al. 2003). Staining the wild type clumps with a rhodamine-labelled lectin specific for galactose showed presence of cell envelope-associated galactose (Fig. 6.4b). The galactose presented more densely in some parts, indicating perhaps a stochastic distribution due to bistability (Veening et al. 2008).

6.2 Studying the Ecophysiology of Soil *Bacillus*

B. cereus display a distinct life cycle in soil (Fig. 6.5). *Bacillus* species are readily culturable in liquid soil extract (SESOM), opening the door to addressing questions on the ecophysiology of this omnipresent group in soil. Technical advances that will aid these efforts include the growing number of genome sequences, the advancement of analytical instruments for studying single cells such as HISH–SIMS (Musat et al. 2008) and Raman microscopy (Huang et al. 2009), but also

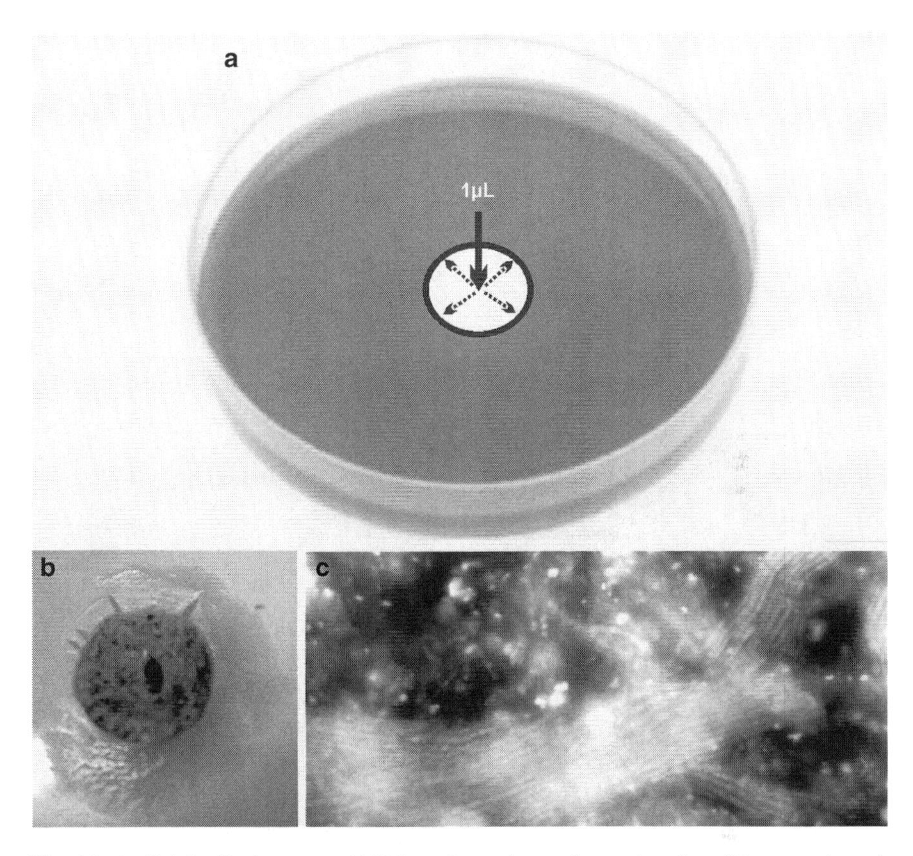

Fig. 6.3 Artificial soil microcosms (ASM) can be used to study translocation of *B. cereus* through soil. Artificially reconstituted soil is placed in wells in SESOM agar and point-inoculated at the centre with 1 μL of spore suspension (**a**). Upon germination, the population spreads through the soil matrix as seen by outgrowth onto the surrounding agar surface (**b**), and by fluorescence microscopy of BacLight Live/Dead stained samples of ASM (**c**)

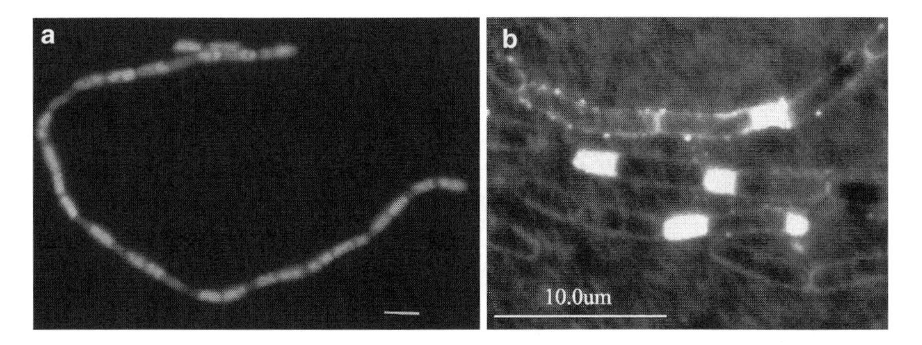

Fig. 6.4 Fluorescence microscopy of BacLight-stained late stationary phase culture of the *galE* mutant, showing formation of chains but not clumps (**a**). Binding of the rhodamine-tagged galactose-specific lectin GSL I to wild type revealed the uneven distribution of galactose both at envelopes and between cells of the wild type (**b**). Size bar: 10.0 μm

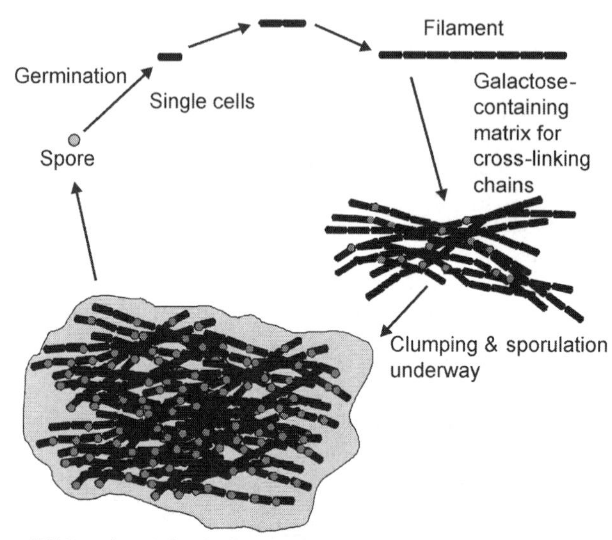

Fig. 6.5 Model of the life cycle of *B. cereus* in SESOM. Spores are able to germinate and outgrow as vegetative cells. The vegetative cells grow as filaments, form clumps, produce extracellular polymeric substance (EPS), and subsequently sporulate. With time, clumps become encased in EPS (as indicated by a *light grey colour* surrounding the clump), and mature spores are released. Mutants insertionally inactivated in the purA and galE genes are unable form filaments and clumps respectively

fluorescence and laser scanning confocal microscopy in combination with a plethora of specific fluorescent reporters. Below we list questions that require attention and that could well be addressed using the basic culturing tools described in this chapter:

- What are the roles played by the various *Bacillus* species in carbon cycling in soil?
- Why do these rapid growing species generally make up a rather small proportion of the soil community?
- What is the mechanism by which *Bacillus* species translocate through soil?
- What is the ratio of vegetative cells to spores in soils?
- Is there a stationary phase state of *Bacillus* as an alternative to sporulation in the soil environment?

6.3 Protocols

Outlined below are several protocols used in our laboratory for studying the ecophysiology of *B. cereus*.

6.3.1 *Culture Environments*

6.3.1.1 Preparation of SESOM

SESOM is a liquid extract of soil prepared without autoclaving in order to avoid chemical alterations at the high temperature of autoclaving. It can be prepared from a variety of soils, but growth will depend on the concentrations and spectrum of organic constituents present.

- Air-dry a volume of soil sufficient for the series of experiments planned, and store at −20°C.
- For 400 mL of SESOM, pre-warm 500 mL 3-[*N*-Morpholino] propanesulfonic acid (MOPS) buffer (10 mM, pH 7) to 50°C.
- To a 1 L Erlenmeyer flask, add 100 g of air-dried soil of choice, and then the pre-warmed buffer.
- Shake in a shaker at 200 rpm for 1 h.
- Remove the larger particulate matter by filtering using filter paper. We suggest using fluted conical paper filters in funnels.
- Sequentially remove finer particles by filtering through hydrophilic PVDF membranes with 5 μm pore size, and then 0.45 μm pore size (Millipore). This can be performed using single filter discs (47 mm diameter) and a re-usable filter device as used for filtering 100 mL water samples.
- After filtration, adjust the pH to 6.5, 7.0 or desired pH.
- Sterilize by filtration (0.22 μm diameter pore size).
- Check for sterility of each batch by depositing 5 μL of SESOM onto R2A or LB agar, and incubating at 30°C for 24 h.
- For SESOM+, SESOM was supplemented with glucose (100 mg/L), casein hydrolysate (1 mg/L, Sigma) and yeast extract (1 mg/L, Fisher Scientific). MOPS buffer (10 mM) supplemented as described above, and designated MOPS+, was used as a negative control. For SESOM agar, 80 mL SESOM+ was mixed with 20 mL of tempered 7.5% agar.

6.3.1.2 Preparation of SESOM Agar

SESOM agar can be prepared without exposing the chemical constituents in the SESOM to autoclave temperatures by mixing the extract with fivefold concentrated agar. SESOM agar may be prepared by adding to 400 mL SESOM (see protocol above) warmed to 40°C, 100 mL of 7.5% agar (7.5 g in 100 mL dH_2O, then autoclaved for 20 min at 121°C). Add the concentrated agar while hot, mix well with the SESOM and leave to stand to allow bubbles to dissipate. Pour into plates.

6.3.1.3 Preparation of Artificial Soil Microcosms

ASMs can be used to study growth of pure cultures under simulated soil conditions. The ASM consists of four parts acid washed, autoclaved sand and one part montmorillonite, made homoionic for calcium (Stotzky and Burns 1982) before autoclaving, and then saturated with SESOM (see protocol above).

- Wash sand in 0.1 M HCl repeatedly until no bubbles appear, or purchase clean sand.
- Autoclave the sand in a glass beaker or easily accessible, sealable container, such as a screw-cap jar for 15 min at 121°C.
- Prepare Ca-Montmorillonite homo-ionic for calcium by washing three times with $CaCl_2$. This entails suspending bentonite (Sigma) in twice the volume of $CaCl_2$ (100 mM, pH 7.0), and removing the free liquid by centrifugation (2,000 $\times g$, 5 min) and decanting.
- Repeat the $CaCl_2$ treatment two more times, for a total of three times.
- Wash the Ca-Montmorillonite with two volumes of dH_2O and remove by centrifugation as above.
- Was at least three times with dH_2O until no free chloride ions can be detected in the supernatant.
- Detect free chloride ions in the supernatant using the silver nitrate test. Add 100 μL of $AgNO_3$ (100 mM) to 1 mL supernatant. Formation of a precipitate indicates presence of residual chloride ions.
- Remove free water by centrifugation and removal of the supernatant.
- Autoclave in an easily accessible sealable container with wide neck, such as a screw-cap jar, for 15 min at 121°C.
- After cooling, add four parts sand to one part Ca-Montmorillonite and mix using a sterile spatula.
- Add two volumes of SESOM, and mix by shaking.
- Transfer to a centrifuge tube, and remove free liquid by centrifugation and removal of the supernatant.
- The soil-like material is now ready for use.
- For ASM, punch wells (8 mm in diameter) using a sterile cork borer into an agar of choice, such as SESOM agar, and remove the agar plug. If the plug is not readily removed, use a sterile needle.
- Using two sterile spatulas, prepare a small ball of soil and place it into the well in the agar. For 8 mm diameter wells add about 0.35 mL soil.
- Inoculate the ASM by applying 1 μL cell suspension of choice into the centre of the ASM by inserting a micropipettor tip about 1 mm under the surface.
- Incubate for the desired time period.
- For extended incubation take measures to prevent drying, such as placing Petri dishes in a plastic bag or wrapping with Parafilm.

6.3.2 Microscopic Analysis

Cells from liquid SESOM may be viewed microscopically, but should be concentrated before staining due to the low cell density obtained in most SESOM (Fig. 6.1).

- Concentrate cells by centrifugation.
- Place 1 mL culture into a 1.5 mL microfuge tube and centrifuge at 10,000 $\times g$ for 3 min.
- Remove the supernatant very carefully in the case of *Bacillus* as pellets tend to be much less cohesive than those of *Escherichia coli*, especially when grown in SESOM.
- Resuspend the pellet in phosphate buffer (10 mM, pH 7.0).
- Add stains of choice, e.g.
  - BacLight Live/Dead stain (Syto 9 and propidium iodide).
  - DAPI or PoPro for nucleic acid
  - FM4-64 for membrane staining
  - Fluorescently conjugated lectins for specific surface carbohydrates
- To a glass microscope slide apply 20 μL suspension.
- Cover with a coverslip.
- Place slide on absorbent paper, cover with absorbent paper and press down firmly to render a single cell layer. This takes some practise as pressing too firmly causes air pockets, and too little pressure removes too little liquid, allowing cells to swim during observation.
- Paint clear nail varnish around the edges of the cover slip to prevent drying of the sample.
- View by fluorescence or laser scanning confocal microscopy.

6.3.3 Proteomic Analysis by Two-Dimensional Gel Electrophoresis

Before undertaking two-dimensional electrophoresis, decide on key parameters such as the pH range (e.g. 4–7, 3–10 non-linear, etc.) and the resolution required. Immobilized pH gradient strips for the first dimension are available from GE Healthcare (previously Amersham), and BioRAD in various lengths. Also consider the width of gel accommodated by the electrophoresis equipment available.

6.3.3.1 Preparation of Cytosolic Proteins

- Grow bacteria to desired stage.
- Add phosphate buffer (10 mM, pH 7.0) to one quarter of the volume of each tube, to centrifuge tubes and freeze, laying them horizontally to get flat layers of

ice down the sides of the tubes. For 50 mL conical tubes, we would add 10 mL buffer for freezing.

- Before adding culture samples, crush the ice by banging the centrifuge tubes against an old Styrofoam container to get crushed ice but without cracking the tube.
- Add the desired volume of culture to the crushed ice, replace the cap, and mix by inversion. That way the samples coming out of the incubator are cooled down very quickly.
- Harvest the cells by centrifugation at 10,000 $\times g$ for 5 min.
- Wash cells in cold 10 mM phosphate buffer (pH 7.0) or Tris (10 mM, pH 8).
- Resuspend in small volume of ice-cold Tris (10 mM, pH 8.0). Cells should be very concentrated for ultrasonication, preferably in a 1.5 mL microfuge tubes.
- Add 0.01 volume PMSF (phenylmethanesulphonylfluoride; 100 mM stock in 100% ethanol), i.e., final concentration of 1 mM and mix. *Careful*: PMSF is toxic. PMSF has a short half-life in water, but is stable once bound to protease.
- Disrupt cells by ultrasonication – depends on your instrument but we do 4 × 30 s while on ice with 2 min to cool in between applications.
- Centrifuge 10 min at 10,000 $\times g$ to remove unbroken cells and particulate debris
- Transfer supernatant and freeze.

6.3.3.2 Preparation of Extracellular Proteins

- Grow bacteria to desired stage.
- Prepare cell-free supernatant by centrifugation of 200 mL of culture at 1,000$\times g$ for 20 min.
- Transfer the supernatant to a clean 250 mL centrifuge tube, adding 25 mL 1 M Trifluoroacetic acid (TCA) for a final TCA concentration of 10% (v/v).
- Mix well and incubate overnight at 4°C.
- Harvest proteins by centrifugation at 8,000$\times g$ for 60 min at 4°C
- Discard the supernatant.
- Wash the pellet eight times with ethanol as follows:
  - Add 10 mL 90% ethanol
  - Resuspend the pellet (sticky) using a clean stainless steel spatula
  - Transfer the suspension into a smaller centrifuge tube, such as a 15 mL conical tube
  - Harvest the pellet by centrifugation at 10,000$\times g$ for 5 min at 4°C
  - Repeat the wash step seven times, for a total of eight times
- Resuspend the pellet in TE buffer (10 mM TRIS, 1 mM EDTA, pH 8.0) and store frozen.

6.3.3.3 First Dimension

- Determine the protein concentration in the sample using either the Bradford assay or Amersham 2D Quant kit
  - Set up a standard curve using bovine serum albumen as standard
  - Determine protein concentration of samples by preparing various dilutions
- Determine volume required for 500 μg if staining with Coomassie R250, 200 μg if staining with colloidal Coomassie, and 50 μg if staining with a silver staining procedure. If the volume of sample required is greater than 20 μL, concentrate by vacuum centrifugation.
- Aliquot out volume of protein to deliver 500, 200 or 50 μg.
- Prepare fresh rehydration buffer by adding dithiothreitol (DTT) to the stock, and make up to 400 μL with rehydration buffer.
- Mix by pipetting, avoiding formation of bubbles (*do not vortex*).
- Leave for 30 min at room temperature to allow proteins to solubilize.
- Centrifuge at 4°C for 30 min.
- Set clean dry re-swelling cassette to level and remove lid.
- Transfer supernatant well in re-swelling cassette, avoiding formation of bubbles
- Remove clear protective cover from strip and insert strip gel side down into well.
- Cover with oil and slide lid back in.
- Leave overnight in cool (*not cold*) place. Urea at this concentration crystallizes below 18°C.
- Place strip briefly in dH_2O to remove oil.
- Dry by tipping sideways onto Kimwipe. Do not allow gel side to cling to the paper wipe.
- Place gel side up onto IPG Phor surface, with acidic end (+) towards anode (red, at cooling outlet).
- Cut 2× filter paper strips 110 mm long to width covering all strips.
- Hydrate with 500 μL dH_2O and pat dry (they should be moist but not wet).
- Place on both ends across strips.
- Insert electrodes.
- Cover strips with oil.
- Place lid on.
- Resolve proteins as follows:
  - 150 V for 1 h
  - 350 V for 1 h
  - 500 V for 4 h
  - 750 V for 1 h
  - 1 kV for 1 h
  - 1.5 kV for 1 h
  - 3.5 kV for 11 h (3 mA max) for a total of ~44 kVh
- Pour off oil into receptacle through filter paper (to re-use).
- Remove strips and place gel side up onto pre-labelled clear plastic sheath.
- Cover with clear plastic and freeze. Freeze briefly, even if the second dimension is to take place directly following the first.

6.3.3.4 Second Dimension

- Assemble dry plates (wiped with ethanol and dust-free) with narrow (IEF) spacers in BioRAD casting stand, set to level
- Pour separating gel until 1 cm from top
- Overlay immediately with isobutanol and leave to set for at least 2 h or overnight
- Remove isobutanol by washing with dH_2O
- Remove residual water between plates – dry with Kimwipe
- Set level
- Pour in stacking gel, cover with isobutanol
- Leave to set for 30 min
- Remove isobutanol by washing with dH_2O
- Equilibrate strips for 15 min in equilibration solution A, shaking slowly
- Transfer to equilibration solution B and leave for 15 min, shaking slowly
- Remove excess fluid by tipping sideways onto Kimwipe
- Dip briefly into tank buffer to lubricate the strip
- Insert strip in between the glass plates, plus at right
- Resolve at 10 mA per gel for 1 h through the stacking gel
- Resolve at 20 mA per gel with the maximum voltage limited to 350 V, until the buffer front reaches the bottom
- Transfer gel to fixing solution or stain of choice

6.3.3.5 Solutions Required

1. Rehydration buffer:

Urea (7 M)	10.5 g
Thiourea (2 M)	3.8 g
CHAPS*	1 g
Bromophenol blue stock	50 μL
H_2O	to 25 mL

*3-[(3-Cholamidopropyl)dimethylammonio]-1-propanesulfonate

Add fresh before use:

Ampholyte (pH 3–10)	500 μL (or 20 μL per mL of rehydration buffer)
DTT	70 mg (or 2.8 mg per mL rehydration buffer)

2. Bromophenol Blue stock solution

Tris base:	60 mg
Bromophenol Blue	100 mg
dH_2O	to 10 mL

3. Equilibration buffer stock:

Tris HCl (pH 6.8), 50 mM	10 mL
Urea	72 g
SDS	4 g
H_2O	to 120 mL

Dissolve, then add:

Glycerol	69 mL
H_2O	to 200 mL

4. Equilibration buffer A:

Equilibration buffer stock	15 mL
DTT	300 mg

5. Equilibration buffer B:

Equilibration buffer stock	15 mL
Iodoacetamide	375 mg
Bromophenol Blue stock	30 μL

6. Upper tank buffer:

Trizma base	3 g
SDS	1 g
Glycine	14 g
dH_2O	to 1 L

7. Lower tank buffer:

Trizma base	120 g
SDS	5 g
Glycine	20 g
H_2O	to 5 L

8. Fixing solution:

dH_2O	380 mL
Ethanol	500 mL
Acetic acid	120 mL
Formaldehyde (37%)	500 μL

9. Stacking gel buffer (pH 6.8):

Tris –HCl	15.0078 g
Trizma base	0.5768 g
dH_2O	to 100 mL
Confirm pH to be 6.8	

10. Separating gel buffer (pH 8.9):

Tris HCl	2.1561 g
Trizma base	10.4567 g
dH_2O	to 100 mL
Confirm pH to be 8.9	

Composition of separating gel for second dimension

Separating gel	Solution	Final conc.	For 35 mL (for one slab)	For 150 mL
Acrylamide	40%	12.175%	10.65 mL	45.795
Bis-acrylamide	2%	0.325%	5.69 mL	24.470

(*continued*)

Separating gel	Solution	Final conc.	For 35 mL (for one slab)	For 150 mL
TRIS pH 8.9	1 M	0.375 M	13.13 mL	56.460
dH_2O			5.23 mL	22.490
Degass for 10 min (apply vacuum carefully)				
SDS	10%	0.10%	350.0 μL	1.505
TEMED (6.6 M)	76.7%	0.06%	273.7 μL	1.177
NH_4 persulphate	10%	0.03%	105.0 μL	0.451
Total volume			35.2537 mL	151.59

Composition of stacking gel for second dimension

Stacking gel 4.67%	Solution	Final conc.	For 4 gels
Acrylamide	40%	4.549%	0.923 mL
Bis-acrylamide	2%	0.121%	0.491 mL
TRIS pH 6.8	1 M	0.125 M	1.015 mL
dH_2O			5.583 mL
Degass for 10 min (apply vacuum carefully)			
SDS	10%	0.10%	81.2 μL
TEMED (6.6 M)	76.7%	0.25%	26.5 μL
NH_4 persulphate	10%	0.05%	40.6 μL
Total volume			8.12 mL

6.3.4 Screening for Genes Contributing to Self-Assembling Biofilm

When growing in SESOM, *B. cereus* ATCC 14579 switches to a multicellular, self-assembling biofilm (Fig. 6.2). This is mirrored by formation of ramificate or rhizoidal colonies on SESOM agar, similar to colonies of the related *B. mycoides* (Di Franco et al. 2002). Genes contributing to the multicellular phenotype can therefore be screened for by a change in colony morphotype of mutants on SESOM agar. The approach taken in our laboratory is outlined below:

- Transform *B. cereus* with a transposon delivery plasmid such as pLTV1 (Tn917) (Camilli et al. 1990) or pAW016 (mini Tn10) (Wilson et al. 2007) by electroporation (Bone and Ellar 1989)
- Perform transposon mutagenesis as described previously (Clements and Moir 1998, or Wilson et al. 2007)
- Select for transposon mutants using the prescribed temperature and antibiotic
- Replica-plate mutants together with a wild-type control onto SESOM agar, by spot-inoculating using sterile tooth picks
- Incubate at 30°C for 5 days
- Inspect colonies for changes in the ramificate appearance. This could include non-ramificate and hyper-ramificate colony morphotypes
- Verify candidate mutants by spot-inoculating in triplicate
- Inspect the colonies microscopically, directly on the agar surface using 200X magnification
- Map the respective mutations by plasmid rescue as described (Wilson et al. 2007)

References

Bone EJ, Ellar DJ (1989) Transformation of *Bacillus thuringiensis* by electroporation. FEMS Microbiol Lett 49:171–177

Branda SS, Gonzalez-Pastor JE, Ben-Yehuda S, Losick R, Kolter R (2001) Fruiting body formation by *Bacillus subtilis*. Proc Natl Acad Sci USA 98:11621–11626

Camilli A, Portnoy DA, Youngman P (1990) Insertional mutagenesis of *Listeria monocytogenes* with a novel Tn917 derivative that allows direct cloning of DNA flanking transposon insertions. J Bacteriol 172:3738–3744

Chattopadhyay A, Bhatnagar NB, Bhatnagar R (2004) Bacterial insecticidal toxins. Crit Rev Microbiol 30:33–54

Clements MO, Moir A (1998) Role of the gerI operon of *Bacillus cereus* 569 in the response of spores to germinants. J Bacteriol 180:6729–6735

Costerton JW, Lewandowski Z, Caldwell DE, Korber DR, Lappin-Scott HM (1995) Microbial biofilms. Annu Rev Microbiol 49:711–745

Daniel R (2004) The soil metagenome–a rich resource for the discovery of novel natural products. Curr Opin Biotechnol 15:199–204

Davey ME, O'Toole GA (2000) Microbial biofilms: from ecology to molecular genetics. Microbiol Mol Biol Rev 64:847–867

Davis KE, Joseph SJ, Janssen PH (2005) Effects of growth medium, inoculum size, and incubation time on culturability and isolation of soil bacteria. Appl Environ Microbiol 71:826–834

Di Franco C, Beccari E, Santini T, Pisaneschi G, Tecce G (2002) Colony shape as a genetic trait in the pattern-forming *Bacillus mycoides*. BMC Microbiol 2:1–15

Dragon DC, Rennie RP (1995) The ecology of anthrax spores: tough but not invincible. Can Vet J 36:295–301

Dragon DC, Bader DE, Mitchell J, Woollen N (2005) Natural dissemination of *Bacillus anthracis* spores in Northern Canada. Appl Environ Microbiol 71:1610–1615

Fierer N, Jackson RB (2006) The diversity and biogeography of soil bacterial communities. Proc Natl Acad Sci USA 103:626–631

Fierer N, Bradford MA, Jackson RB (2007) Toward an ecological classification of soil bacteria. Ecology 88:1354–1364

Garbeva P, Van Veen JA, Van Elsas JD (2003) Predominant *Bacillus* spp. in agricultural soil under different management regimes detected via PCR-DGGE. Microb Ecol 45:302–316

Guggenberger G, Zech W (1993a) Dissolved organic-matter (Dom) dynamics in spruce forested sites – examinations by analytical dom fractionation. Zeitschrift Fur Pflanzenernahrung Und Bodenkunde 156:341–347

Guggenberger G, Zech W (1993b) Dissolved organic carbon control in acid forest soils of the Fichtelgebirge (Germany) as revealed by distribution patterns and structural composition analyses. Geoderma 59:109–129

Handelsman J, Stabb EV (1996) Biocontrol of soilborne plant pathogens. Plant Cell 8:1855–1869

Helgason E, Tourasse NJ, Meisal R, Caugant DA, Kolstø AB (2004) Multilocus sequence typing scheme for bacteria of the *Bacillus cereus* group. Appl Environ Microbiol 70:191–201

Henrichsen J (1972) Bacterial surface translocation: a survey and a classification. Bacteriol Rev 36:478–503

Henrici AT (1934) The biology of bacteria. D.C. Heath, Boston

Honerlage W, Hahn D, Zeyer J (1995) Detection of mRNA of *nprM* in *Bacillus megaterium* ATCC 14581 grown in soil by whole-cell hybridization. Arch Microbiol 163:235–241

Huang Y, Eglinton G, Van Der Hage ERE, Boon JJ, Bol R, Ineson P (1998) Dissolved organic matter and its parent organic matter in grass upland soil horizons studied by analytical pyrolysis techniques. Eur J Soil Sci 49:1–15

Huang WE, Ferguson A, Singer AC, Lawson K, Thompson IP, Kalin RM, Larkin MJ, Bailey MJ, Whiteley AS (2009) Resolving genetic functions within microbial populations: in situ analyses using rRNA and mRNA stable isotope probing coupled with single-cell raman-fluorescence in situ hybridization. Appl Environ Microbiol 75:234–241

Ivanova N, Sorokin A, Anderson I, Galleron N, Candelon B, Kapatral V, Bhattacharyya A, Reznik G, Mikhailova N, Lapidus A, Chu L, Mazur M, Goltsman E, Larsen N, D'Souza M, Walunas T, Grechkin Y, Pusch G, Haselkorn R, Fonstein M, Ehrlich SD, Overbeek R, Kyrpides N (2003) Genome sequence of *Bacillus cereus* and comparative analysis with *Bacillus anthracis*. Nature 423:87–91

Jensen GB, Hansen BM, Eilenberg J, Mahillon J (2003) The hidden lifestyles of *Bacillus cereus* and relatives. Environ Microbiol 5:631–640

Jones DL, Kemmitt SJ, Wright D, Cuttle SP, Bol R, Edwards AC (2005) Rapid intrinsic rates of amino acid biodegradation in soils are unaffected by agricultural management strategy. Soil Biol Biochem 37:1267–1275

Kaiser K, Guggenberger G, Haumaier L, Zech W (2001) Seasonal variations in the chemical composition of dissolved organic matter in organic forest floor layer leachates of old-growth Scots pine (Pinus sylvestris L.) and European beech (Fagus sylvatica L.) stands in northeastern Bavaria, Germany. Biogeochemistry 55:103–143

Kalbitz K, Solinger S, Park JH, Michalzik B, Matzner E (2000) Controls on the dynamics dissolved organic matter in soils: a review. Soil Sci 165:277–304

Kalbitz K, Schwesig D, Schmerwitz J, Kaiser K, Haumaier L, Glaser B, Ellerbrock R, Leinweber P (2003) Changes in properties of soil-derived dissolved organic matter induced by biodegradation. Soil Biol Biochem 35:1129–1142

Liebeke M, Brozel VS, Hecker M, Lalk M (2009) Chemical characterization of soil extract as growth media for the ecophysiological study of bacteria. Appl Microbiol Biotechnol 83:161–173

Luo Y, Vilain S, Voigt B, Albrecht D, Hecker M, Brozel VS (2007) Proteomic analysis of *Bacillus cereus* growing in liquid soil organic matter. FEMS Microbiol Lett 271:40–47

Lynch JM (1982) Limits to microbial-growth in soil. J Gen Microbiol 128:405–410

Margulis L, Jorgensen JZ, Dolan S, Kolchinsky R, Rainey FA, Lo SC (1998) The *Arthromitus* stage of *Bacillus cereus*: intestinal symbionts of animals. Proc Natl Acad Sci USA 95:1236–1241

Mishustin EN (1972) Microflora of soils in the northern and central USSR. Israel Program for Scientific Translations, Jerusalem, Israel

Musat N, Halm H, Winterholler B, Hoppe P, Peduzzi S, Hillion F, Horreard F, Amann R, Jorgensen BB, Kuypers MM (2008) A single-cell view on the ecophysiology of anaerobic phototrophic bacteria. Proc Natl Acad Sci USA 105:17861–17866

Olsen RA, Bakken LR (1987) Viability of soil bacteria: optimization of plate-counting technique and comparison between total counts and plate counts within different size groups. Microb Ecol 13:59–74

Oosthuizen MC, Steyn B, Theron J, Cosette P, Lindsay D, Von Holy A, Brözel VS (2002) Proteomic analysis reveals differential protein expression by *Bacillus cereus* during biofilm formation. Appl Environ Microbiol 68:2770–2780

O'Toole G, Kaplan HB, Kolter R (2000) Biofilm formation as microbial development. Annu Rev Microbiol 54:49–79

Pace NR (1997) A molecular view of microbial diversity and the biosphere. Science 276:734–740

Pizzeghello D, Zanella A, Carletti P, Nardi S (2006) Chemical and biological characterization of dissolved organic matter from silver fir and beech forest soils. Chemosphere 65:190–200

Priest FG, Barker M, Baillie LW, Holmes EC, Maiden MC (2004) Population structure and evolution of the *Bacillus cereus* group. J Bacteriol 186:7959–7970

Rappe MS, Giovannoni SJ (2003) The uncultured microbial majority. Annu Rev Microbiol 57:369–394

Rickard AH, Gilbert P, High NJ, Kolenbrander PE, Handley PS (2003) Bacterial coaggregation: an integral process in the development of multi-species biofilms. Trends Microbiol 11:94–100

Rihani M, Cancela Da Fonseca JP, Kiffer E (1995) Decomposition of beech leaf litter by microflora and mesofauna. II. Food preferences and action of oribatid mites on different substrates. Eur J Soil Biol 31:67–79

Saggar S, Parshotam A, Hedley C, Salt G (1999) ^{14}C-labelled glucose turnover in New Zealand soils. Soil Biol Biochem 31:2025–2037

Saleh SM, Harris RF, Allen ON (1970) Fate of *Bacillus thuringiensis* in soil: effect of soil pH and organic amendment. Can J Microbiol 16:677–680

Schloss PD, Handelsman J (2006) Toward a census of bacteria in soil. PLoS Comput Biol 2:e92

Schneckenberger K, Demin D, Stahr K, Kuzyakov Y (2008) Microbial utilization and mineralization of [^{14}C]glucose added in six orders of concentration to soil. Soil Biol Biochem 40:1981–1988

Schoeni JL, Wong AC (2005) *Bacillus cereus* food poisoning and its toxins. J Food Prot 68:636–648

Stabb EV, Jacobson LM, Handelsman J (1994) Zwittermicin A-producing strains of *Bacillus cereus* from diverse soils. Appl Environ Microbiol 60:4404–4412

Stoodley P, Dodds I, Boyle JD, Lappin-Scott HM (1999) Influence of hydrodynamics and nutrients on biofilm structure. J Appl Microbiol Suppl 85:19S–28S

Stoodley P, Sauer K, Davies DG, Costerton JW (2002) Biofilms as complex differentiated communities. Annu Rev Microbiol 56:187–209

Stotzky G, Burns RG (1982) The soil environment: Clay – Humus – Microbe interactions. In: Burns RG and Slater JH (eds) Experimental microbial ecology. Blackwell Scientific, Oxford

Strobel BW (2001) Influence of vegetation on low-molecular-weight carboxylic acids in soil solution: a review. Geoderma 99:169–198

Sutherland IW (2001) The biofilm matrix – an immobilized but dynamic microbial environment. Trends Microbiol 9:222–227

Tiunov AV, Scheu S (2004) Carbon availability controls the growth of detritivores (Lumbricidae) and their effect on nitrogen mineralization. Oecologia 138:83–90

Torsvik V, Øvreås L (2002) Microbial diversity and function in soil: from genes to ecosystems. Curr Opin Microbiol 5:240–245

Torsvik V, Goksoyr J, Daae FL (1990) High diversity in DNA of soil bacteria. Appl Environ Microbiol 56:782–787

Van Hees PAW, Jones DL, Finlay R, Godbold DL, Lundstromd US (2005) The carbon we do not see – the impact of low molecular weight compounds on carbon dynamics and respiration in forest soils: a review. Soil Biol Biochem 37:1–13

Van Ness GB (1971) Ecology of anthrax. Science 172:1303–1307

Veening JW, Smits WK, Kuipers OP (2008) Bistability, epigenetics, and bet-hedging in bacteria. Annu Rev Microbiol 62:193–210

Vilain S, Brozel VS (2006) Multivariate approach to comparing whole-cell proteomes of *Bacillus cereus* indicates a biofilm-specific proteome. J Proteome Res 5:1924–1930

Vilain S, Luo Y, Hildreth MB, Brözel VS (2006) Analysis of the life cycle of the soil saprophyte *Bacillus cereus* in liquid soil extract and in soil. Appl Environ Microbiol 72:4970–4977

Vilain S, Pretorius JM, Theron J, Brozel VS (2009) DNA as an adhesin: *Bacillus cereus* requires extracellular DNA to form biofilms. Appl Environ Microbiol 75:2861–2868

von Stetten F, Mayr R, Scherer S (1999) Climatic influence on mesophilic *Bacillus cereus* and psychrotolerant *Bacillus weihenstephanensis* populations in tropical, temperate and alpine soil. Environ Microbiol 1:503–515

Waksman SA (1932) Principles of soil microbiology. Williams and Wilkins, Baltimore

Wilson AC, Perego M, Hoch JA (2007) New transposon delivery plasmids for insertional mutagenesis in *Bacillus anthracis*. J Microbiol Methods 71:332–335

Wolfaardt GM, Lawrence JR, Robarts RD, Caldwell SJ, Caldwell DE (1994) Multicellular organization in a degradative biofilm community. Appl Environ Microbiol 60:434–446

Chapter 7
Dispersal of Aerobic Endospore-forming Bacteria from Soil and Agricultural Activities to Food and Feed

Marc Heyndrickx

7.1 Introduction

There is a clear association between soil-borne endospore-forming bacteria and food contamination. The spore-formers implicated belong both to the strictly anaerobic and to the aerobic phylogenetic groups of microorganisms. However, in the last few decades, it has been especially the aerobic endospore-forming bacteria belonging to the genus *Bacillus* and related genera that have caused specific problems for the food industry. Several reasons can be proposed to explain this phenomenon and most are related to the characteristics of the spores: their ubiquitous presence in soil, their resistance to heat in common industrial processes such as pasteurization and the adhesive characters of particular spores that facilitate their attachment to processing equipment (Andersson et al. 1995). Probably of more recent concern is the possibly increasing tolerance, adaptation or resistance of spores or vegetative cells of particular spore-forming species to conditions or treatments that were previously presumed to stop growth (low temperatures and low pH) or to inactivate all living material, such as ultrahigh heat treatment (UHT) and commercial sterilization. The food industry seems to be increasingly confronted with tolerant or resistant spore-formers that might be side effects of the use of new ingredients and the application of new processing and packaging technologies.

Aerobic endospore-formers cause two kinds of problems in the food industry. In the first place, there are some food-borne pathogens such as *Bacillus cereus*. Secondly, there is the reduction of shelf life and food spoilage (in't Veld 1996). Microbial spoilage of food is usually indicated by changes in texture or the development of off-flavours. Food defects, however, can also simply be caused by unwanted microbial growth in commercially sterile products. It is estimated that

M. Heyndrickx
Institute for Agricultural and Fisheries Research (ILVO), Technology and Food Science Unit, Brusselsesteenweg 370, 9090 Melle, Belgium
e-mail: Marc.Heyndrickx@ilvo.vlaanderen.be

N.A. Logan and P. De Vos (eds.), *Endospore-forming Soil Bacteria*, Soil Biology 27,
DOI 10.1007/978-3-642-19577-8_7,

food spoilage constitutes significant financial losses despite the contributions of modern food technology and preservation techniques.

In this chapter, two aerobic endospore-forming species present in soil, *B. cereus* and *Alicyclobacillus acidoterrestris*, are discussed in detail. Also described is how agricultural activities in general can introduce other unwanted spore-formers (e.g. *Bacillus sporothermodurans*) either directly or indirectly (via the feed) into the food chain.

7.2 *Bacillus cereus* in Soil

B. cereus is a ubiquitous organism present in many types of soil, sediments, dust and plants. It is has long been believed that this organism has a saprophytic life cycle in soil with the presence of spores that only germinate and grow upon contact with soil-associated organic matter (i.e. nutrient-rich conditions). However, in laboratory experiments with liquid soil extract and artificial soil microcosms, it was observed that *B. cereus* (as well as other soil-isolated *Bacillus* species) displayed a complete life cycle (germination, growth and sporulation) and adapted to translocate in soil by switching from a single-cell phenotype to a multicellular one with the formation of filaments and clumps that encased the ensuing spores in an extracellular matrix (Vilain et al. 2006). The multicellular filamentous stage of *B. cereus sensu lato* has also been observed in the gut of soil-dwelling arthropod species, in which this symbiotic intestinal stage was initially described as *Arthromitus* (Margulis et al. 1998). In addition to a full life cycle in soil, *B. cereus* is also adapted to a lifestyle in the animal or human gut, where it can behave as a pathogen or as a part of the intestinal microbiota, as well as to growth in food and feed. *Bacillus anthracis,* the mammalian pathogen of the *B. cereus* group, proliferates in lymphatic vessels of the host and is released by shedding into the soil where sporulation occurs (so-called storage areas; Dragon and Rennie 1995). *Bacillus thuringiensis* being an insect pathogen of the *B. cereus* group is also a ubiquitous soil microorganism, but it is also found on the phylloplane and in insects. Presumably, *B. thuringiensis* proliferates in the guts of insects and is then released into soil where it can subsequently proliferate under favourable nutrient conditions (Jensen et al. 2003). However, *B. thuringiensis* seems to be less capable of adapting to soil habitats than *B. cereus*, as in the study of Yara et al. (1997) *B. cereus* grew even when indigenous soil bacteria were added at high density to sterile soil, whereas *B. thuringiensis* sporulated immediately without any growth. A high genetic diversity among *B. cereus/B. thuringiensis* isolates from Norwegian soils has been reported, probably resulting from extensive genetic recombination (Helgason et al. 1998). Less is known on the ecology of the other members of the *B. cereus* group (*Bacillus mycoides*, *Bacillus pseudomycoides*, *Bacillus weihenstephanensis*), except that they have been isolated from a wide variety of environmental niches such as soils, sludge, arthropods, earthworms and rhizospheres (Jensen et al. 2003). In soil from a tropical habitat only mesophilic *B. cereus* was found, in soil from an alpine habitat only the psychrotolerant species *B. weihenstephanensis* was commonly found, while in soil from a temperate habitat mesophilic *B. cereus*, *B. weihenstephanensis* and "intermediate thermal types" of the *B. cereus* group were found (von Stetten et al. 1999). The latter are probably reflections of adaptation based on genetic recombination in temperate habitats with high temperature fluctuations, resulting in a high population diversity, especially amongst the psychrotolerant bacteria.

7.3 Transfer of *Bacillus cereus* and Other Aerobic Endospore-Formers from Soil and Agricultural Environments to Food and Feed

7.3.1 Transfer to Food in General

B. cereus is found in a wide range of habitats, with air and soil probably being the primary sources of food contamination. Soil, especially, is heavily contaminated with *B. cereus* spores (see Sect. 7.3.2). Being a soil resident, *B. cereus* is part of the microbiota of plant raw materials, attached as vegetative cells or spores. At harvest, this plant raw material can be used for direct human consumption as fresh produce, as ingredients for food or feed production or directly as animal feed. Dairy cows consuming such feed will excrete *B. cereus* spores in the faeces, and these spores will contaminate the raw milk (see Sect. 7.3.2). However, as *B. cereus* is not dominant in the microbiota of plant raw materials, fresh produce for direct human consumption (fresh vegetables and fruits, fresh herbs and spices, fresh potatoes) has not been reported in association with outbreaks or cases of food-borne illness caused by *B. cereus* (Bassett and McClure 2008). Nevertheless, recent investigations in Denmark have shown that *B. thuringiensis* strains, in some cases indistinguishable from the commercial strains used in microbial bioinsecticides, were present on fresh vegetables (e.g. fresh cucumbers and tomatoes) in Danish retail shops to levels that may exceed 10^4 cfu/g (Frederiksen et al. 2006). Since these strains harboured genes encoding enterotoxins, increased concern regarding the residual amount of *B. thuringiensis* insecticide on vegetable products after harvest may be imminent. Despite these observations, problems of *B. cereus* food poisoning and/or spoilage are mainly restricted to pasteurized or dried products manufactured in food processing units. A combination of the specific attributes of the spores and of the resulting vegetative cells not only gives *B. cereus* a huge advantage over non-spore-formers, but also over other spore-formers, and they explain why this organism is a special threat to the food processing industry. Firstly, like all bacterial spores, *B. cereus* spores are resistant to heat and desiccation and they will thus survive food-processing steps, such as pasteurization or thermization and dehydration or drying, that all eliminate or reduce vegetative cells. de Vries (2006) reported decimal reduction times at a heating temperature of 95°C (D_{95}-values) varying from around

5 to as high as 80 min for spores of naturally occurring *B. cereus* strains. As a result, a final product may be contaminated with spores that face little or no competition from vegetative cells that would otherwise outgrow *B. cereus*.

A second aspect is that *B. cereus* spores, which are introduced into the food production chain via plant ingredients or via milk or milk powder, can adhere to equipment surfaces and in pipelines. The ability of *B. cereus* spores, especially, to adhere and act as initiation stages for biofilm formation on a wide variety of materials commonly encountered in food processing plants is well known (Peng et al. 2001). The strong adhesion properties of *B. cereus* spores have been attributed to the hydrophobic character of the exosporium and to the presence of appendages on the surfaces of the spores (Husmark and Rönner 1992). According to Wijman et al. (2007), a *B. cereus* biofilm may particularly develop in a partly filled industrial storage or a piping system and such a biofilm acts as a nidus for formation of spores that can subsequently be dispersed by release into the food production system. Spores embedded in biofilms are protected against disinfectants (Ryu and Beuchat 2005).

A third aspect is the use of extended refrigeration in food production and distribution, as well as in the kitchen to increase the shelf life of processed foods. A particular category of thermally processed food products of high organoleptic quality that rely on refrigeration for their safety and quality are the Refrigerated Processed Foods of Extended Durability, or REPFEDs, such as vegetable purées and ready-to-eat meals with shelf lives up to and exceeding 10 days. These REPFED food items, as well as dairy products (milk and desserts) with a shorter shelf life (<10 days), are specific niches for psychrotolerant endospore-forming bacteria. An important feature of the *B. cereus* strains, especially diarrhoeal and food-environment strains (but, however, virtually none of the emetic strains), is psychrotolerance; that is to say growing at temperatures ≤7°C (Carlin et al. 2006). Moreover, it is important to note that the majority of *B. cereus* strains are able to start growing from 10°C, which represents mild temperature abuse conditions (Guinebretière et al. 2008). Emphasizing the importance of the psychrotolerant *B. cereus* group strains, a separate species, *B. weihenstephanensis*, was proposed for them (Lechner et al. 1998). Later on, it was observed that not all psychrotolerant strains of the *B. cereus* group should be classified as *B. weihenstephanensis*, and that some still belong to *B. cereus*/*B. thuringiensis* (Stenfors and Granum 2001).

Due to its ubiquitous presence in soil, and on plant material which is used for a variety of purposes, and in food processing environments, as well as its special characteristics described earlier, the presence of *B. cereus* (and other species of the *B. cereus* group) seems to be inevitable in many types of foods and ingredients. These include raw foods such as fresh vegetables and fruits, including sprouted seeds, raw herbs and raw milk, ingredients such as pasteurized liquid egg and milk powder, flour, dried spices and herbs and all kinds of processed foods such as rice, pasta, cereals, dried potato products, meat products, sauces, dehydrated soups, salads, cut and pre-packed vegetables and prepacked sprouts, dehydrated mushrooms, chilled ready-to-eat and ready-to-cook meals, Chinese meals, vegetable and potato purées, pasteurized milk and dairy products, bakery products, desserts and custards, powdered infant formulae and chocolate (EFSA (European Food Safety Authority) 2004; Arnesen et al. 2008). The level of *B. cereus* in raw foods and in processed foods before storage is usually very low (<100 spores/g or ml) and poses no direct health or spoilage concern. Some samples of dried herbs and spices, however, may contain >1,000 spores/g (see also Sect. 7.3.3). Upon storage of processed foods, or the use of contaminated ingredients in complex foods or use of contaminated herbs and spices to season foods, conditions may allow germination and outgrowth of spores to levels that present hazards for consumers. Although no microbiological criterion exists in the European Union for *B. cereus* in foodstuffs (except for a process hygiene criterion for dried infant formulae), some countries have specific criteria or guidelines for this food-borne pathogen; these are generally in the range of $>10^4$–10^5 cfu/g. As shown in Table 7.1, it seems that starchy cooked products such as desserts and rice, heat-treated products such as pasteurized milk and ready-to-eat meals, cold and fresh dishes as well as precut and packed vegetables and fruits are especially associated with the highest prevalence (>1% of samples analysed) of unacceptable levels. This is mostly due to temperature abuse or to overlong storage of the product.

As a result, outbreaks of *B. cereus* food-borne illness have been associated with a diversity of foods such as take-away meals and lunch boxes, oriental meals, cold dishes, minced meats, pita, merguez (North African sausage), chicken, sprouts, meals with rice, boiled and fried rice, pasta salad, pastry, spaghetti, noodles, spices, sauces, soups, stew, quiche, puddings and cream pastries, pasteurized milk and milk products, mashed potatoes and potato salad with mayonnaise, vegetable purée, salad, fish, orange juice and onion powder (EFSA (European Food Safety Authority) 2004; Arnesen et al. 2008; K. Dierick personal communication).

7.3.2 Transfer into Milk from Soil and Agricultural Practises

The dairy production line selects for spore-formers through several heating processes, and selects for psychrophiles or psychrotolerant organisms during cold storage. *B. cereus*, *B. circulans* and *B. mycoides* are the predominant psychrophilic (better described as psychrotolerant) species, with *B. cereus* being the most common (reviewed by Heyndrickx and Scheldeman 2002). The incidence of psychrotolerant spore-formers, and of *B. cereus* in particular, seems to be highest in the summer and autumn. In recent years, the importance of the psychrotolerant aerobic endospore-formers for the keeping quality of milk has increased significantly, owing to extended refrigerated storage of raw milk on the farm before processing higher pasteurization temperatures, reduction of post-pasteurization contamination and prolonged shelf-life requirements of the consumer product. It should be remembered that pasteurization activates spore germination, and thus enhances vegetative cell growth. It has been estimated that 25% of all shelf-life problems of pasteurized milk and cream in the USA are caused by psychrotolerant spore-formers (Sorhaug and Stepaniak 1997).

Table 7.1 Unacceptable *B. cereus* prevalence in different food types at distribution or retail. Food products with prevalence of 1% or greater are indicated in bold

Country (year or period)	Microbiological guideline for unacceptable levels	Food category or item	Number of samples or batches analysed	*B. cereus* prevalence in unacceptable levels (%)	Reference
UK (1999)	$\geq 10^4$ cfu/g	Ready-to-eat foods with added spices	1,946	0.9	Little et al. (2003)
		Dried individual spice	250	**2**	
		Dried spice mix	175	**1.9**	
		Spice paste	243	**4.5**	
		Other spice ingredient (sauce)	82	0	
UK	$\geq 10^4$ cfu/g	**Precooked stored rice for reheating**	2,190	**2**	Nichols et al. (1999)
		Ready-to-eat rice at point-of-sale	1,972	0.5	
Denmark (2000–2003)	$>10^4$ cfu/g	**Cucumbers, tomatoes**	38	**2.6**	Rosenquist et al. (2005)
		Other fresh vegetables	367	0.3	
		Ready-prepared dishes	14,393	0.4	
		Sauces	4,288	0.4	
		Soups	1,723	0.7	
		Vegetables (heat-treated)	428	0.9	
		Sausages	1,666	0.3	
		Meat for open sandwiches	4,215	0.1	
		Pasta	2,216	0.3	
		Rice	1,070	**1.3**	
		Pasta salad	593	0.3	
		Dressings	696	0.1	
		Vegetable/meat/fish mayonnaise	1,748	0.1	
		Desserts with milk and flour	2,350	**1.0**	
		Desserts with milk and rice	223	**3.1**	
		Cake custard	1,601	**1.2**	
		Ice cream with milk products	4,751	0.2	
		Cream cakes	4,948	0.8	

Wales, UK (2003–2005)	$\geq 10^5$ cfu/g	Cream cakes	433	0.5	Meldrum et al. (2006)
		Pasty-meat	515	0.4	
UK (2004)	Production[a]: $n = 5$, $c = 1$, $m = 10^3$ cfu/g, $M = 10^4$ cfu/g	**Spice production batch**	109	**4.6**	Sagoo et al. (2009)
		Herb production batch	23	**8.7**	
	Retail: $\geq 10^4$ cfu/g	Spice retail	2,090	0.9	
		Herb retail	743	0.3	
The Netherlands (2007)	10^5 cfu/g	Tapenade	125	0.8	VWA[b]
		Shoarma meat	324	0.3	
		Whipped cream	475	0.2	
Belgium (2001–2006)	10^3 cfu/g	**Ready-to-eat meals for infants (6–18 months)**	155	**4.6**	Data from FASFC[c] (K. Dierick personal communication)
		Pasteurized milk	419	**2.7**	
	10^4 cfu/g	**Ready-to-eat dishes**	484	**2.5**	
		Precut packed fruit and vegetables including sprouted seeds	252	**2.6**	
		Herbs and spices	323	0.5	
		Dehydrated mushrooms	114	**3.7**	
Belgium (2007)	10^3 cfu/g	**Ready-to-eat meals for infants**	119	**4.0**	FASFC[c]
	10^4 cfu/g	**Fresh aromatic herbs**	52	**4.0**	
		Fresh precut packed fruit and vegetables	59	**12**	
		Cold dishes	87	**6**	

[a]The status of a production batch is considered satisfactory when counts for *B. cereus* in all five sample units (n) are m or less, acceptable when a maximum of c sample units has a count of *B. cereus* between m and M, and unacceptable when one or more sample unit has a count equal to or above M or when more than c sample units have a count between m and M

[b]data from Voedsel en Waren Autoriteit, the Netherlands

[c]data from Belgian Federal Agency for the Safety of the Food Chain

The defects caused by psychrotolerant aerobic endospore-formers are off-flavours and structural defects caused by proteolytic, lipolytic and/or phospholipase enzymes (reviewed by Heyndrickx and Scheldeman 2002). *Paenibacillus polymyxa*, which causes a sour, yeasty and gassy defect in milk, has been found as the main Gram-positive spoilage organism of Swedish and Norwegian pasteurized milks stored at 5 or 7°C (Ternstrom et al. 1993). Two other possible problems for the cheese industry, associated with this and other milk endospore-formers, are fermentative growth with gas production, which causes blowing of soft and semi-hard cheeses (Quiberoni et al. 2008), and vigorous denitrification that diminishes the anti-clostridial effect of added nitrate. The structural defects are sweet curdling in fluid milk and bitty cream, mainly caused by psychrotolerant *B. cereus* strains, although sweet curdling may also be caused by *Brevibacillus laterosporus* or *Bacillus subtilis* strains. Sweet curdling is rather common and mostly affects containers undergoing prolonged refrigerated storage.

B. cereus is also the causative agent of two distinct types of food poisoning, the emetic and diarrhoeal syndrome (reviewed by Arnesen et al. 2008). The diarrhoeal type of *B. cereus* food poisoning has been related to heat-labile protein complexes (haemolysin BL (Hbl) and non-haemolytic enterotoxin (Nhe)) or a single protein (cytotoxin K (CytK)) as causative agents. *B. cereus* diarrhoeal cases or outbreaks associated with milk or dairy products are scarce, although *B. cereus* is commonly isolated from pasteurized milk (Christiansson et al. 1989). This may be due to the fact that in the cold dairy chain, selection occurs for psychrotolerant members of the *B. cereus* group, which are mostly strains of *B. weihenstephanensis*. It has been shown that these strains are less toxigenic than mesophilic *B. cereus* strains and are thus less likely to cause illness (Svensson et al. 2007). In the period 2002–2006, one diarrhoeal *B. cereus* outbreak affecting 70 persons, and caused by milk (containing 3.9×10^5 cfu *B. cereus*/ml) in a scouting camp, was reported in Belgium (K. Dierick personal communication). The causative agent of emesis (food poisoning characterized by vomiting 1–5 h after ingestion of the incriminated food) is a small ring-structured dodecadepsipeptide named cereulide, which is extremely resistant to heat, pH and proteolytic activity of pepsin and trypsin. Food poisoning caused by cereulide is an intoxication; the toxin is formed and released in food, as a consequence of the proliferation of cereulide-producing strains of *B. cereus* under favourable conditions. In high concentrations, cereulide has been reported to cause fatal liver failure and respiratory distress leading to a patient's death (e.g. Dierick et al. 2005). Milk is a suitable substrate for emetic toxin production under optimal conditions (e.g. Finlay et al. 2000). Emetic food poisoning caused by dairy products has been reported (see references in Svensson et al. 2006). Although the overall prevalence of emetic strains in the dairy production chain is very low (<1%), colonization and proliferation of emetic *B. cereus* was observed at particular farms with saw dust bedding as a likely contamination source for the raw milk and in a silo tank at a dairy plant (Svensson et al. 2006). As the general level of spores of emetic toxin producing strains in milk is low (<1,000 cfu/l), there is no risk associated with consumer milk provided that the milk is properly refrigerated.

If contaminated milk is used for milk powder, a risk to the consumer exists if this milk powder is used in baby food or as raw ingredient in foodstuffs, where proliferation of *B. cereus* may occur, and there have been reports of emetic food poisoning episodes where contaminated milk powder was a likely source (see references in Svensson et al. 2006).

From the above, it is not surprising that *B. cereus* is a major concern of the dairy industry, because it seems impossible to avoid its presence in milk completely. In Sweden, there is a legal limit of 1,000 *B. cereus*/ml pasteurized milk after storage at 8°C until the "best before" date (Christiansson et al. 1999). *B. cereus* was found in 56% of samples of Danish pasteurized milk cartons, and both prevalence and counts per ml (after storage at 7°C for 8 days) were significantly higher in the summer than in the winter (Larsen and Jorgensen 1997). In the Netherlands, a maximum *B. cereus* spore limit in farm tank milk of 3 log spores/l is necessary to achieve a shelf life for pasteurized milk of at least 7 days (Vissers et al. 2007a). During recent decades, several possible contamination routes for *B. cereus* in pasteurized milk have been described, with either raw milk or post-pasteurization contamination of the pasteurized milk being the sources (Fig. 7.1). As for the raw milk route of contamination, evidence has been obtained using sensitive detection methods, as *B. cereus* levels in raw milk are usually very low (e.g. average of 1.2 log spores/l) (Vissers et al. 2007a; see Table 7.2), and by molecular typing. Raw milk in the farm tank is contaminated with *B. cereus* spores via the exterior of the cow's teats and through improperly cleaned milking equipment. Contamination of the exterior of the teats occurs when they are contaminated with dirt; during the grazing season of the cows, this dirt is mainly soil, while during the housing season the attached dirt is mainly faeces and bedding material. During milking, this dirt is rinsed off and spores present in the dirt can contaminate the raw milk. As a result, soil, feed (through excretion of spores in faeces) and bedding material are the major sources of contamination of raw milk with *B. cereus*, but the relative contribution of these

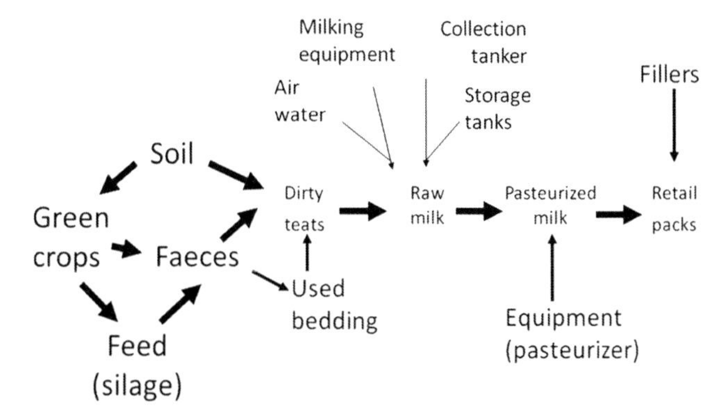

Fig. 7.1 Contamination routes of *B. cereus* spores in the dairy production chain. The importance of contamination sources and routes is indicated by the size of type and *arrow*

Table 7.2 Presence of *B. cereus* spores in soil, agricultural samples and raw milk

Main sample category	Sample type	Number of samples	Total aerobic spore count (log cfu/g or ml, unless otherwise indicated)		*B. cereus* spore prevalence and count (log cfu/g or ml, unless otherwise indicated)			Reference
			Average[a] or median[c]	Range[b] or standard deviation	Prevalence (%)	Average[a] or median[c]	Range[b]	
Soil	Soil						2–5	te Giffel et al. (1995)
		20	5.52[c]	0.64	65		2–5	Slaghuis et al. (1997)
		72			100	3.1[c]–3.9[a]	2.5–5.2	Christiansson et al. (1999)
		38[i]–14[h]				4.9[a,i]–5.0[a,h]	4.1[i]–6.0[i]	Vissers et al. (2007a)
Feed	Grass						1–3	te Giffel et al. (1995)
		20	3.80[c]	0.93	50		<1–3	Slaghuis et al. (1997)
		51			18	<2[c]–1.9[a]	<2–3.3	Christiansson et al. (1999)
	Silage	69–70	4.71[c]	1.18	67		<1–3	Slaghuis et al. (1997)
							<1–4	te Giffel et al. (1995)
		9			78	2[c]–2.1[a]	<2–2.3	Christiansson et al. (1999)
		66[g]–19[h]				2.2[a,g]–2.8[a,k]	<1.0[g]–3.7[h]	Vissers et al. (2007a)
	Concentrate						<1–4	te Giffel et al. (1995)
		82	4.10[c]	1.08	24		<1–3	Slaghuis et al. (1997)
		6			50	<2.2[c]–2.9[a]	<2–3.7	Christiansson et al. (1999)
	Hay	15	4.16		53	<2[c]–3.5[a]	<2–4.7[d]	Christiansson et al. (1999)
Faeces	Faeces				61		2–4	te Giffel et al. (1995)
		87	5.20[c]	1.02	63		<1–3	Slaghuis et al. (1997)
		56			37.5	<2[c]–2.2[a]	<2 – 2.9	Christiansson et al. (1999)
				4.7[b,e]–6.2[b,f]				Wu et al. (2007)
		66[g]–14[h]				2.0[a,g]–2.5[a,h]	1.0[g]–3.8[h]	Vissers et al. (2007a)
Water	Drinking water						<1	te Giffel et al. (1995)
	Rinse water	32			22	<0.7[c] – 0.5[a]/l	<0.7–1/l	Christiansson et al. (1999)
Bedding	Unused bedding						<1–1	te Giffel et al. (1995)
		73	4.08[c]	0.91	6		<1–2	Slaghuis et al. (1997)
	Used bedding						1–4	te Giffel et al. (1995)
		78	5.15[c]	0.83	53		<1–4	Slaghuis et al. (1997)
		66[g]–38[i]				2.7[a,g]–3.1[a,i]	1.0[g]–4.0[i]	Vissers et al. (2007a)
	Used deep-straw bedding	4				6.4[a]		Magnusson et al. (2007)

	Used sand bedding	2				2.3[a]		Magnusson et al. (2007)
	Used sawdust bedding	4				4.3[a]		Magnusson et al. (2007)
	Used straw bedding	4				3.9[a]		Magnusson et al. (2007)
Udder	Teats	86	2.82/swab[c]	0.79	11		<0.6–1/swab	Slaghuis et al. (1997)
Raw milk	First milk, individual cow	91–94	1.61[c]	0.86	21		<0.3–1/10 ml	Slaghuis et al. (1997)
	First milk, installation	59–60	1.48[c]	0.73	15		<0.3–1/10 ml	Slaghuis et al. (1997)
	Raw bulk tank milk	288–296	1.20[c]	0.65	5		<0.3–1/10 ml	Slaghuis et al. (1997)
		18	3.74[a]					Scheldeman et al. (2005)
		66[g]–38[i]				1.0[a,g]–1.5[a,i]/l	<1.0[g]–2.9[i]/l	Vissers et al. (2007a)
	Milk, individual cows	144			69	1[c]–1.6[a]/l	<1–2.9/l	Christiansson et al. (1999)

[a]Average concentration as indicated in table
[b]Minimum and maximum value
[c]Median concentration as indicated in table
[d]One extreme value
[e]Prior to feedlotting
[f]At day 76 of feedlotting
[g]Samples analysed for *B. cereus* count for cows housed in barn all day in housing period (November to April)
[h]Samples analysed for *B. cereus* count for cows at pasture 24h per day during grazing period (May to October)
[i]Samples analysed for *B. cereus* count for cows at pasture during daytime during grazing period (May to October)

sources will be different according to the housing conditions of the cows (i.e. outdoor grazing versus indoor housing in stables) (Fig. 7.1).

A higher incidence of *B. cereus* in raw milk when cows were at pasture (23%) compared with milk from cows housed (4%) during the same period in the summer (Slaghuis et al. 1997) was evidence that soil is the initial contamination source in the grazing season (called "a pasturing effect"). This pasturing effect was also seen in a comparative study of raw milks from organic dairy farms (with obligatory grazing during the summer and autumn) and conventional farms (with less grazing during summer and autumn); higher incidences of *B. cereus* were found in milks from organic farms (Coorevits et al. 2008). Soil is very frequently contaminated with *B. cereus* (65–100%) and can contain up to 10^6 *B. cereus* spores/g, but high variations (up to 3 log) can occur between sampling sites or times (Table 7.2); even higher concentrations of vegetative *B. cereus* cells up to 10^7 cfu/g have been reported (te Giffel et al. 1995). A very high diversity of genetic *B. cereus* types was found in soil, but identical or nearly identical types were found in soil, milk from cows and in the bulk tank milk, which proves the importance of soil as major contamination source of raw milk during the grazing season (Christiansson et al. 1999). The milk contained more spores when the water content of the soil was high, indicating the influence of weather conditions, and the spore content decreased when teats were cleaned before milking, indicating the importance of correct milking routines.

Another important source of contamination of raw milk with *B. cereus* spores, identified by different studies, is cows' faeces or dung. According to Table 7.2, highly variable levels of *B. cereus* spores (from below the detection limit up to 10^5 cfu/g) can be present in different samples of faeces, but the average concentration seems to be around 2 log cfu/g. It is assumed that *Bacillus* spp. are not commensals of the cow's gastrointestinal tract but are introduced via the feed and subsequently shed in the faeces (Wu et al. 2007). Thus, *B. cereus* spores present in faeces probably originate from feed and from soil, either directly (during grazing on the pasture when soil is also taken up by the cows) or indirectly (through the green crops used as feed or feed ingredients). This is corroborated by the fact that the majority of the strains in faeces (57%–66%), on grass (63%–97%), in silage (39%), on hay (33%–67%) and in soil (61%–66%) were psychrotolerant (te Giffel et al. 1995; Christiansson et al. 1999). Raw milk contained a comparable level of psychrotolerant strains (30%–55%). In contrast, feed concentrates did not contain psychrotolerant strains and thus do not contribute to *B. cereus* contamination of milk. Using molecular typing, silage has been shown to be a significant source of contamination of raw milk with spores, including those of *B. cereus* that may occur in levels up to 10^4 cfu/g (Table 7.2). The spores in silage principally arise from soil and farmyard manure. In maize silage, lower concentrations of total spores of aerobes were found than in grass silage (10–10^3 cfu/g versus 10^2 to $>10^5$ cfu/g), which can be explained by the fact that grass silage is more contaminated with soil or faeces than maize silage (te Giffel et al. 2002). Grass samples taken close to soil or faeces contained levels of 10^1–10^3 *B. cereus* spores/g, while this was <10 spores/g for grass without soil. Nowadays, ensiled grass and maize form the major part of the feed ration of cattle in Europe and North America during the whole milking year. The main principles of preservation of crops by ensilage are a rapid achievement of a low pH by lactic acid bacteria and the maintenance of anaerobic conditions. From the moment the silage is used for feeding, it becomes exposed to air and aerobic deterioration takes place, initiated by acid-tolerant yeasts. During the later stages of aerobic spoilage of silage, growth of aerobic endospore-formers (including *B. cereus*) occurs, leading to elevated levels of spores (especially in the surface layers) compared to the initial levels on the green crops. Therefore, it is important to prevent outgrowth of spores in silage by application of special cultures of lactic acid bacteria or chemical additives to improve aerobic stability of the silage (Driehuis and Elferink 2000); however, these measures are not always used on the farm. In the housing period of the cows, feed is the only source of spores and teats become contaminated mainly through the bedding material that is contaminated with faeces. It was found that an increase in the spore contents of the bedding material exceeding 10^4 spores/g increased the risk for elevated spore concentrations in the bulk tank milk (Magnusson et al. 2007). Used bedding material may contain high levels of *B. cereus* spores – up to 10^6 cfu/g (Table 7.2) – which indicates that there is active growth and sporulation of *B. cereus* in bedding material, especially at the back parts of deep sawdust beds that come into contact with the udders.

Using a predictive modelling approach, it was found that when soil was the source of *B. cereus* spores, 33% of simulated spore concentrations in farm tank milk in the Netherlands were above the maximum spore limit of 3 log cfu/l, while this was only 2% when feeds (through faeces and bedding) were the sources of spores (Vissers et al. 2007b). During the grazing period, a 99% reduction of the average concentration of *B. cereus* spores in raw milk might be achieved by minimization of contamination of teats with soil and optimization of teat cleaning efficiency at all farms in the Netherlands. Another control option would be to house cows on damp and rainy days during summer, when the concentration of *B. cereus* spores in soil is high. During such a housing period, care would have to be taken that the initial spore concentration in feeds is <3 log cfu/g and that the pH of the ration offered to the cows is below 5. On the basis of a survey on 24 farms in the Netherlands (none of them exceeding the maximum spore limit in the farm tank milk), it was found that transmission of *B. cereus* spores to farm tank milk occurred predominantly via faeces the whole year round (Vissers et al. 2007a). The higher spore concentrations in milk observed during summer, compared with the rest of the year (a difference of 0.5 log spores/l), coincided with similar higher spore concentrations in faeces, bedding material and mixed silage (but not in soil), and were probably caused by higher temperatures. Nevertheless, the authors pointed to the fact that under certain circumstances, especially wet conditions, transmission of substantial amounts of soil (>1 mg/l milk) can occur, leading to spore levels in milk exceeding the maximum spore limit. Hence, for the dairy industry, control of the soil contamination pathway is most important.

7.3.3 Transfer to Vegetables, Fruit, Ready-to-Eat Meals, Herbs and Spices from Soil and by Agricultural Practises

Contamination of vegetables and fruit frequently occurs either directly through soil or indirectly through agricultural practises such as irrigation with polluted water or fertilization with manure or sewage sludge. Also, wild and domestic animals can contaminate crops or irrigation water; and farm workers and other human handlers further on in the food chain have important roles in the microbiological safety of produce given the possibility of cross-contamination. Although *B. cereus* is frequently present on fresh and intact vegetables (e.g. prevalence of 20%, $n = 81$, in Japan), this is usually at a low level (2–2.7 log cfu/g) (Kaneko et al. 1999). As a result, and as already outlined in Sect. 7.3.1, fresh crops have not yet been reported as sources of *B. cereus* food-borne outbreaks and incidents. Notable exceptions are sprouted seeds, for which a few outbreaks have been reported (reviewed by Bassett and McClure 2008); these products are usually pre-packed for retail or are home-grown. Prepared salads, with a variety of other ingredients besides vegetables and/or fruit, and which are sold in ready-to-use or ready-to-eat formats and rely upon refrigeration for preservation, have been increasingly associated with *B. cereus* or *Bacillus* spp. food poisoning in recent years. In England and Wales in the period 1992–2006, 4% of the outbreaks associated with prepared salads were caused by *Bacillus* spp. (Little and Gillespie 2008). In general, for outbreaks associated with prepared salads, cross-contamination is the most commonly reported fault, followed by an infected food handler (probably of no importance for *Bacillus* spp.) and inappropriate storage (of high importance for *Bacillus* spp., see further).

Besides prepared salads, cooked-chilled foods, also known as REPFEDs (see Sect. 7.3.1), are increasingly popular. They are produced with a mild heat treatment and without additives or preservatives, so achieving a high organoleptic quality, and rely on refrigeration and (frequently) modified-atmosphere packaging for extended durability (from a few days up to 3 months). To this category of foods belong the vegetable purées, which are prepared by steam cooking of washed vegetables (10–50 min. at 80–100°C), mixing with other ingredients (UHT-cream, milk proteins, starch, salt) and pasteurization in their packaging (10 min at 80–90°C). Although these products are relatively stable up to 21 days when stored at 4°C, the processing conditions will not, completely, eliminate the spore-formers that can develop during cold storage. Vegetable purées were the source of a severe fatal *B. cereus* outbreak in a French nursing home for elderly persons, and this episode led to the discovery of a new *B. cereus* diarrhoeal toxin, cytotoxin K (Hardy et al. 2001). According to data in the literature, in the period 1969–1998, food-borne outbreaks after the consumption of products containing vegetables, or with a high probability of containing vegetables, have been caused by *B. cereus* (38 outbreaks), *B. subtilis* (11 outbreaks), *B. licheniformis* (24 outbreaks) and *B. pumilus* (Carlin et al. 2000a). In cooked, pasteurized and chilled vegetable purées (broccoli, carrot, courgette, leek, potato, split pea) aerobic endospore-formers are the dominant microbiota (albeit at low numbers of $\leq$2 log cfu/g before storage) and an increase in their numbers (up to 7–8 log cfu/g) depends on the storage temperature and the type of vegetable (Carlin et al. 2000b). Courgette purée (also called zucchini purée) consistently supported the most rapid microbial growth at low temperature and was the product most prone to spoilage (pack swelling and/or off-odours). In zucchini purées stored at 10°C, spoilage occurred after 21 days, concomitant with the dominance of *B. pumilus* and a new *Paenibacillus* species (Guinebretière et al. 2001). This indicates that at the temperature conditions (around 10°C) that often occur during distribution or in consumers' refrigerators, spoilage may already develop in this product before the end of the specified shelf-life, while strict compliance with the recommended storage temperature of 4°C should prevent spoilage and growth of *B. cereus*. On raw and processed zucchini, *B. cereus* counts were $\leq$0.4 log cfu/g, and in packaged purées *B. cereus* counts were $\leq$0.2 log cfu/g; however, this low level seemed to be sufficient to permit bacterial growth in the stored product after 21 days at 10°C (up to 4.6 log cfu/g, exceeding the upper limit of 4 log cfu/g recommended for processed vegetables in France), while at 4°C no significant growth could be detected. By means of molecular typing it could be established that this outgrowth at 10 or 7°C is of psychrotolerant strains that originated in the zucchini, while the strains from the dehydrated texturing agents (milk proteins and starch) were mesophilic strains that were genetically different from the zucchini strains and which probably contributed less to the growth at 10°C (Guinebretière and Nguyen-The 2003). Furthermore, it was found that the initial source of contamination for the zucchini, with psychrotolerant *B. cereus,* was soil. These data indicate that processing and storage of vegetables leads to selection of *B. cereus* strains, and that this is enabled by the fact that soil as the initial source of contamination is a reservoir of a large diversity of strains. The importance of this genetic diversity was illustrated by a quantitative exposure assessment that took into account the heterogeneous population of *B. cereus* in courgette purée (Afchain et al. 2008). The model predicted that at the time of consumption, *B. cereus* may be present in all purée packages, but that not all *B. cereus* genetic groups (I to VII as defined by Guinebretière et al. 2008) were present in all packages. The highest food poisoning risk in REPFED foods was predicted to be associated with members of the psychrotolerant *B. cereus* group II because of their predicted prevalence (16.8% of the *B. cereus* population per package) and concentrations at the time of consumption (2.5% of the packages contaminated with >6.7 log cfu/g). Although some strains implicated in *B. cereus* food poisoning belong to psychrotolerant genetic group II, it must be noted that most strains related to cases or outbreaks belong to mesophilic genetic groups III and IV. This may explain why most reported cases have been associated with incidents of storage temperature abuse.

Finally, herbs and spices are a potential hazard if they are used at the end of cooking or added to prepared foods without further cooking, although they are not major contributors to food poisoning. Usually, levels of *B. cereus* and of other spore-formers are below 10^3 cfu/g, but these low levels may be sufficient to cause food poisoning if the seasoned food is inappropriately handled. A likely microbiological criterion for unacceptable retail samples of dried herbs and spices is $\geq 10^4$ *B. cereus* cfu/g. According to recent monitoring programmes, between

<1% and 2% of sampled dried herbs and spices in retail or ready-to-eat food premises, respectively, and up to 4.5% of spice pastes used in ready-to-eat food premises was of unacceptable quality (Table 7). The corresponding ready-to-eat foods, to which these spices with unacceptable levels of contamination were added, contained similarly high levels of spore-formers (Little et al. 2003). The prevention of contamination of dried herbs and spices lies in the application of good hygienic practises during growing, harvesting and processing, as well as in effective subsequent decontamination processes such as steam or dry heat treatments.

7.4 Transfer of *Alicyclobacillus* Species from Soil to Fruit Juices

In 1992, the new genus *Alicyclobacillus* was created to accommodate the species *A. acidocaldarius*, *A. acidoterrestris* and *A. cycloheptanicus*, three species possessing high amounts of ω-alicyclic fatty acids as major lipids in the cell membrane, a unique phenotype amongst the aerobic endospore-formers (Wisotzkey et al. 1992). Initial interest in these organisms was purely academic, but there had been a report from Germany in 1984 (Cerny et al. 1984) of large-scale spoilage, following an exceptionally long, hot summer, of shelf-stable apple juice by unusual bacilli able to grow at pH values as low as 2.5. This was followed in the 1990s by observations of episodic spoilage of fruit juice and similar products by *A. acidoterrestris* in Europe, the USA and Japan (Jensen 1999). Until then, it was believed that the high acidity of fruit juices (pH <4), combined with pasteurization in the fruit beverage industry (hot-fill and hold for 2 min at 88–96°C), was sufficient to render these-products commercially sterile without the use of preservatives. More recently, three new species *A. acidiphilus*, which is closely related to *A. acidoterrestris*, and *A. pomorum* were isolated from spoiled juice products, and *A. herbarius* was isolated from spoiled herbal tea (Goto et al. 2002, 2003; Matsubara et al. 2002).

Spoilage (long before the expiration date) is sometimes associated with a slight increase in turbidity and white sediment at the bottom of packages, but the most important fault is taint of strong medicinal or antiseptic flavour caused mainly by the production of guaiacol (2-methoxyphenol) at levels up to 100 ppb (the taste threshold in water is 2 ppb) (Pettipher et al. 1997), but also by 2,6 dibromophenol and 2,6-dichlorophenol at levels up to 20 ppt (taste thresholds 0.5 ppt and 30 ppt in juice, respectively) (Jensen and Whitfield 2003). Affected products are pasteurized fruit juices (mainly apple and orange) and fruit juice blends, but there have also been reports of spoiled carbonated fruit drinks, berry juice containing iced tea and diced canned tomatoes (Walker and Phillips 2008). Although spoilage is incidental, requiring a combination of adequate conditions such as available oxygen (determined by the head space in the package or container), nutrient availability, taint precursors and high temperature for a long time, all fruit juice manufacturers have started quality assurance programmes to monitor and control levels of *Alicyclobacillus* species in raw materials and fruit concentrates. Between 14.7% and 83% of fruit juices, fruit concentrates or fruit juice products have been found positive for

Alicyclobacillus (Walker and Phillips 2008). As *A. acidoterrestris* is mainly soil borne, unwashed or insufficiently washed raw fruit, which is contaminated on the surface with soil, and poor hygiene measures in fruit juice concentrate production are the most likely contamination sources. More than one third of all fruits sampled at two juice-processing facilities were contaminated with presumptive *Alicyclobacillus* species (Parish and Goodrich 2005). *A. acidoterrestris* has also been isolated from process water (Mcintyre et al. 1995) and *Alicyclobacillus* strains have been found in liquid sugar (Heyndrickx and Scheldeman 2002).

A. acidoterrestris survives the hot-fill and hold processes because of the heat resistance of the spores (*D*-values of 65.6 min at 85°C and 11.9 min at 91°C in orange juice; Silva et al. 1999), which germinate following heat shock, and the organism starts to grow in the product up to a level of 10^5–10^6 cfu/ml because of its unique physiological characteristics (growth from pH 2.5 to 6.0 with optima around pH 3.5–5.0, and from 20 to 60°C with optima at 42–53°C). Spore numbers in raw materials and processed products may be very low (<1 per 100 g), requiring sensitive detection procedures, including heat shock to induce germination and enrichment or membrane filtration to test larger volumes, but even batches of materials with such low numbers should be rejected (Jensen 1999). Storage of pasteurized fruit juices below 20°C would be sufficient to prevent outgrowth of spores, but these juices are usually distributed under ambient conditions.

Recently, both *A. acidoterrestris* and *A. acidocaldarius* have been isolated from orchard soil (pH 5.9–6.7) collected from an apple and pear farm in South Africa (Groenewald et al. 2008) and *A. acidocaldarius* was isolated from pre-pasteurized pear purée and vinegar flies (Groenewald et al. 2009). The roles of these specific strains for fruit juice spoilage remain to be established.

7.5 Transfer of *Bacillus sporothermodurans* from Concentrated Feeds into Milk

UHT or sterilization combined with aseptic filling are intended to eliminate all viable microorganisms from products and result in commercially sterile liquid milk products, having long shelf-lives without refrigeration. According to the former Milk Hygiene EC-directive 92/46, the colony count at 30°C after incubation of unopened packages for 15 days at 30°C should be <10 cfu per 0.1 ml. However, massive occurrences of contamination with the newly recognized, highly heat-resistant endospore-former *B. sporothermodurans* at levels above this threshold were first reported in Italy and Austria in 1985, and this problem later spread to other European (France, Benelux, Spain) and extra-European countries (Mexico, USA) (reviewed by Scheldeman et al. 2006). Milk products affected included whole, skimmed, evaporated or reconstituted UHT-milk, UHT-cream and chocolate milk; milk powders were also found to be contaminated. Real spoilage defects in consumer milks are only rarely noticed – as slight pink colour changes, off-flavours and coagulation, especially in containers with low oxygen barriers (e.g. plastic bottles), but the main problem caused by this organism is non-sterility of commercially sterile milk and milk products, and currently this contamination continues at a sporadic level.

It seems that UHT-milk represents a special ecological niche, for which several contamination sources have been suggested. The most important route of contamination has been identified as reprocessing of contaminated lots of UHT-milk in dairy factories (Guillaume-Gentil et al. 2002). Reprocessing of one contaminated package can contaminate a considerable fraction of the whole day's production at a level of 1 spore/l, which is regarded as the common contamination level. Contamination is also possible via contaminated milk powder.

B. sporothermodurans has also been isolated from raw farm milk, although at very low frequency and contamination levels (Scheldeman et al. 2006). As already explained in Sect. 7.3.2, contamination of raw milk at the farm is possible via the feed through excretion in faeces. *B. sporothermodurans* has been isolated from silage and (usually sugar beet) pulp, but has mainly been isolated from feed concentrates (Scheldeman et al. 2005). There are also claims of isolation from compost and cattle faeces (Wu et al. 2007; Zhang et al. 2002). *B. sporothermodurans* thus seems to be associated with agricultural environments, but there are no reports of isolation from soil as yet. Feed concentrates often contain tropical ingredients (e.g. coconut, citrus pulp, manioc, cacao), and it might therefore be speculated that *B. sporothermodurans* has a (sub)tropical origin.

7.6 Conclusions

In many types of food and feed, soil can be considered as the initial contamination source for aerobic spore-formers. Usually, when direct transfer from soil is involved, levels of these spore-formers in foods, ingredients or feeds are too low to cause problems. However, because of the complexity of the food chain, particular spore forming species or types may encounter niches where proliferation occurs. This can happen on the primary production level (e.g. silage, bedding material), in the processing line (e.g. storage tanks), during distribution (e.g. temperature abuse during refrigerated storage) or in the final product (e.g. complex foods, fruit juice). These proliferation steps enable the endospore-former such as *B. cereus* or *A. acidoterrestris* either to enter as a contaminant into a next step of the production chain or to provoke food quality or safety problems. For the food industry, it is a challenge to gain insight into the whole contamination flow of endospore-formers originating in soil.

References

Afchain AL, Carlin F, Nguyen-the C, Albert I (2008) Improving quantitative exposure assessment by considering genetic diversity of *B. cereus* in cooked, pasteurised and chilled foods. Int J Food Microbiol 128:165–173

Andersson A, Ronner U, Granum PE (1995) What problems does the food industry have with the spore-forming pathogens *Bacillus cereus* and *Clostridium perfringens*? Int J Food Microbiol 28:145–155

Arnesen LPS, Fagerlund A, Granum PE (2008) From soil to gut: *Bacillus cereus* and its food poisoning toxins. FEMS Microbiol Rev 32:579–606

Bassett J, McClure P (2008) A risk assessment approach for fresh fruits. J Appl Microbiol 104:925–943

Carlin F, Girardin H, Peck MW, Stringer SC, Barker GC, Martinez A, Fernandez A, Fernandez P, Waites WM, Movahedi S, van Leusden F, Nauta M, Moezelaar R, Torre MD, Litman S (2000a) Research on factors allowing a risk assessment of spore-forming pathogenic bacteria in cooked chilled foods containing vegetables: a FAIR collaborative project. Int J Food Microbiol 60:117–135

Carlin F, Guinebretière MH, Choma C, Pasqualini R, Braconnier A, Nguyen-The C (2000b) Spore-forming bacteria in commercial cooked, pasteurised and chilled vegetable purees. Food Microbiol 17:153–165

Carlin F, Fricker M, Pielaat A, Heisterkamp S, Shaheen R, Salonen MS, Svensson B, Nguyen-The C, Ehling-Schulz M (2006) Emetic toxin-producing strains of *Bacillus cereus* show distinct characteristics within the *Bacillus cereus* group. Int J Food Microbiol 109:132–138

Cerny G, Hennlich W, Poralla K (1984) Fruchtsaftverderb durch Bacillen: Isolierung und Charakterisierung des Verderbniserregers. Z Lebens Unters Forsch 179:224–227

Christiansson A, Naidu AS, Nilsson I, Wadstrom T, Pettersson HE (1989) Toxin production by *Bacillus cereus* dairy isolates in milk at low temperatures. Appl Environ Microbiol 55:2595–2600

Christiansson A, Bertilsson J, Svensson B (1999) *Bacillus cereus* spores in raw milk: factors affecting the contamination of milk during the grazing period. J Dairy Sci 82:305–314

Coorevits A, De Jonghe V, Vandroemme J, Reekmans R, Heyrman J, Messens W, De Vos P, Heyndrickx M (2008) Comparative analysis of the diversity of aerobic spore-forming bacteria in raw milk from organic and conventional dairy farms. Syst Appl Microbiol 31:126–140

de Vries YP (2006) *Bacillus cereus* spore formation, structure, and germination. PhD Thesis. Wageningen University, Wageningen, the Netherlands

Dierick K, Van Coillie E, Swiecicka I, Meyfroidt G, Devlieger H, Meulemans A, Hoedemaekers G, Fourie L, Heyndrickx M, Mahillon J (2005) Fatal family outbreak of *Bacillus cereus*-associated food poisoning. J Clin Microbiol 43:4277–4279

Dragon DC, Rennie RP (1995) The ecology of anthrax spores – tough but not invincible. Can Vet J Rev Vet Can 36:295–301

Driehuis F, Elferink SJWH (2000) The impact of the quality of silage on animal health and food safety: a review. Vet Qtly 22:212–216

EFSA (European Food Safety Authority) (2004) Opinion of the Scientific Panel on Biological Hazards on *Bacillus cereus* and other *Bacillus* spp. in foodstuffs. EFSA J 175:1–49

Finlay WJJ, Logan NA, Sutherland AD (2000) *Bacillus cereus* produces most emetic toxin at lower temperatures. Lett Appl Microbiol 31:385–389

Frederiksen K, Rosenquist H, Jorgensen K, Wilcks A (2006) Occurrence of natural *Bacillus thuringiensis* contaminants and residues of *Bacillus thuringiensis*-based insecticides on fresh fruits and vegetables. Appl Environ Microbiol 72:3435–3440

Goto K, Matsubara H, Mochida K, Matsumura T, Hara Y, Niwa M, Yamasato K (2002) *Alicyclobacillus herbarius* sp nov., a novel bacterium containing omega-cycloheptane fatty acids, isolated from herbal tea. Int J Syst Evol Microbiol 52:109–113

Goto K, Mochida K, Asahara M, Suzuki M, Kasai H, Yokota A (2003) *Alicyclobacillus pomorum* sp nov., a novel thermo-acidophilic, endospore-forming bacterium that does not possess omega-alicyclic fatty acids, and emended description of the genus *Alicyclobacillus*. Int J Syst Evol Microbiol 53:1537–1544

Groenewald WH, Gouws PA, Witthuhn RC (2008) Isolation and identification of species of *Alicyclobacillus* from orchard soil in the Western Cape, South Africa. Extremophiles 12:159–163

Groenewald WH, Gouws PA, Witthuhn RC (2009) Isolation, identification and typification of *Alicyclobacillus acidoterrestris* and *Alicyclobacillus acidocaldarius* strains from orchard soil and the fruit processing environment in South Africa. Food Microbiol 26:71–76

Guillaume-Gentil O, Scheldeman P, Marugg J, Herman L, Joosten H, Heyndrickx M (2002) Genetic heterogeneity in *Bacillus sporothermodurans* as demonstrated by ribotyping and repetitive extragenic palindromic-PCR fingerprinting. Appl Environ Microbiol 68:4216–4224

Guinebretière MH, Nguyen-The C (2003) Sources of *Bacillus cereus* contamination in a pasteurized zucchini puree processing line, differentiated by two PCR-based methods. FEMS Microbiol Ecol 43:207–215

Guinebretière MH, Berge O, Normand P, Morris C, Carlin F, Nguyen-The C (2001) Identification of bacteria in pasteurized zucchini purees stored at different temperatures and comparison with those found in other pasteurized vegetable purees. Appl Environ Microbiol 67:4520–4530

Guinebretière MH, Thompson FL, Sorokin A, Normand P, Dawyndt P, Ehling-Schulz M, Svensson B, Sanchis V, Nguyen-The C, Heyndrickx M, De Vos P (2008) Ecological diversification in the *Bacillus cereus* Group. Environ Microbiol 10:851–865

Hardy SP, Lund T, Granum PE (2001) CytK toxin of *Bacillus cereus* forms pores in planar lipid bilayers and is cytotoxic to intestinal epithelia. FEMS Microbiol Lett 197:47–51

Helgason E, Caugant DA, Lecadet MM, Chen YH, Mahillon J, Lovgren A, Hegna I, Kvaloy K, Kolsto AB (1998) Genetic diversity of *Bacillus cereus – B. thuringiensis* isolates from natural sources. Curr Microbiol 37:80–87

Heyndrickx M, Scheldeman P (2002) Bacilli associated with spoilage in dairy products and other food. In: Berkeley RCW, Heyndrickx M, Logan NA, De Vos P (eds) Applications and systematics of *Bacillus* and relatives. Blackwell Science, Oxford, UK, pp 65–82

Husmark U, Rönner U (1992) The influence of hydrophobic, electrostatic and morphologic properties on the adhesion of *Bacillus* spores. Biofouling 5:335–344

in't Veld JHJH (1996) Microbial and biochemical spoilage of foods: an overview. Int J Food Microbiol 33:1–18

Jensen N (1999) *Alicyclobacillus* – a new challenge for the food industry. Food Aust 51:33–36

Jensen N, Whitfield FB (2003) Role of *Alicyclobacillus acidoterrestris* in the development of a disinfectant taint in shelf-stable fruit juice. Lett Appl Microbiol 36:9–14

Jensen GB, Hansen BM, Eilenberg J, Mahillon J (2003) The hidden lifestyles of *Bacillus cereus* and relatives. Environ Microbiol 5:631–640

Kaneko K, Hayashidani H, Ohtomo Y, Kosuge J, Kato M, Takahashi K, Shiraki Y, Ogawa M (1999) Bacterial contamination of ready-to-eat foods and fresh products in retail shops and food factories. J Food Protect 62:644–649

Larsen HD, Jorgensen K (1997) The occurrence of *Bacillus cereus* in Danish pasteurized milk. Int J Food Microbiol 34:179–186

Lechner S, Mayr R, Francis KP, Pruss BM, Kaplan T, Wiessner-Gunkel E, Stewartz GSAB, Scherer S (1998) *Bacillus weihenstephanensis* sp. nov. is a new psychrotolerant species of the *Bacillus cereus* group. Int J Syst Bacteriol 48:1373–1382

Little CL, Gillespie IA (2008) Prepared salads and public health. J Appl Microbiol 105:1729–1743

Little CL, Omotoye R, Mitchell RT (2003) The microbiological quality of ready-to-eat foods with added spices. Int J Environ Health Res 13:31–42

Magnusson M, Christiansson A, Svensson B (2007) *Bacillus cereus* spores during housing of dairy cows: factors affecting contamination of raw milk. J Dairy Sci 90:2745–2754

Margulis L, Jorgensen JZ, Dolan S, Kolchinsky R, Rainey FA, Lo SC (1998) The *Arthromitus* stage of *Bacillus cereus*: intestinal symbionts of animals. Proc Natl Acad Sci USA 95:1236–1241

Matsubara H, Goto K, Matsumura T, Mochida K, Iwaki M, Niwa M, Yamasato K (2002) *Alicyclobacillus acidiphilus* sp nov., a novel thermo-acidophilic, omega-alicyclic fatty acid-containing bacterium isolated from acidic beverages. Int J Syst Evol Microbiol 52:1681–1685

McIntyre S, Ikawa JY, Parkinson N, Haglund J, Lee J (1995) Characteristics of an acidophilic *Bacillus* strain isolated from shelf-stable juices. J Food Protect 58:319–321

Meldrum RJ, Smith RMM, Ellis P, Garside J (2006) Microbiological quality of randomly selected ready-to-eat foods sampled between 2003 and 2005 in Wales, UK. Int J Food Microbiol 108:397–400

Nichols GL, Little CL, Mithani V, de Louvois J (1999) The microbiological quality of cooked rice from restaurants and take-away premises in the United Kingdom. J Food Protect 62:877–882

Parish ME, Goodrich RM (2005) Recovery of presumptive *Alicyclobacillus* strains from orange fruit surfaces. J Food Protect 68:2196–2200

Peng JS, Tsai WC, Chou CC (2001) Surface characteristics of *Bacillus cereus* and its adhesion to stainless steel. Int J Food Microbiol 65:105–111

Pettipher GL, Osmundson ME, Murphy JM (1997) Methods for the detection and enumeration of *Alicyclobacillus acidoterrestris* and investigation of growth and production of taint in fruit juice and fruit juice-containing drinks. Lett Appl Microbiol 24:185–189

Quiberoni A, Guglielmotti D, Reinheimer J (2008) New and classical spoilage bacteria causing widespread blowing in Argentinean soft and semihard cheeses. Int J Dairy Technol 61:358–363

Rosenquist H, Smidt L, Andersen SR, Jensen GB, Wilcks A (2005) Occurrence and significance of *Bacillus cereus* and *Bacillus thuringiensis* in ready-to-eat food. FEMS Microbiol Lett 250:129–136

Ryu JH, Beuchat LR (2005) Biofilm formation and sporulation by *Bacillus cereus* on a stainless steel surface and subsequent resistance of vegetative cells and spores to chlorine, chlorine dioxide, and a peroxyacetic acid-based sanitizer. J Food Protect 68:2614–2622

Sagoo SK, Little CL, Greenwood M, Mithani V, Grant KA, McLauchlin J, de Pinna E, Threlfall EJ (2009) Assessment of the microbiological safety of dried spices and herbs from production and retail premises in the United Kingdom. Food Microbiol 26:39–43

Scheldeman P, Pil A, Herman L, De Vos P, Heyndrickx M (2005) Incidence and diversity of potentially highly heat-resistant spores isolated at dairy farms. Appl Environ Microbiol 71:1480–1494

Scheldeman P, Herman L, Foster S, Heyndrickx M (2006) *Bacillus sporothermodurans* and other highly heat-resistant spore formers in milk. J Appl Microbiol 101:542–555

Silva FM, Gibbs P, Vieira MC, Silva CLM (1999) Thermal inactivation of *Alicyclobacillus acidoterrestris* spores under different temperature, soluble solids and pH conditions for the design of fruit processes. Int J Food Microbiol 51:95–103

Slaghuis BA, Giffel MCT, Beumer RR, Andre G (1997) Effect of pasturing on the incidence of *Bacillus cereus* spores in raw milk. Int Dairy J 7:201–205

Sorhaug T, Stepaniak L (1997) Psychrotrophs and their enzymes in milk and dairy products: quality aspects. Trends Food Sci Technol 8:35–41

Stenfors LP, Granum PE (2001) Psychrotolerant species from the *Bacillus cereus* group are not necessarily *Bacillus weihenstephanensis*. FEMS Microbiol Lett 197:223–228

Svensson B, Monthan A, Shaheen R, Andersson MA, Salkinoja-Salonen M, Christiansson A (2006) Occurrence of emetic toxin producing *Bacillus cereus* in the dairy production chain. Int Dairy J 16:740–749

Svensson B, Monthan A, Guinebretière MH, Nguyen-The C, Christiansson A (2007) Toxin production potential and the detection of toxin genes among strains of the *Bacillus cereus* group isolated along the dairy production chain. Int Dairy J 17:1201–1208

te Giffel MC, Beumer RR, Slaghuis BA, Rombouts FM (1995) Occurrence and characterization of (psychrotrophic) *Bacillus cereus* on farms in the Netherlands. Netherlands Milk Dairy J 49:125–138

te Giffel MCT, Wagendorp A, Herrewegh A, Driehuis F (2002) Bacterial spores in silage and raw milk. Antonie Leeuwenhoek 81:625–630

Ternstrom A, Lindberg AM, Molin G (1993) Classification of the spoilage flora of raw and pasteurized bovine-milk, with special reference to *Pseudomonas* and *Bacillus*. J Appl Bacteriol 75:25–34

Vilain S, Luo Y, Hildreth MB, Brozel VS (2006) Analysis of the life cycle of the soil saprophyte *Bacillus cereus* in liquid soil extract and in soil. Appl Environ Microbiol 72:4970–4977

Vissers MMM, Giffel MCT, Driehuis F, De Jong P, Lankveld JMG (2007a) Minimizing the level of *Bacillus cereus* spores in farm tank milk. J Dairy Sci 90:3286–3293

Vissers MMM, Giffel MCT, Driehuis F, De Jong P, Lankveld JMG (2007b) Predictive modeling of *Bacillus cereus* spores in farm tank milk during grazing and housing periods. J Dairy Sci 90:281–292

von Stetten F, Mayr R, Scherer S (1999) Climatic influence on mesophilic *Bacillus cereus* and psychrotolerant *Bacillus weihenstephanensis* populations in tropical, temperate and alpine soil. Environ Microbiol 1:503–515

Walker M, Phillips CA (2008) *Alicyclobacillus acidoterrestris*: an increasing threat to the fruit juice industry? Int J Food Sci Technol 43:250–260

Wijman JGE, de Leeuw PPLA, Moezelaar R, Zwietering MH, Abee T (2007) Air-liquid interface biofilms of *Bacillus cereus*: formation, sporulation, and dispersion. Appl Environ Microbiol 73:1481–1488

Wisotzkey JD, Jurtshuk P, Fox GE, Deinhard G, Poralla K (1992) Comparative sequence analyses on the 16S ribosomal-RNA (rDNA) of *Bacillus acidocaldarius*, *Bacillus acidoterrestris*, and *Bacillus cycloheptanicus* and proposal for creation of a new genus, Alicyclobacillus gen. nov. Int J Syst Bacteriol 42:263–269

Wu XY, Walker M, Vanselow B, Chao RL, Chin J (2007) Characterization of mesophilic bacilli in faeces of feedlot cattle. J Appl Microbiol 102:872–879

Yara K, Kunimi Y, Iwahana H (1997) Comparative studies of growth characteristic and competitive ability in *Bacillus thuringiensis* and *Bacillus cereus* in soil. Appl Entomol Zool 32:625–634

Zhang YC, Ronimus RS, Turner N, Zhang Y, Morgan HW (2002) Enumeration of thermophilic *Bacillus* species in composts and identification with a random amplification polymorphic DNA (RAPD) protocol. Syst Appl Microbiol 25:618–626

Chapter 8
Biological Control of Phytopathogenic Fungi by Aerobic Endospore-Formers

Alejandro Pérez-García, Diego Romero, Houda Zeriouh, and Antonio de Vicente

8.1 Introduction

Plant diseases are of paramount importance to humans. It is conservatively estimated that plant diseases interfere with the production of 14% of all the crops produced throughout the world, resulting in a total annual worldwide crop loss of about $220 billion. To this should be added the 6–12% of crops that are lost after harvest, which is a substantial problem in developing, tropical countries where training and resources such as refrigeration are generally lacking (Agrios 2005). The increasing demand for a steady, healthy food supply to feed the burgeoning human population requires efficient control of the main plant diseases that reduce crop yield. Current practises for controlling plant diseases are based largely on genetic resistance in the host plant, management of the plant and its environment and the application of synthetic pesticides. Nevertheless, excessive use of agrochemicals in conventional crop management has caused serious environmental and health problems. Therefore, there is a demand for new methods to supplement the existing disease control strategies, and thereby, to achieve better disease control.

Of the plant disease agents known to humankind, phytopathogenic fungi are the most prevalent. Most of the more than 100,000 known fungal species live on dead organic matter and thereby help decompose it (i.e., they are strictly saprophytic). However, about 100 species cause diseases in animals (including humans), and more than 10,000 species may induce diseases in plants. Fungi can destroy crops, and the economic consequences of this have been enormous throughout human history. The

A. Pérez-García (✉)
Departamento de Microbiología, Universidad de Málaga, Instituto de Fruticultura Subtropical y Mediterránea (IHSM-UMA-CSIC), Bulevar Louis Pasteur-Campus Universitario de Teatinos s/n, 29071 Málaga, Spain
e-mail: aperez@uma.es

D. Romero, H. Zeriouh, and A. de Vicente
Grupo de Microbiología y Patología Vegetal, Departamento de Microbiología, Facultad de Ciencias, Universidad de Málaga, Bulevar Louis Pasteur-Campus Universitario de Teatinos s/n, 29071 Málaga, Spain

N.A. Logan and P. De Vos (eds.), *Endospore-forming Soil Bacteria*, Soil Biology 27,
DOI 10.1007/978-3-642-19577-8_8, © Springer-Verlag Berlin Heidelberg 2011

endless variety and complexity of the many diseases of plants caused by fungi have led to the development of a correspondingly large number of approaches for their control. For many fungal diseases, however, the most effective (and sometimes the only) method available for their control is the application of fungicides. Unfortunately, just as human pathogens can become resistant to antibiotics, several plant pathogens have developed resistance to certain fungicides (Agrios 2005).

Biological control, i.e., the use of microorganisms to suppress plant disease, offers an interesting alternative to the use of fungicides. The rich diversity of the microbial world provides a seemingly endless resource for this purpose. The complexity of the interactions between organisms, the existence of numerous mechanisms of disease suppression by a single microorganism, and the adaptation of most biological control agents to the environment, have all contributed to the belief that biological control will be more durable than fungicides (Emmert and Handelsman 1999). In this chapter, we seek to highlight the potential of aerobic endospore-forming bacteria as biological control agents for fungal plant diseases. To this end, we provide examples of the successful control of fungal plant diseases by these bacteria, we describe the mechanisms involved in this effective means of disease control and we provide protocols and tools for further study. And finally, we discuss future prospects for aerobic endospore-former-based biological control products in the integrated pest management (IPM) systems of twenty-first century agriculture.

8.2 Aerobic Endospore-forming Bacteria in Agroecosystems and Plant Health

Aerobic endospore-forming bacteria are ubiquitous in agricultural systems. Common metabolic traits important to their survival include the production of a multi-layered cell wall, the formation of endospores, and the secretion of a myriad of antibiotics, signal molecules, and extracellular enzymes. In addition, variability in other key traits such as nutrient utilization, motility, and physicochemical growth optima, allow for these bacteria to inhabit diverse niches in agroecosystems (McSpadden Gardener 2004), and they are common soil inhabitants (see Mandic-Mulec and Prosser 2011). Multiple *Bacillus* spp. strains can be readily cultured from both bulk and rhizospheric soils on solid media, and cultivable counts of these bacteria generally range from 10^3 to 10^6 cells per gram, fresh weight, of soil. The most abundant species are *Bacillus subtilis* and *Bacillus cereus*, but multiple isolates of phenetically and phylogenetically related species can also be recovered. The most morphologically distinctive of these is *Bacillus mycoides*, which often confounds attempts to accurately enumerate cultured populations by virtue of its rapid mycelium-like growth patterns on solid media (McSpadden Gardener 2004). In addition, cultivation-independent analyses of soil DNA have demonstrated the presence of a diversity of uncultured *Bacillus*-related lineages (see De Vos 2011) and have also revealed significant differences between the soil and rhizosphere communities of aerobic endospore-forming bacteria (Smalla et al. 2001).

Over the past century, numerous examples have accumulated of susceptible plants remaining virtually disease-free despite ample exposure to virulent soil-borne pathogens. Soils in which pathogens fail to establish disease are known as disease-suppressive soils. Two classical types of soil disease suppressiveness are known. General suppression is related to the total microbial biomass in soil, which either competes with the pathogen for resources or causes inhibition through more direct forms of antagonism. General suppression is often enhanced by the addition of organic matter (e.g., compost or green manure) and/or other agronomic practises that increase soil microbial activity (Hoitink and Boehm 1999). No single microorganism is responsible for general suppression, and the suppressiveness is not transferable between soils. By contrast, specific suppression is due largely to the activity of an individual (or select group) of microorganisms that inhibits some stage in the life cycle of the pathogen. Specific suppression is transferable between soils, so it is also known as transferable suppression. Suppressive organically amended soils undoubtedly owe their activity to a combination of general and specific suppression. The two mechanisms function as a continuum in soil, although they may be affected differently by edaphic, climatic, and agronomic conditions (Weller et al. 2002).

Several studies have suggested an active role for aerobic endospore-forming bacteria in soil suppressiveness. Early reports on suppression of *Fusarium* wilt and take-all decline implicated the *Bacillus* genus in this phenomenon (Kim et al. 1997). Similarly, *Bacillus* spp. have been identified as antagonistic bacteria responsible for the disease control effect in compost-amended substrates (Hoitink et al. 1997); for example, turf grass diseases such as brown patch are particularly suppressed by compost that has been inoculated with *B. subtilis* (Nakasaki et al. 1998). Diverse populations of aerobic endospore-forming bacteria are denizens of agricultural fields and may directly or indirectly contribute to crop productivity. The biological mechanisms by which *Bacillus* species suppress pathogens and/or promote plant growth are beginning to be understood, and recent insights will be discussed later.

8.3 Biological Control of Plant Fungal Diseases by Aerobic Endospore-forming Bacteria

While diverse bacteria may contribute to the biological control of plant pathogens, most of the research and development efforts have focused on two genera, *Bacillus* and *Pseudomonas*. The Gram-positive bacteria have received less attention than the fluorescent pseudomonads in the literature on biocontrol. In part, this is because the Gram-positive organisms have been less tractable for genetic study and less is known about the mechanisms by which they suppress disease. However, their efficacy is striking. Many surveys of soil bacteria have identified *Bacillus* spp. strains as potential biocontrol agents. Thus it is probably not a coincidence that members of the genus *Bacillus* were among the first successful biopesticides. *Bacillus thuringiensis* is the best known and most famous example of a bioinsecticide (see Ohba 2011). The success of these organisms is due in part to the ease of formulation and storage of the products. The Gram-positive spore offers a product formulation that has been selected over billions of years of evolution for its robustness and durability. In this part of the chapter, we present examples of the successful use of various *Bacillus* spp. strains for the control of fungal diseases that affect plants and plant products.

8.3.1 Biological Control of Soil-Borne Diseases

Soil-borne fungal diseases cause worldwide economically significant diseases that affect important crops, including cereals, grapevine and many different horticultural and ornamental crops. Furthermore, soil-borne fungal diseases are the limiting factors in the productivity of many crop systems. Controlling soil-borne fungal pathogens has always been very difficult. Fungicide drenches are expensive and impractical, and they have undesirable effects on the environment. Physical methods, such as soil sterilization by heat and soil solarization, are sometimes useful for reducing the pathogen inoculums. However, these techniques can have undesirable side effects on the plants and on beneficial microflora. Among the major soil-borne fungal pathogens are *Rhizoctonia* and the oomycete *Pythium*, which are especially problematic in disinfected soil and soilless substrates, where the microbial diversity and biological buffering present in natural soils are lacking. Moreover, sclerotia-producing fungi such as *Botrytis*, *Sclerotinia* and *Sclerotium* are very difficult to eradicate from the soil by existing methods. In this scenario, the use of antagonistic rhizospheric microorganisms would be an interesting alternative means of control. In fact, the biocontrol of soil-borne fungal pathogens has probably been the subject of more research than any other form of plant disease biological control.

Several biocontrol products containing various aerobic endospore-forming bacterial species have been marketed especially for the control of soil-borne fungal pathogens (Table 8.1). *Bacillus* species such as *B. amyloliquefaciens*, *B. licheniformis*, *B. megaterium*, *B. pumilus* and *B. subtilis* have been commercially formulated and registered for use against soil-borne pathogens such as *Pythium*, *Rhizoctonia*, *Fusarium*, and others that affect field, ornamental and vegetable crops, as well as tree seedlings. Several reports have described the effects of these species against root diseases such as rhizoctonia crown and root rot of sugar beet, stem canker and black scurf of potato, damping-off, fusarium crown and root rot of tomato or avocado *Dematophora* root rot (Kiewnick et al. 2001; Brewer and Larkin 2005; Omar et al. 2006; Cazorla et al. 2007; Jayaraj et al. 2009). However, most of these reports highlight the need to integrate aerobic endospore-forming biocontrol agents with fungicides and other control strategies to optimize disease management.

Table 8.1 Some commercial formulations of *Bacillus*-based plant disease biological control products

Product name	Biocontrol organism	Target pathogen/disease	Crop	Formulation	Manufacturer
Avogreen	*B. subtilis*	*Cercospora* spot	Avocado	Liquid	Stimuplant
Ballad Plus	*B. pumilus*	Rust, powdery mildew, cercospora, brown spot	Soybean	Liquid	AgraQuest
Companion	*B. subtilis*	*Rhizoctonia, Pythium, Fusarium, Phytophthora, Sclerotinia*	Greenhouse, nursery and ornamental crops	Liquid	Growth Products
EcoGuard	*B. licheniformis*	Dollar spot, anthracnose	Turf	Liquid	Novozymes
HiStick[a]	*B. subtilis*	*Fusarium, Rhizoctonia, Aspergillus*	Soybean, peanuts	Powder	Becker Underwood
Pythium	*B. subtilis*	*Rhizoctonia, Fusarium, Phytium, Aspergillus*	Cotton, legumes, soybean and vegetable crops	Powder	Bayer CropScience
Rhapsody	*B. subtilis*	*Rhizoctonia, Fusarium, Pythium, Phytophthora*	Turf and ornamental, vegetable and fruit greenhouse crops	Liquid	AgraQuest
Serenade	*B. subtilis*	Rusts, powdery mildews, *Botrytis, Sclerotinia*	Vegetable, wine, nut and fruit crops	Liquid or powder	AgraQuest
Sonata	*B. pumilus*	Rusts, powdery and downy mildews	Vegetable and fruit crops	Liquid	AgraQuest
Subtilex	*B. subtilis*	*Rhizoctonia, Fusarium, Aspergillus*	Field, ornamental and vegetable crops	Powder	Becker Underwood
Taegro	*B. amyloliquefaciens*	*Rhizoctonia, Fusarium*	Tree seedlings, ornamentals and shrubs	Granules	Novozymes

[a]This formulation is composed by *B. subtilis* and rhizobial cells

8.3.2 *Biological Control of Foliar Diseases in Greenhouse Systems*

Like pathogens, biocontrol agents are very sensitive to environmental conditions. The failure and/or inconsistent performance of biological controls in the field have frequently been attributed to environmental factors. It is now well established that biological control agents require specific environmental conditions (such as high relative humidity) for optimal suppressive activity, and it is presumed that they should perform better under greenhouse conditions than in the field. Therefore, the greenhouse environment represents a unique situation that is more favourable for diseases but also more favourable for the management of foliar diseases with biological control agents (Paulitz and Bélanger 2001). The logistics and economics of this crop system also favour the use of biocontrol agents in the greenhouse. Greenhouse crops have a high economic value, and therefore can absorb higher disease-control cost inputs. Moreover, there is increasing societal concern over the environmental and health effects of fungicides. A pesticide-free vegetable or floral product may give greenhouse growers a market advantage. Therefore, agriculture in greenhouses and protected structures offers a unique niche for the development and use of biological control agents such as aerobic endospore-forming bacteria.

The primary pathogens in most greenhouses are fungal diseases. Among these are foliar pathogens such as the grey mould fungus *Botrytis cinerea* and powdery mildews. Both of these are very destructive diseases responsible for important losses in many horticultural crops and ornamental plants. Species such as *B. subtilis* and *B. pumilus* have been formulated and registered for use on vegetables, grapes and ornamentals against foliar diseases such as rusts, powdery and downy mildews, grey mould and others (Table 8.1). Several reports have described the management of foliar fungal diseases with aerobic endospore-forming bacteria (Elad et al. 1996; Collins and Jacobsen 2003; Romero et al. 2007c), but their implementation is still very limited. Biological agents are becoming an increasingly attractive alternative to chemicals, but recent studies revealed that they did not perform as expected in all crop conditions. Integration with other control strategies may be required to increase the performance and consistency of biocontrol agents in the phyllosphere.

Foliar diseases can also be suppressed by the use of so-called plant growth-promoting rhizobacteria (PGPR). PGPR are rhizosphere inhabitants that can promote plant growth as well as suppress diseases. PGPR can confer on the plant an enhanced defensive capacity against a broad spectrum of plant diseases by means of a phenomenon known as induced systemic resistance (ISR) (see Sect. 8.4.2) (Van Loon et al. 1998). An increasing number of reports have described the effects on systemic disease protection of several *Bacillus* species. Among them, *B. mycoides*, *B. pasteurii* (now *Sporosarcina pasteurii*), *B. pumilus*, *B. sphaericus* (now *Lysinibacillus sphaericus*) and *B. subtilis* have been described as reducing the severity of fungal diseases such as sugar beet *Cercospora* leaf spot, tobacco blue mould, tomato late blight, pine fusiform rust and cucumber anthracnose (Kloepper et al. 2004). This illustrates the potential of *Bacillus* spp. to control foliar diseases by means of ISR. Interestingly, this mechanism of disease suppression protects the plant not only from fungal diseases but also from pathogenic bacteria, viruses, nematodes and insects. Thus, aerobic endospore-forming bacteria are very promising weapons for IPM systems.

8.3.3 Biological Control of Postharvest Diseases

Despite the development of modern fungicides and improved storage technologies, more than 25% of the total commercial harvest of fruits is at risk from fungal attack, and very often the losses exceed 50% of the total harvest of certain sensitive fruits (Spadaro and Gullino 2004). Over the past 20 years, biological control has emerged as an effective alternative for combating the post-harvest decay of fruits and vegetables. The substantial progress made in this field can be attributed to the uniqueness and relative simplicity of the postharvest system. Wounds incurred during harvesting and fruit handling can be protected from invading pathogens with a single postharvest application of the antagonist directly to the wounds using existing fungicide delivery systems. Environmental conditions such as temperature and relative humidity can be managed to favour antagonist survival. Also, biotic interference is minimal, so the antagonist encounters little competition from indigenous microorganisms. Consequently, biocontrol of postharvest diseases tends to be more consistent than biocontrol in the field, and given the high retail value of fresh fruits, the application of high concentrations of antagonists to fruit surfaces is economically viable (Janisiewicz and Korsten 2002). Since the large-scale introduction to the market in 1996 of the pioneering biocontrol product BioSave, many other microbial products have become commercially available, including *Bacillus*-based biofungicides (Table 8.1).

The greatest progress in biological control of postharvest diseases of fruits has been made against common wound-invading necrotrophic postharvest pathogens such as *Penicillium expansum* and *B. cinerea* (pome fruits), *Penicillium italicum* and *Penicillium digitatum* (citrus fruits), the wound-invading phase of *Monilinia fructicola* (stone fruits), and latent infections caused by *Colletotrichum* spp. (avocado, mango) and *B. cinerea* (strawberries). The *Bacillus* species most frequently noted for their potential as biocontrol agents for these postharvest diseases are *B. amyloliquefaciens, B. licheniformis, B. pumilus* and *B. subtilis* (Pusey and Wilson 1984; Huang et al. 1992; Korsten et al. 1995; Mari et al. 1996; Leibinger et al. 1997; Korsten and Jeffries 2000; Korsten 2004; Govender et al. 2005; Jamalizadeh et al. 2008). The progress made in this area during the past two decades has been remarkable; if this pace continues, the use of biological control to combat postharvest diseases will be greatly expanded in the future.

8.3.4 Biocontrol Assays for the Screening and Selection of Biocontrol Agents

Bacillus spp. are well known for their ability to control several plant pathogens. Isolating bacteria from the same environment in which they will be used may increase the chances of finding effective biocontrol strains. Endospore-forming bacteria are usually selectively isolated by heat treatment of the sample (80°C, 10 min). In order to test a collection of isolates of aerobic endospore-forming bacteria for their ability to control plant fungal pathogens, a number of relatively simple tests can be undertaken. Some of these assays are described below (Fig. 8.1).

Fig. 8.1 Bioassays for the screening and selection of potential biological control agents. (**a**) Antagonism assay. Dual culture analysis technique showing the growth inhibition of the grey mould fungus *Botrytis cinerea* caused by two strains of *B. subtilis*. (**b**) Detached leaf assay. Picture illustrating the double Petri plate system. (**c**) Seedling biocontrol assay for a foliar pathogen. An untreated melon leaf showing typical symptoms of cucurbit powdery mildew (*left*) and a leaf treated with *B. subtilis* showing no disease symptoms (*right*). (**d**) Seedling biocontrol assay for a soilborne pathogen. An untreated avocado seedling showing wilting symptoms due to Dematophora root rot (*left*) and a disease-protected plant after treatment with *B. subtilis* (*right*)

8.3.4.1 Antagonism Assays

Bacterial isolates can be screened for their ability to inhibit fungal growth on agar plates using a technique known as dual culture analysis (Fig. 8.1a). Bacterial inocula are prepared from cultures grown on nutrient agar plates for 24 h, whereas fungal cultures are grown on potato dextrose agar (PDA) plates for several days. A plug containing the mycelium of a target fungus (1 cm in diameter) is placed at the centre of a dual PDA plate, and single bacterial colonies are patched along the perimeter of the plate at a distance of 4 cm from the fungal block. Plates are incubated for 15 days in the dark at 25°C, and inhibition of fungal growth is assessed by measuring the diameters of inhibition zones, in mm (Romero et al. 2004).

8.3.4.2 Detached Leaf Assays

Potentially antagonistic isolates can be tested against foliar fungal pathogens using detached leaves maintained in vitro (Fig. 8.1b). Prior to inoculation, fully expanded leaves are detached from young seedlings. Bacterial inocula are prepared from fresh liquid cultures. Bacterial cells are washed in distilled water and cell suspensions are adjusted to 10^8 cfu ml^{-1}. The bacterial suspension is spread onto a leaf until it runs-off and, after the whole leaf surface has completely air-dried (4 h), the leaf is spread with a conidial suspension of the fungal pathogen (10^5 spores ml^{-1}) in a similar way (Romero et al. 2004). The detached leaves are then placed in double Petri dishes. Their petioles are immersed in a nutrient solution (e.g., 50% Hoagland solution) in the lower plate and the leaf blades are deposited in the upper plate. After incubation, the plates are then incubated at 25°C and 95% relative humidity (RH) under a 16 h photoperiod (Romero et al. 2003). The respective disease symptoms are then conveniently evaluated and disease reduction indexes are calculated.

8.3.4.3 Seedling Assays

To select candidate bacteria for the biological control of foliar fungal pathogens, growth chamber experiments can be carried out using plant seedlings (Fig. 8.1c). Bacterial candidates and fungal pathogens are applied essentially as previously described for the detached leaf assay. The inoculated seedlings are then maintained under the same conditions of light, temperature and relative humidity (Romero et al. 2004), and disease symptoms and disease reduction indexes are estimated. For soil-borne pathogens, soil inoculation is usually carried out by thoroughly mixing spore suspensions of the fungal pathogen with potting soil to a final concentration of 10^6 spores kg^{-1}. In this case, bacteria can be applied to seeds by dipping the seeds in a mixture of 1% methylcellulose and 10^9 cfu ml^{-1} bacteria. Alternatively, the roots of seedlings previously grown in perlite can be immersed in the bacterial suspension (10^9 cfu ml^{-1}) prior to being placed in the infested soil. The seedlings are then incubated in a growth chamber at 25°C and 70% RH under a 16 h photoperiod (Cazorla et al. 2007). The number of diseased seedlings is usually determined 21 days after bacterization. Plants are removed from the soil and washed, and the roots are examined for disease symptoms (crown or root rot). Alternatively, aerial symptoms (yellowing and wilting) can be recorded (Fig. 8.1d). In any case, after the disease symptoms are quantified, disease reduction indexes are calculated. To select bacterial isolates able to induce plant systemic resistance, bacteria are applied to the root system as previously indicated, and plants are watered weekly with bacterial suspensions. After 2 weeks of incubation, the plants are inoculated with a low dose of a foliar pathogen. For fungal pathogens such as powdery mildews, leaves are spread with conidial suspensions of 10^3 conidia ml^{-1}. Disease symptoms and disease reduction indexes are then calculated.

8.3.4.4 Fruit Assays

To test bacteria for antagonism to fungal pathogens responsible for postharvest spoilage, fruits are disinfected by immersion for 1 min in a diluted solution of sodium hypochlorite (1% active chlorine), and are then rinsed with sterile distilled water and air-dried prior to use. Fruits are then prepared for inoculation by creating wells 6 mm wide by 3 mm deep with a sterile cork borer. As a preventive treatment, fruits are treated with the antagonist by adding 50 μl of a bacterial suspension (10^8 cfu ml^{-1}) to each wound 24 h prior to challenge with the pathogen. After this incubation, the fungal pathogen is applied to the wound by adding the same volume of a suspension containing 10^5 conidia. Treated fruits are incubated at 22°C and 70% RH. To assay as a curative treatment, the antagonist and the pathogen are applied as mentioned above, but in this case the pathogen is added first and the antagonist 24 h later. Disease symptoms are evaluated 6, 15 and 21 days after pathogen challenge, and disease reduction indexes are calculated on the basis of the diameter of the disease lesions that have developed around the infected site (Touré et al. 2004).

8.4 Biocontrol Mechanisms of *Bacillus* spp.

Bacillus species possess several advantages that make them good candidates for use as biological control agents. First, they produce several different types of antimicrobial compounds, such as antibiotics (e.g., bacilysin, iturin and mycosubtilin) and siderophores (Shoda 2000). Second, they induce growth and defence responses in the host plant (Raupach and Kloepper 1998). Furthermore, *Bacillus* spp. are able to produce spores that are resistant to UV light and desiccation, and that allow them to resist adverse environmental conditions and permit easy formulation for commercial purposes (Emmert and Handelsman 1999). In this part of the chapter we describe the biological mechanisms by which *Bacillus* species suppress fungal plant pathogens, and thereby promote plant health (Fig. 8.2).

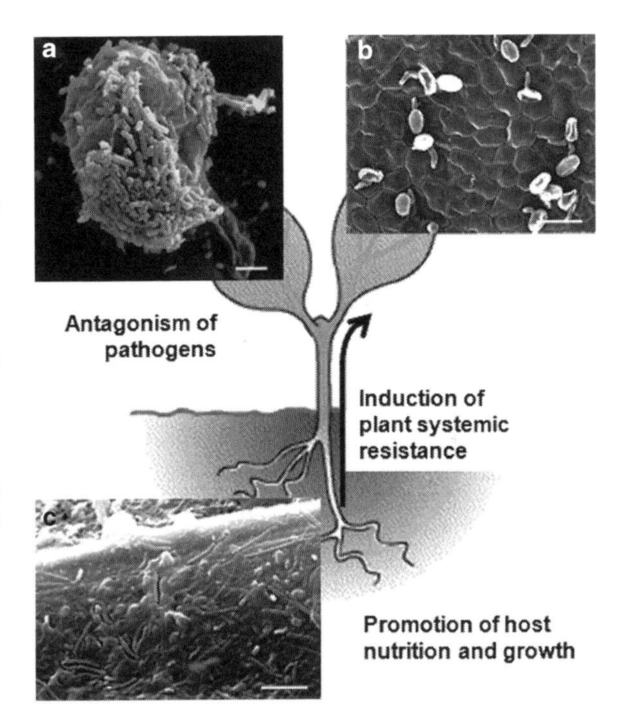

Fig. 8.2 Schematic representation of the main biocontrol mechanisms of *Bacillus* spp. illustrated by SEM micrographs. (**a**) Antagonism of pathogens. Attachment of *B. subtilis* cells to a germinated conidium of the foliar fungal pathogen of cucurbits *Podosphaera fusca*. (**b**) Induction of plant systemic resistance. Collapsed germinated conidia of *P. fusca* as a consequence of elicitation of ISR by *B. cereus*. (**c**) Promotion of host nutrition and growth. Colonization of a tomato root by *B. subtilis* promoting plant health by stimulating plant host nutrition. Scale bars represent 5 μm (plate **c**), 10 μm (plate **a**) and 50 μm (plate **b**)

8.4.1 Antagonism to Pathogens

Several species of the genus *Bacillus* are microbial factories that produce a vast array of biologically active molecules, potentially inhibitory to the growth and/or activities of plant pathogens (Fig. 8.2a). One of the most thoroughly studied species in this regard is *B. subtilis*. In addition, a number of studies have reported direct antagonism by several other species, including *B. amyloliquefaciens*, *B. cereus*, *B. licheniformis*, *B. megaterium*, *B. mycoides* and *B. pumilus*, as well as various unidentified *Bacillus* species. Cell wall-degrading enzymes (such as chitinases, glucanases and proteases), peptide antibiotics and other small secondary metabolites (such as volatile organic compounds) are all secreted by various species, and many of these contribute to pathogen suppression (Shoda 2000). As an illustration of this, it is noteworthy that 4–8% of the genomes of *B. subtilis* and *B. amyloliquefaciens* are devoted to antibiotic synthesis, giving them the potential to produce more than two dozen structurally diverse antimicrobial compounds (Stein 2005; Chen et al. 2009).

Among these antimicrobial compounds, peptide antibiotics predominate. They exhibit highly rigid, hydrophobic and/or cyclic structures with unusual constituents such as D-amino acids, and they are generally resistant to hydrolysis by peptidases and proteases. Furthermore, the cysteine residues of peptide antibiotics are either oxidized to disulphides or modified to characteristic intramolecular C–S (thioether) linkages, making them insensitive to oxidation (Katz and Demain 1977). These unusual non-proteinaceous compounds are divided into two classes according to the biosynthetic pathways that allow their incorporation, namely lantibiotics and lipopeptides. Lantibiotics (or lanthionine-containing antibiotics) are peptide antibiotics with inter-residual thioether bonds as a unique feature. Lanthionine formation occurs through the post-translational modification of ribosomally synthesized precursor peptides. These modifications include the dehydration of serine and threonine residues and the subsequent addition of neighbouring cysteine thiol groups (McAuliffe et al. 2001). Lipopeptides are oligopeptides synthesized in a non-ribosomal manner by large multienzyme complexes. These amphiphilic compounds share a common cyclic structure consisting of a β-amino or β-hydroxy fatty acid integrated into a peptide moiety. Lipopeptides are classified into three families depending on their amino acid sequence and fatty acid branching: the iturin, fengycin and surfactin lipopeptide families (Stein 2005).

Lipopeptides are the antibiotic compounds most frequently produced by *Bacillus* spp. The iturin family, represented by iturin A, mycosubtilin and bacillomycin, are heptapeptides with a β-amino fatty acid, and they exhibit strong antifungal activity (Magnet-Dana and Peypoux 1994). The fengycin family (which includes plipastatin) are decapeptides with a β-hydroxy fatty acid. Members of this family have unusual properties, such as the presence of ornithine in the peptide portion, and they also have antifungal activity, although they are more specific for filamentous fungi (Vanittanakom et al. 1986). The surfactin family consists of heptapeptides containing a β-hydroxy fatty acid with anywhere between 3 and 15 carbon atoms. Surfactins are possibly the most powerful biosurfactants described, and although only slightly antifungal, they show a strong synergistic antifungal action in combination with iturin A (Magnet-Dana et al. 1992). The ability of various *Bacillus* strains to control soilborne pathogens such as *Rhizoctonia solani* (Asaka and Shoda 1996) and *Pythium aphanidermatum* (Leclère et al. 2005), and foliar and postharvest diseases such as cucurbit powdery mildew (Romero et al. 2007b) and grey mould (Ongena et al. 2005), has been attributed to iturins and fengycins. The amphiphilic structure of lipopeptides allows them to antagonistically interact with biological membranes and induce the formation of pores (Magnet-Dana and Peypoux 1994). Such alterations of plasma membrane integrity promote internal osmotic imbalance and widespread disorganization of the cytoplasm in fungal cells (Romero et al. 2007a).

8.4.2 Induction of Plant Systemic Resistance

Unlike animals, plants cannot defend themselves against microbial attacks by producing circulating antibodies or specialized cells. Instead, they respond to pathogen infection through physical and chemical defences that may be either preformed (e.g., the cuticle and cell wall) or induced after pathogen penetration. The timing of this defence response is critical and makes the difference between coping and succumbing to pathogen challenge. If defence mechanisms are triggered by a stimulus prior to infection by a plant pathogen, disease can be reduced. Selected strains of plant growth-promoting rhizobacteria (PGPR) can suppress plant diseases by inducing a systemic resistance in plants to both root and foliar pathogens (van Loon 2007). This phenomenon, termed "induced systemic resistance" (ISR), was first described in carnation and cucumber (van Peer et al. 1991; Wei et al. 1991). Since then, it has been documented in many more plant species. Several bacterial groups with potential to induce ISR have been described. *Pseudomonas*-mediated ISR has been extensively studied in different plant models, including *Arabidopsis*. In addition to *Pseudomonas* strains, ISR is induced by *Bacillus* spp. (Fig. 8.2b). Published results show that several specific strains of the species *B. amyloliquefaciens*, *B. subtilis*, *B. pasteurii*, *B. cereus*, *B. pumilus*, *B. mycoides*, and *B. sphaericus* elicit significant reductions in the incidence and/or severity of various diseases in a diversity of hosts (Kloepper et al. 2004; Choudhary et al. 2007).

ISR enhances the plant's defences against a broad spectrum of fungal, bacterial and viral diseases, and once induced, it is maintained for prolonged periods (van Loon et al. 1998). In response to pathogenic challenge, ISR fortifies cell walls and alters host physiology and metabolism to enhance the synthesis of defence chemicals. Cell wall strengthening is due to the deposition of callose and phenolic compounds at the site of pathogen attack (Benhamou et al. 1996, 1998). Physiological changes in plants include the accumulation of pathogenesis-related (PR) proteins such as chitinases, β-1,3-glucanases and peroxidases. However, certain PGPR do not induce PR proteins but rather increase the accumulation of plant antimicrobial compounds termed phytoalexins (Ramamoorthy et al. 2001). Regarding signalling pathways, ISR is typically independent of salicylic acid (SA) and mostly dependent on jasmonate (JA) and/or ethylene (ET) signalling (Verhagen et al. 2004). However, certain PGPR also appear to activate an SA-dependent pathway during ISR, indicating that different signalling pathways may be operable when ISR is elicited (Ryu et al. 2003).

Bacterial determinants involved in the induction of systemic resistance by Gram-negative PGPR include lipopolysaccharides present in the outer membrane, siderophores, flagella, the antibiotic 2,4-diacetylphloroglucinol, and the *N*-acylhomoserine lactones involved in quorum sensing (Van Loon 2007). Much remains to be discovered about the molecular aspects of *Bacillus*-mediated ISR. The sole determinants of ISR elicitation identified from *Bacillus* spp. so far are volatile organic compounds (in particular 2,3-butendiol) and the lipopeptides fengycin and surfactin (Ryu et al. 2004a; Ongena et al. 2007). In *Arabidopsis*, ISR triggered by volatiles is independent of the SA and JA signalling pathways, and more or less mediated by ET. However, ISR activation by certain strains is independent of these signalling pathways (Ryu et al. 2004a), raising the possibility that additional volatiles might trigger ISR via alternative pathways. In bean and tomato, a role for surfactins and fengycins in plant defence induction was demonstrated by the similarity between the protective activities of the purified compounds with that of the producing strain. More conclusively, a significant protective effect was gained by treating plants with various lipopeptide-overproducing derivatives generated from the wild-type *B. subtilis* 168, which is not able to synthesize these compounds and is not active on plants (Ongena et al. 2007). How these compounds are perceived by plant cells as signals to activate ISR has yet to be determined.

8.4.3 Promotion of Host Nutrition and Growth

Bacillus populations may also promote plant health by stimulating plant host nutrition as mutualistic symbionts (Fig. 8.2c). Phosphate-solubilizing *B. subtilis* strains have been reported to synergistically increase plant nitrogen and phosphate accumulation when co-inoculated with mycorrhiza (Toro et al. 1997). In addition, the production of extracellular phytases under conditions of phosphate limitation and in the presence of phytate seems to contribute to the plant-growth promoting activity of *B. amyloliquefaciens* (Idris et al. 2002). Phytases are enzymes that sequentially remove phosphate groups from *myo*-inositol 1,2,3,4,5,6-hexakisphosphate (phytate), the main storage form of phosphate in plants. Besides their ability to make phytate phosphorus available, another beneficial effect of the extracellular phytase activities of *Bacillus* spp. is the elimination of chelate-forming phytate, which is known to bind nutritionally important minerals (such as Zn^{2+}, Fe^{2+} and Ca^{2+}) (Kerovuo et al. 1998). In *B. amyloliquefaciens* the *phyC* (phytase) gene is a member of the phosphate starvation-inducible Pho regulon and its expression is controlled by the PhoP–PhoR two-component signal transduction system (Makarewicz et al. 2006).

In *Bacillus*, other mechanisms of growth promotion involve the modulation of plant regulatory mechanisms through the production of phytohormones that influence plant development. Various *Bacillus* species are capable of producing auxins, gibberellins and cytokinins, which are compounds that positively influence cell enlargement, cell division, root initiation and root growth. The production of gibberellins has been confirmed in *B. licheniformis* and *B. pumilus* strains that enhance the growth of pine plants (Gutiérrez-Mañero et al. 2001; Probanza et al. 2002). In *B. subtilis*, the production of cytokinins (such as zeatin riboside) seems to stimulate the growth of lettuce by increasing the levels of cytokinins and other hormones (Arkhipova et al. 2005). Similarly, a recent study has shown that cytokinin signalling is also involved in plant growth promotion by *B. megaterium* (Ortíz-Castro et al. 2008). Regarding the production of auxins, in *B. amyloliquefaciens* the biosynthesis of indole-3-acetic acid (IAA) affects its ability to promote plant growth. This ability is dependent on the presence of tryptophan, which is one of the main compounds present in several plant root exudates (Idris et al. 2007). Interestingly, a recent study has shown that volatile organic compounds from *B. subtilis* trigger growth promotion in *Arabidopsis* by regulating auxin homeostasis and cell expansion, thus providing a new paradigm as to how rhizobacteria promote plant growth (Zhang et al. 2007).

In order for a biocontrol strain to induce systemic resistance, promote host nutrition and growth, and efficiently protect the root system from phytopathogenic microorganisms, it has to establish itself in the rhizosphere. Bacterial properties such as chemotaxis, motility and growth play essential roles in this process (de Weert and Bloemberg 2007). The properties of *Bacillus* biocontrol strains that are involved in rhizosphere competence and the role of root colonization in biological control are poorly understood. In this regard, a recent report has highlighted the role of the formation of a stable, extensive biofilm and of the secretion of surfactin that occurs upon root colonization by *B. subtilis* in the protection of plants against pathogenic attack (Bais et al. 2004). The lack of knowledge in this critical aspect of *Bacillus* ecology should stimulate more intensive research not only on rhizosphere competence and root colonization but also on phyllosphere competence and fruit colonization. In this sense, the isolation of *Bacillus* strains with enhanced colonization capabilities could be another strategy to improve the performance of biofungicides based on aerobic endospore-forming bacteria.

8.4.4 Characterization of Biological Control Mechanisms

In order to develop more effective and consistent biological control products, a good understanding of the mechanisms of action of the biological control agents is essential. Although the main biocontrol mechanisms of *Bacillus* spp. are relatively well characterized, many aspects of the multitrophic interactions between plants, pathogens and biological control agents remain to be analysed in detail. In order to unravel the mechanisms underlying the biological control capabilities of an interesting *Bacillus* strain, a number of basic research tools can be exploited. Some of these research tools are described below.

8.4.4.1 Microscopic Analyses

Microscopic examination can be very useful for studying the interactions between the biocontrol agent and the pathogen. Scanning electron microscopy (SEM) is commonly used to examine interactions on the plant surface. For SEM analysis, plant material (leaf discs, root samples or fruit pieces) is first fixed in 0.1 M sodium cacodylate (pH 7.2), 50 mM L-lysine and 2.5% glutaraldehyde for 12–24 h at 4°C, and then dehydrated in an ethanol series from 50 to 100% ethanol. Prior to examination, the material is critical-point dried with carbon dioxide, mounted on aluminium stubs, and coated with gold (Pérez-García et al. 2001). To specifically visualize a particular biocontrol agent, confocal laser scanning microscopy (CLSM) is nowadays the better choice. For this analysis, a *Bacillus* strain tagged with green fluorescent protein (GFP) is needed. Several plasmids containing the *gfp* marker gene have been developed for *Bacillus* (e.g., pGFP4412) (Ji et al. 2008), but their use in plant-associated bacteria is still very limited. To analyse the plant surface, samples are directly examined under a confocal laser scanning microscope. To study internal proliferation, tissue fragments are first fixed and then embedded in agarose (Pliego et al. 2009). Subsequently, semithin tissue sections are obtained and observed by CLSM.

8.4.4.2 Thin Layer Chromatography Analyses

In order to determine the pattern of lipopeptide antibiotics produced by a particular *Bacillus* strain, thin layer chromatography (TLC) analyses can be undertaken. The *Bacillus* strains are grown in a medium optimal for lipopeptide production at 37°C (Ahimou et al. 2000). After 5 days of incubation, cells are removed by centrifugation and the supernatants are extracted with *n*-butanol. Once the butanol layers completely evaporate, the remaining residues are dissolved in methanol (Romero et al. 2007b). The methanolic fractions are then analysed by TLC on normal-phase silica gel 60 sheets and developed at room temperature with chloroform–methanol–water (65:25:4 by volume) (Razafindralambo et al. 1993), using purified iturin A, fengycin and surfactin as standards or cell-free filtrates of previously characterized *Bacillus* strains. Once the solvent front has migrated to within 1 cm of the top of the chromatogram, the sheets are air-dried. The presence of organic compounds is revealed by spraying with a solution of 10% H_2SO_4, whereas the specific visualization of lipopeptides is revealed simply by spraying with water. For the detection of antifungal activity associated with lipopeptide bands, a simple bioassay can be conducted. TLC sheets developed under identical conditions are placed in appropriate Petri plates, each is overlaid with a thin layer of PDA amended with conidia of the target fungus (10^3 spores ml^{-1}), and is inspected for areas of inhibition of fungal growth after incubation.

8.4.4.3 Transformation Methods

The genetic manipulation of the biocontrol agent is an essential tool for gaining insight into the molecular mechanisms involved in biological control. The most common methods used to transform *Bacillus* strains are natural competence and the highly versatile electrotransformation. However, these strategies have been developed using reference or culture collection strains, well adapted to laboratory conditions, and they do not always work with undomesticated strains. Recently, a method combining the use of protoplasts and electroporation has been developed to transform recalcitrant wild strains of *B. subtilis* (Romero et al. 2006). Briefly, the method involves obtaining protoplasts in the presence of lysozyme and mutanolysin. These are then mixed with plasmid DNA and electroporated at 25 μF, 400 Ω and 0.7 kV. Transformants are recovered in the presence of appropriate antibiotics and 0.5 M mannitol or 0.5 M sorbitol as osmotic support.

8.4.4.4 Detached Leaf, Seedling and Fruit Assays

To identify bacterial traits involved in biological control (e.g., antagonism or induction of systemic resistance), transposon-mediated (random) or site-directed transformants have to be tested for reduction and/or suppression of their biocontrol capabilities. For this, the previously mentioned bioassays with detached leaves, seedlings or fruits (see Sect. 8.3.4), can be used. In these experiments, the parental wild-type strain must always be included as a reference. In order to prove the role of a given *Bacillus* gene (through insertion or deletion) in the biocontrol activity, a decrease in the disease suppression ability relative to the wild-type strain must be observed.

8.4.4.5 *Arabidopsis* Assays

Although *Arabidopsis thaliana* is a model organism for research in plant biology, it can be also be used to study various aspects of biological control, especially those related to the induction of plant systemic resistance and growth promotion by PGPR. As previously mentioned, several *Bacillus* species have been shown to elicit ISR and promote growth in *Arabidopsis*. Therefore, it is possible to use the tools developed for this plant to characterize these types of responses. To study effects on plant growth, it is particularly useful growing *Arabidopsis* seedlings on agarized medium. To study the induction of plant systemic resistance, ISR bioassays can be performed using oomycete (*Peronospora parasitica*), fungal (*Alternaria brassicicola*), bacterial (*Pseudomonas syringae* pv. *tomato*) or viral (cucumber mosaic virus) pathogens as challenge inocula (van Wees et al. 1997; Ton et al. 2002; Iavicoli et al 2003; Ryu et al. 2004b). Furthermore, to characterize the signal transduction pathways elicited by a bacterial strain, several *Arabidopsis* mutants can be tested in order to determine whether the induced systemic resistance is SA-dependent or JA/ET-dependent (Iavicoli et al. 2003). Moreover, *Arabidopsis* microarrays are commercially available, allowing the transcriptional response to a particular rhizobacterial strain to be characterized in detail.

8.5 Development of *Bacillus*-Based Biofungicides

For the commercial development of a microbial pesticide, the biocontrol agent must be produced on an industrial scale (by fermentation), preserved, stored and formulated. Aerobic endospore-forming bacteria offer biological solutions to the formulation problems that have plagued biocontrol agents because they produce heat- and desiccation-resistant spores that can readily be formulated into stable products (Emmert and Handelsman 1999). Effective formulations of *Bacillus* spp. biomass for the biocontrol of plant diseases are currently on the marketplace (Table 8.1), but details of the specific compositions and production formulations of commercial biocontrol agents are largely held as proprietary information by the companies. Fortunately, there is a substantial base of industrial experience with *B. thuringiensis* (see Ohba 2011), which has been extensively used for four decades in biopesticidal formulations (Brar et al. 2006). This experience has obviously served as a reference for the use of related species such as *B. subtilis* for biocontrol, helping to overcome obstacles that have arisen during the development of commercial *Bacillus*-based biofungicides.

The production of *Bacillus* biomass is usually carried out by submerged fermentation. In most instances, fermentation protocols are designed to maximize the production of spores rather than vegetative cells (Schisler et al. 2004). The aim is to achieve the highest yield possible with the lowest cost in culture medium, and this is usually achieved with a fed-batch fermentation strategy (Lee et al. 1999). Various raw materials such as agricultural and industrial by-products, as well as wastes such as wastewater and wastewater sludge, have successfully replaced costly synthetic media (Yezza et al. 2004; El-Bendary 2006). In developing countries, however, the use of submerged fermentation may not be economically feasible due to the high cost of the equipment. Accordingly, solid state fermentation methodology offers an alternative approach. Unfortunately, little information has so far been published on the use of this technology for the production of *Bacillus* (El-Bendary 2006).

The problem of the stability of biocontrol agents both during storage and after application has historically stalled biopesticide development (Brar et al. 2006). Formulation therefore plays a significant role in determining the final efficacy of a *Bacillus*-based biological control product (Schisler et al. 2004). A formulated microbial product is a product composed of biomass of the biocontrol agent together with other ingredients that improve the survival and effectiveness of the product. The most common adjuvants/additives consist of wetting and dispersal agents, nutrients, and agents that protect against ultraviolet light and osmotic stress. These help the microbial cells to survive under field conditions (Boyetchko et al. 1998). The selection of adjuvants/additives is governed by the type of formulation desired. In general, there are two types of formulations, dry solid and liquid, and these include encapsulated forms. The current commercial formulations of *Bacillus* spp. for biocontrol of plant diseases include liquids (aqueous suspensions and flowables), wettable powders and granules (Table 8.1). The commercial formulations of *B. thuringiensis* products have undergone a wide range of transitions over the years. With continued research, similar improvements in the formulations of *Bacillus*-based plant disease biological control products can be expected.

8.6 Conclusions and Future Perspectives

Synthetic fungicides are the primary means of controlling plant fungal diseases. Biological control has emerged as one of the most promising alternatives to chemicals. During the last 20 years, several aerobic endospore-forming bacterial species have been extensively investigated for their ability to protect a variety of crops from various pathogens. Many biological control mechanisms have been implicated, including antagonism, induction of systemic resistance, and promotion of host nutrition and growth. *Bacillus*-based biological control agents have great potential for use in IPM systems (Fig. 8.3); however, relatively little work has been

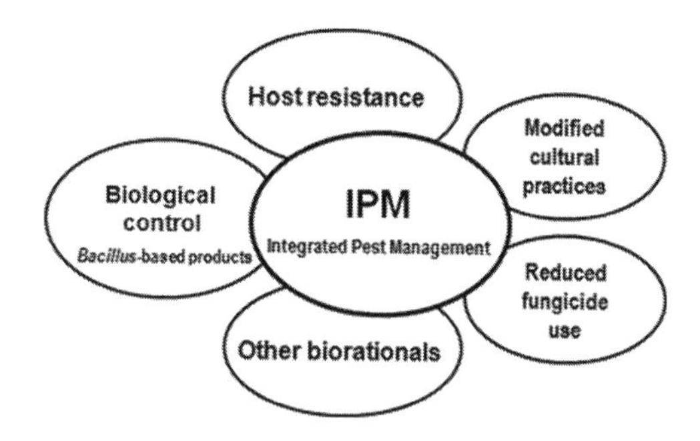

Fig. 8.3 The main components of integrated pest management systems (IPMs)

published on their integration with other IPM management tools. In order to extend the use of these biofungicides, it will be important to study how they interact with other aspects of disease control, including host resistance, chemicals, other biological control agents, cultural practises, and other means of protection. The formulation and the method of application are key issues influencing the efficacy and successful of commercial products. Genetic engineering may provide a useful tool for the enhancement of biological control efficacy. Whole genome sequences are now available for several *Bacillus* species. With this information, high-throughput studies can be undertaken to gain prior knowledge about the plant protection properties and rhizosphere/phyllosphere competence of these biocontrol agents. Since biofungicides are usually not as effective as their chemical counterparts, this approach should be viewed as an important component of an integrated disease management scheme. A significant and permanent reduction in fungicide use is our main goal, and this may also be very helpful to fungicide resistance management programmes. As noted by Jacobsen et al. (2004), it is critical that the research base be expanded so that science-based recommendations for the integration of *Bacillus*-based products into IPM systems can be made.

Acknowledgements The authors gratefully acknowledge past and ongoing support for their work on aerobic endospore-forming biological control agents from Plan Nacional de I+D+I of the Ministerio de Ciencia e Innovación, Spain (AGL2001-1837; AGL2004-0656; AGL2007-65340).

References

Agrios GN (2005) Plant pathology. Elsevier, Amsterdam

Ahimou F, Jacques P, Deleu M (2000) Surfactin and iturin A effects on *Bacillus subtilis* surface hydrophobicity. Enz Microbiol Technol 27:749–754

Arkhipova TN, Veselov SU, Melentiev AI, Martynenko EV, Kudoyarov GR (2005) Ability of bacterium *Bacillus subtilis* to produce cytokinins and to influence the growth and endogenous hormone content of lettuce plants. Plant Soil 272:201–209

Asaka O, Shoda M (1996) Biocontrol of *Rhizoctonia solani* damping-off of tomato with *Bacillus subtilis* RB14. Appl Environ Microbiol 62:4081–4085

Bais HP, Fall R, Vivanco JM (2004) Biocontrol of *Bacillus subtilis* against infection of *Arabidopsis* roots by *Pseudomonas syringae* is facilitated by biofilm formation and surfactin production. Plant Physiol 134:307–319

Benhamou N, Kloepper JW, Quadt-Hallmann A, Tuzun S (1996) Induction of defense-related ultrastructural modifications in pea root tissues inoculated with endophytic bacteria. Physiol Plant Pathol 112:919–929

Benhamou N, Kloepper JW, Tuzun S (1998) Induction of resistance against Fusarium wilt of tomato by combination of chitosan with an endophytic bacterial strain: ultrastructure and cytochemistry of the host response. Planta 204:153–168

Boyetchko S, Pedersen E, Punja Z, Reddy M (1998) Formulations of biopesticides. In: Hall FR, Barry JW (eds) Biopesticides: use and delivery (Methods in Biotechnology, vol 5). Humana, Totowa, NJ, pp 487–508

Brar SK, Verma M, Tyagi RD, Valéro JR (2006) Recent advances in downstream processing and formulations of *Bacillus thuringiensis* based biopesticides. Process Biochem 41:323–342

Brewer MT, Larkin RP (2005) Efficacy of several potential biocontrol organisms against *Rhizoctonia solani* on potato. Crop Prot 24:939–950

Cazorla FM, Romero D, Pérez-García A, Lugtenberg BJJ, de Vicente A, Bloemberg G (2007) Isolation and characterization of antagonistic *Bacillus subtilis* strains from the avocado rhizoplane displaying biocontrol activity. J Appl Microbiol 103:1950–1959

Chen XH, Koumoutsi A, Scholz R, Schneider K, Vater J, Süssmuth R, Piel J, Borriss R (2009) Genome analysis of *Bacillus amyloliquefaciens* FZB42 reveals its potential for biocontrol of plant pathogens. J Biotechnol 140:27–37

Choudhary DK, Prakash A, Johri BN (2007) Induced systemic resistance (ISR) in plants: mechanism of action. Indian J Microbiol 47:289–297

Collins DP, Jacobsen BJ (2003) Optimizing a *Bacillus subtilis* isolate for biological control of sugar beet cercospora leaf spot. Biol Control 26:153–161

De Vos P (2011) Studying the bacterial diversity of the soil by culture-independent approaches. In: Logan NA, de Vos P (eds) Endospore forming soil bacteria (Soil Biology 27). Springer, Heidelberg, doi:10.1007/978-3-642-19577-3

de Weert S, Bloemberg GV (2007) Rhizosphere competence and the role of root colonization in biocontrol. In: Gnanamanickam SS (ed) Plant-associated bacteria. Springer, Dordrecht, pp 317–333

Elad Y, Malathrakis NE, Dik AJ (1996) Biological control of *Botrytis* incited diseases and powdery mildews in greenhouse crops. Crop Prot 15:229–240

El-Bendary MA (2006) *Bacillus thuringiensis* and *Bacillus sphaericus* biopesticides production. J Basic Microbiol 46:158–170

Emmert EAB, Handelsman J (1999) Biocontrol of plant disease: a (Gram-) positive perspective. FEMS Microbiol Lett 171:1–9

Govender V, Korsten L, Sivakumar D (2005) Semi-commercial evaluation of *Bacillus licheniformis* to control mango postharvest diseases in South Africa. Postharv Biol Technol 38:57–65

Gutiérrez-Mañero F, Ramos-Solano B, Probanza A, Mehouachi J, Tadeo FR, Talon M (2001) The plant-growth-promoting rhizobacteria *Bacillus pumilus* and *Bacillus licheniformis* produce high amounts of physiologically active gibberellins. Physiol Plant 111:206–211

Hoitink H, Boehm M (1999) Biocontrol within the context of soil microbial communities: a substrate-dependent phenomenon. Annu Rev Phytopathol 37:427–446

Hoitink HAJ, Stone AG, Han DY (1997) Suppression of plant disease by composts. Hortscience 32:184–187

Huang Y, Wild BL, Morris SC (1992) Postharvest biological control of *Penicillium digitatum* decay on citrus fruit by *Bacillus pumilus*. Ann Appl Biol 120:367–372

Iavicoli A, Boutet E, Buchala A, Metraux JP (2003) Induced systemic resistance in *Arabidopsis thaliana* in response to root inoculation with *Pseudomonas fluorescens* CHA0. Mol Plant Microbe Interact 16:851–858

Idris EE, Makarewicz O, Farouk A, Rosner K, Greiner R, Bochow H, Ritcher T, Borriss R (2002) Extracellular phytase activity of *Bacillus amyloliquefaciens* FZB45 contributes to its plant-growth promoting effect. Microbiology 148:2097–2109

Idris EE, Iglesias DJ, Talon M, Borriss R (2007) Tryptophan-dependent production of indole-3-acetic acid (IAA) affects level of plant growth promotion by *Bacillus amyloliquefaciens* FZB42. Mol Plant Microbe Interact 20:619–626

Jacobsen BJ, Zidack NK, Larson BJ (2004) The role of *Bacillus*-based biological control agents in integrated pest management systems: plant diseases. Phytopathology 94:1272–1275

Jamalizadeh M, Etebarian HR, Alizadeh A, Aminian H (2008) Biological control of gray mold of apple fruits by *Bacillus licheniformis* (EN74-1). Phytoparasitica 36:23–29

Janisiewicz WJ, Korsten L (2002) Biological control of postharvest diseases of fruits. Annu Rev Phytopathol 40:411–441

Jayaraj J, Radhakrishnan NV, Kannan R, Sakthivel K, Suganya D, Venkatesan S, Velazhahan R (2009) Development of new formulations of *Bacillus subtilis* for management of tomato damping-off caused by *Pythium aphanidermatum*. Biocontrol Sci Technol 15:55–65

Ji X, Lu G, Gai Y, Zheng C, Mu Z (2008) Biological control against bacterial wilt and colonization of mulberry by an endophytic *Bacillus subtilis* strain. J Appl Microbiol 65:565–573

Katz E, Demain AL (1977) The peptide antibiotics of *Bacillus*: chemistry, biogenesis, and possible functions. Bacteriol Rev 41:449–474

Kerovuo J, Lauraeus M, Nurminen P, Kalkkinen N, Apajalahti J (1998) Isolation, characterization, molecular gene cloning, and sequencing of a novel phytase from *Bacillus subtilis*. Appl Environ Microbiol 64:2079–2085

Kiewnick S, Jacobsen BJ, Braun-Kiewnick A, Eckhoff JLA, Bergman JW (2001) Integrated control of rhizoctonia crown and root rot of sugar beet with fungicides and antagonistic bacteria. Plant Dis 85:718–722

Kim DS, Cook RJ, Weller DM (1997) *Bacillus* sp. L324-92 for biological control of three root diseases of wheat grown with reduced tillage. Phytopathology 87:551–558

Kloepper JW, Ryu CM, Zhang S (2004) Induced systemic resistance and promotion of plant growth by *Bacillus* spp. Phytopathology 94:1259–1266

Korsten L (2004) Biological control in Africa: can it provide a sustainable solution for control of fruit diseases? S Afr J Bot 70:128–139

Korsten L, Jeffries P (2000) Potential for biological control of diseases caused by *Colletotrichum*. In: Prusky D, Freeman S, Dickman MB (eds) Colletotrichum host specificity, pathology and host-pathogen interaction. APS, St. Paul, pp 266–295

Korsten L, De Jager ES, De Villiers EE, Lourens A, Kotze JM, Wehner FC (1995) Evaluation of bacterial epiphytes isolated from avocado leaf and fruit surfaces for biological control of avocado postharvest diseases. Plant Dis 79:1149–1156

Leclère V, Bechet M, Adam A, Guez JS, Wathelet B, Ongena M, Thonart P, Gancel F, Chollet-Imbert M, Jacques P (2005) Mycosubtilin overproduction by *Bacillus subtilis* BBG100 enhances the organism's antagonistic and biocontrol activities. Appl Environ Microbiol 71:4577–4584

Lee J, Lee SY, Park S, Middelberg AP (1999) Control of fed batch fermentations. Biotechnol Adv 17:29–48

Leibinger W, Breuker B, Hahn M, Mendgen K (1997) Control of postharvest pathogens and colonization of the apple surface by antagonistic microorganisms in the field. Phytopathology 87:1103–1110

Magnet-Dana R, Peypoux F (1994) Iturins, a special class of pore forming lipopeptides: biological and physiological properties. Toxicology 87:151–174

Magnet-Dana R, Thimon L, Peypoux F, Ptak M (1992) Surfactin/iturin A interactions may explain the synergistic effect of surfactin on the biological properties of iturin *A*. Biochemie 74:1047–1051

Makarewicz O, Dubrac S, Msadek T, Borriss R (2006) Dual role of the PhoP P response regulator: Bacillus amyloliquefaciens FZB45 phytase gene transcription is directed by positive and negative interaction with the phyC promoter. J Bacteriol 188:6953–6965

Mandic-Mulec I, Prosser JI (2011) Diversity of endospore-forming bacteria in soil: Characterization and driving mechanisms. In: Logan NA, de Vos P (eds) Endospore forming soil bacteria (Soil Biology 27). Springer, Heidelberg, doi: 10.1007/978-3-642-19577-2

Mari M, Guizzardi M, Pratella GC (1996) Biological control of gray mold in pears by antagonistic bacteria. Biol Control 7:30–37

McAuliffe O, Ross RP, Hill C (2001) Lantibiotics: structure, biosynthesis and mode of action. FEMS Microbiol Rev 25:285–308

McSpadden Gardener BB (2004) Ecology of *Bacillus* and *Paenibacillus* spp. in agricultural systems. Phytopathology 94:1252–1258

Nakasaki K, Hiraoka S, Nagata H (1998) A new operation for producing disease suppressive compost from grass clippings. Appl Environ Microbiol 64:4015–4020

Ohba M (2011) *Bacillus thuringiensis* diversity in soil and phylloplane. In: Logan NA, de Vos P (eds) Endospore forming soil bacteria (Soil Biology 27). Springer, Heidelberg, doi: 10.1007/978-3-642-19577-11

Omar I, O'Neill TM, Rossall S (2006) Biological control of fusarium crown and root rot of tomato with antagonistic bacteria and integrated control when combined with the fungicide carbendazim. Plant Pathol 55:92–99

Ongena M, Jacques P, Tourè Y, Destain J, Jabrane A, Thonart P (2005) Involvement of fengycin-type lipopeptides in the multifaceted biocontrol potential of *Bacillus subtilis*. Appl Microbiol Biotechnol 69:29–38

Ongena M, Jourdan E, Adam A, Paquot M, Brans A, Joris B, Arpigny JL, Thonart P (2007) Surfactin and fengycin lipopeptides of *Bacillus subtilis* as elicitors of induced systemic resistance in plants. Environ Microbiol 9:1084–1090

Ortíz-Castro R, Valencia-Cantero E, López-Bucio J (2008) Plant growth promotion by *Bacillus megaterium* involves cytokinin signaling. Plant Signal Behav 3:263–265

Paulitz TC, Bélanger RR (2001) Biological control in greenhouse systems. Annu Rev Phytopathol 39:103–133

Pérez-García A, Olalla L, Rivera E, del Pino D, Cánovas I, de Vicente A, Torés JA (2001) Development of *Sphaerotheca fusca* on susceptible, resistant, and temperature-sensitive resistant melon cultivars. Mycol Res 105:1216–1222

Pliego C, Kanematsu S, Ruano-Rosa D, de Vicente A, López-Herrera C, Cazorla FM, Ramos C (2009) GFP sheds light on the infection process of avocado roots by *Rosellinia necatrix*. Fungal Genet Biol 46:137–145

Probanza A, García JAL, Palomino MR, Ramos B, Manero FJG (2002) *Pinus pinea* L. seedling growth and bacterial rhizosphere structure after inoculation with PGPR *Bacillus* (*B. licheniformis* CECT 5106 and *B. pumilus* CECT 5105). Appl Soil Ecol 20:75–84

Pusey PL, Wilson CL (1984) Postharvest biological control of stone fruit brown rot by *Bacillus subtilis*. Plant Dis 68:753–756

Ramamoorthy V, Viswanathan R, Raguchander T, Prakasam V, Samiyappan R (2001) Induction of systemic resistance by plant growth promoting rhizobacteria in crop plants against pests and diseases. Crop Prot 20:1–11

Raupach GS, Kloepper JW (1998) Mixtures of plant growth promoting rhizobacteria enhance biological control of multiple cucumber pathogens. Phytopathology 88:1158–1164

Razafindralambo H, Paquot M, Hbid C, Jacques P, Destain J, Thonart P (1993) Purification of antifungal lipopeptides by reversed-phase high performance liquid chromatography. J Chromatogr 639:81–85

Romero D, Rivera ME, Cazorla FM, de Vicente A, Pérez-García A (2003) Effect of mycoparasitic fungi on the development of *Sphaerotheca fusca* in melon leaves. Mycol Res 107:64–71

Romero D, Pérez-García A, Rivera ME, Cazorla FM, de Vicente A (2004) Isolation and evaluation of antagonistic bacteria towards the cucurbit powdery mildew fungus *Podosphaera fusca*. Appl Microbiol Biotechnol 64:263–269

Romero D, Pérez-García A, Veening JW, de Vicente A, Kuipers OP (2006) Transformation of undomesticated strains of *Bacillus subtilis* by protoplast electroporation. J Microbiol Methods 66:556–559

Romero D, de Vicente A, Olmos JL, Dávila JC, Pérez-García A (2007a) Effect of lipopeptides of antagonistic strains of *Bacillus subtilis* on the morphology and ultrastructure of the cucurbit fungal pathogen *Podosphaera fusca*. J Appl Microbiol 103:969–976

Romero D, de Vicente A, Rakotoaly RH, Dufour SE, Veening J-W, Arrebola E, Cazorla FM, Kuipers OP, Paquot M, Pérez-García A (2007b) The iturin and fengycin families of lipopeptides are key factors in antagonism of *Bacillus subtilis* toward *Podosphaera fusca*. Mol Plant Microbe Interact 20:430–440

Romero D, de Vicente A, Zeriouh H, Cazorla FM, Fernández-Oruño D, Torés JA, Pérez-García A (2007c) Evaluation of biological control agents for managing cucurbit powdery mildew on greenhouse-grown melon. Plant Pathol 56:976–986

Ryu CM, Hu CH, Reddy MS, Kloepper JW (2003) Different signalling pathways of induced resistance by rhizobacteria in *Arabidopsis thaliana* against two pathovars of *Pseudomonas syringae*. New Phytol 160:413–420

Ryu CM, Farag MA, Hu CH, Reddy MS, Kloepper JW, Pare PW (2004a) Bacterial volatiles induce systemic resistance in *Arabidopsis*. Plant Physiol 134:1017–1026

Ryu CM, Murphy JF, Mysore KF, Kloepper JW (2004b) Plant growth-promoting rhizobacteria systemically protect *Arabidopsis thaliana* against *Cucumber mosaic virus* by a salicylic acid and NPR1-independent and jasmonic acid-dependent signaling pathway. Plant J 39:381–392

Schisler DA, Slininger PJ, Behle RW, Jackson MA (2004) Formulation of *Bacillus* spp. for biological control of plant diseases. Phytopathology 94:1267–1271

Shoda M (2000) Bacterial control of plant diseases. J Biosci Bioeng 89:515–521

Smalla K, Wieland G, Buchner A, Zock A, Parzy J, Kaiser S, Roskot N, Heuer H, Berg G (2001) Bulk and rhizosphere soil bacterial communities studied by denaturing gradient gel electrophoresis: plant dependent enrichment and seasonal shifts revealed. Appl Environ Microbiol 67:4742–4751

Spadaro D, Gullino ML (2004) State of the art and future prospects of the bological control of postharvest fruit diseases. Int J Food Microbiol 91:185–194

Stein T (2005) *Bacillus subtilis* antibiotics: structures, syntheses and specific functions. Mol Microbiol 56:845–857

Ton J, van Pelt JA, van Loon LC, Pieterse CM (2002) Differential effectiveness of salicylate-dependent and jasmonate/ethylene-dependent induced resistance in Arabidopsis. Mol Plant Microbe Interact 15:27–34

Toro M, Azcon R, Barea JM (1997) Improvement of arbuscular mycorrhiza development by inoculation of soil with phosphate-solubilizing rhizobacteria to improve rock phosphate bioavailability and nutrient cycling. Appl Environ Microbiol 63:4408–4412

Touré Y, Ongena M, Jacques P, Guiro A, Thonart P (2004) Role of lipopeptides produced by *Bacillus subtilis* GA1 in the reduction of grey mould disease caused by *Botrytis cinerea* on apple. J Appl Microbiol 96:1151–1160

van Loon LC (2007) Plant responses to plant growth-promoting rhizobacteria. Eur J Plant Pathol 119:243–254

van Loon LC, Bakker PA, Pieterse CMJ (1998) Systemic resistance induced by rhizosphere bacteria. Annu Rev Phytopathol 36:453–483

van Peer R, Niemann GJ, Schippers B (1991) Induced resistance and phytoalexin accumulation in biological control of fusarium wilt of carnation by *Pseudomonas* sp. strain WCS417r. Phytopathology 81:728–734

van Wees SCM, Pieterse CMJ, Trijssenaar A, van't Westende YAM, Hartog F, van Loon LC (1997) Differential induction of systemic resistance in *Arabidopsis* by biocontrol bacteria. Mol Plant Microbe Interact 10:716–724

Vanittanakom N, Loeffer W, Koch U, Jung G (1986) Fengycin–A novel antifungal lipopeptide antibiotic produced by *Bacillus subtilis* F-29-3. J Antibiot 39:888–901

Verhagen BWM, Glazebrook J, Zhu T, Chang HS, Van Loon LC, Pieterse CMJ (2004) The transcriptome of rhizobacteria-induced systemic resistance in *Arabidopsis*. Mol Plant Microbe Interact 17:895–908

Wei G, Kloepper JW, Tuzun S (1991) Induction of systemic resistance of cucumber to *Colletotrichum orbiculare* by select strains of plant growth-promoting rhizobacteria. Phytopathology 81:1508–1512

Weller DM, Raaijmakers JM, McSpaden Gardener BB, Thomashow LS (2002) Microbial populations responsible for specific soil suppressiveness to plant pathogens. Annu Rev Phytopathol 40:309–348

Yezza A, Tyagi RD, Valéro JR, Surampalli RY (2004) Production of *Bacillus thuringiensis* based biopesticides in batch and fed-batch cultures using wastewater sludge as a raw material. J Chem Technol Biotechnol 80:502–510

Zhang H, Kim MS, Krishnamachari V, Payton P, Sun Y, Grimson M, Farag MA, Ryu CM, Allen R, Melo IS, Paré PW (2007) Rhizobacterial volatile emissions regulate auxin homeostasis and cell expansion in *Arabidopsis*. Planta 226:839–851

Chapter 9
Pasteuria penetrans and Its Parasitic Interaction with Plant Parasitic Nematodes

Alistair H. Bishop

9.1 Introduction

Until quite recently, bacteriology has been the study of culturable organisms. The proportion of unculturable bacteria in the soil is commonly put at over 99% (Sharma et al. 2005). The ecology of many of these will be very difficult to discover (Nannipieri et al. 2003; Riesenfeld et al. 2004) but this is not the case for one extraordinary group of non-culturable, aerobic, spore-forming bacteria: members of the genus *Pasteuria*. The composite species are exclusively obligate parasites of either water fleas or plant parasitic nematodes. These represent the two lineages of the different species of *Pasteuria*. This chapter predominantly focuses on the nematode parasites because of their residence in soil.

The first time that the characteristic spores of this genus were represented in publication were as drawings by Metchnikoff (1888) of what is now called *Pasteuria ramosa* (Ebert et al. 1996), the parasite of the water fleas *Daphnia magna* and *D. pulex*. This is now the type species of the genus *Pasteuria* and was recognized by Metchnikoff (1888) as an endospore-forming bacterium. A *Pasteuria* sp. was first reported in a nematode (*Dorylaimus bulbiferous*), by Cobb (1906), and the suggestion was made that it should be classified as a protozoan. In keeping with this suggestion the nematode parasite was later named *Duboscqia penetrans* by Thorne (1940) and placed in the family *Microsporidiae*.

Until quite recently, developments in taxonomy were based solely upon observations and measurements made from electron micrographs. Mankau (1975) noted that the parasite of the root knot nematode (RKN) *Meloidogyne javanica* divided with the formation of cross walls reminiscent of bacteria, and that it formed spores similar to those of known endospore-formers. He suggested the adoption of a position within the eubacteria with the name *Bacillus penetrans* n. comb. In spite of the outlandish appearance of the highly differentiated vegetative phase (Fig. 9.1)

A.H. Bishop
Detection Department, Defence Science and Technology Laboratory, Building 4, Room 12, Salisbury SP4 0JQ, UK
e-mail: AHBishop@dstl.gov.uk

N.A. Logan and P. De Vos (eds.), *Endospore-forming Soil Bacteria*, Soil Biology 27, DOI 10.1007/978-3-642-19577-8_9, © Springer-Verlag Berlin Heidelberg 2011 and the unique process of sporulation, this placement has subsequently been proved by genetic analysis to be surprisingly accurate. The link between the nematode parasite and that of the cladoceran *Daphnia* reported by Metchnikoff (1888) was suggested by Sayre and Wergin (1977). Based solely upon morphology, such as the elongated and branching septate cells, Sayre et al. (1988) formed the opinion that the nematode parasite was more in keeping with placement in the *Actinomycetales* than the *Bacillaceae*. Indeed, one could imagine the branching, septate hyphae, characteristic of some actinomycetes, to be condensed into the ball mycelia of the vegetative stage (Fig. 9.1). Furthermore, the elaborate differentiation that occurs during sporulation is more reminiscent of conidiospore than endospore formation. The designation *Pasteuria penetrans* (ex Thorne 1940) was finally proposed for the parasite of *Meloidogyne* spp. (Sayre and Starr 1985) based upon the morphology of the vegetative and spore phases and noting the similarity to the parasite of the cladocerans, *P. ramosa*.

An overriding factor that must have contributed to the stuttering start to research into *Pasteuria* spp. and their uncertain early taxonomy was the inability to culture them in vitro. This remains a major obstacle to ongoing study and the ultimate goal of large-scale agricultural exploitation.

9.2 Pathogenesis

The only mechanism by which *P. penetrans* can reproduce is through parasitism of certain phytoparasitic nematodes. Spores remain dormant in the soil until they come into contact with the cuticle of a second-stage juvenile (J2) nematode to which they can bind (Fig. 9.2). The characteristic "flying saucer" shape of the spore allows it to adhere, limpet-like, to the body of its nematode host, even over the uneven lateral fields that run down the length of the body (Fig. 9.3). The crucial event of attachment of spore to nematode cuticle has, understandably, been the most studied event in the bacterium's life cycle. Surrounding the spore is a perisporal "skirt" of proteinaceous fibres which mediates the attachment (Fig. 9.4). Each spore is initially encased in an exosporium that impedes attachment, as the increased levels of attachment following its removal by sonication have demonstrated (Stirling et al. 1986). This takes place naturally in the soil, presumably by processes including enzymatic degradation and abrasion. A close association ensues between the perisporal fibres and the cuticle of an appropriate nematode. The exact nature of this attachment has not yet been fully characterized. The initial recognition is presumed to be between carbohydrates on the surface of the spore, perhaps from glycoproteins containing *N*-acetylglucosamine (Persidis et al. 1991) in the perisporal fibres, and lectin-like proteins on the nematode cuticle. Additionally, it has been suggested that the attachment of spores involves collagen- (Persidis et al. 1991) or fibronectin-like (Mohan et al. 2001; Davies et al. 1996) molecules on the nematode cuticle. At least 13 polypeptides have been identified as components of the perisporal fibres of *P. penetrans* using a polyclonal antibody raised against whole spores (Vaid et al. 2002). Attachment

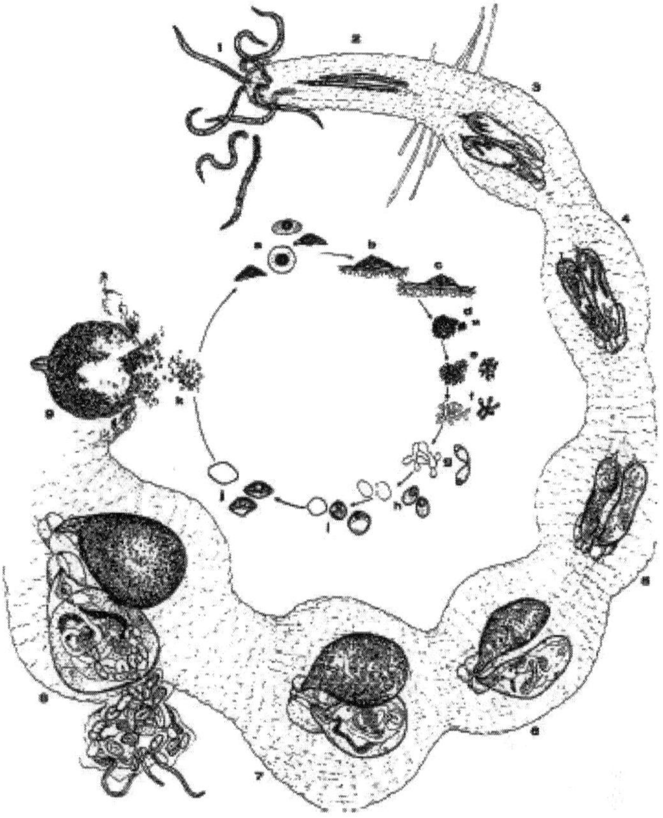

Fig. 9.1 Representation of the life cycle of a root knot nematode, *Meloidogyne* sp. within a root, both uninfected (*outer portion*) and infected with *Pasteuria penetrans* (*inner portion of the diagram*). The *inner circle* of the diagram shows the stages in the life cycle of *P. penetrans* corresponding to those of pathogenesis. (1) Entry of second stage juvenile nematodes (J2s) into the tip of a root, (2) intracellular migration of the J2s to the root cortex, (3) establishment of a feeding site. At this point the spores of *P. penetrans* germinate, (4) moult to third stage juvenile, (5) moult to fourth stage juvenile, (6) development of young females, (7) egg formation in healthy female; little or no egg production in infected females, (8) healthy female releases eggs; the infected female becomes filled with spores of *P. penetrans*, (9) the infected female dies and releases spores of *P. penetrans* into the soil. (a) Mature endospores of *P. penetrans* in the soil, (b) spores attach to the cuticle of *Meloidogyne* juvenile, (c) germination of the spore with formation of a germ tube penetrating the body of the nematode, (d) differentiation into vegetative ball mycelium, (e) release of fragments of the mycelium by lysis of intercalary "suicide" cells, (f) further disintegration of the mycelium and swelling of released fragments, (g) formation of quartets and then doublets of thalli where sporulation ensues at the tips, (h) formation of forespore within the sporangium, (i) engulfment and maturation of the spore with formation of the perisporal fibres, (j) completion of spore development, (k) release of spores into the soil (reproduced with permission from Preston et al. 2003)

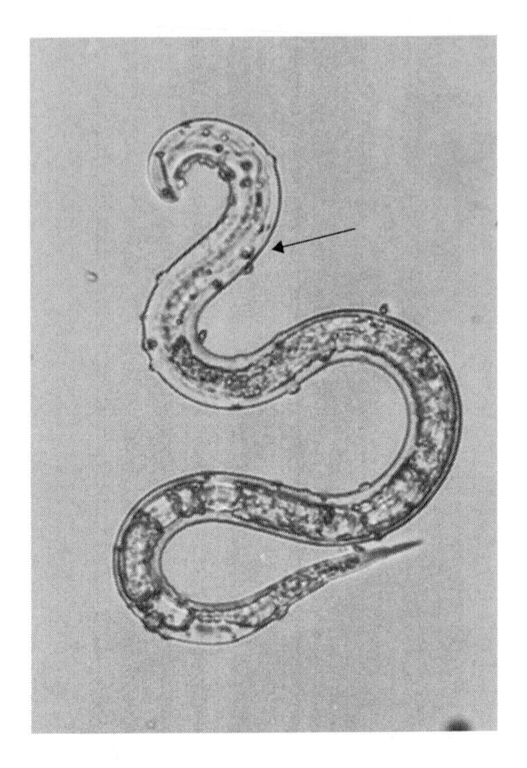

Fig. 9.2 Juvenile root knot nematode encumbered with spores of *Pasteuria penetrans* (*arrowed*) (image kindly supplied by Barbara Pembroke, University of Reading, UK)

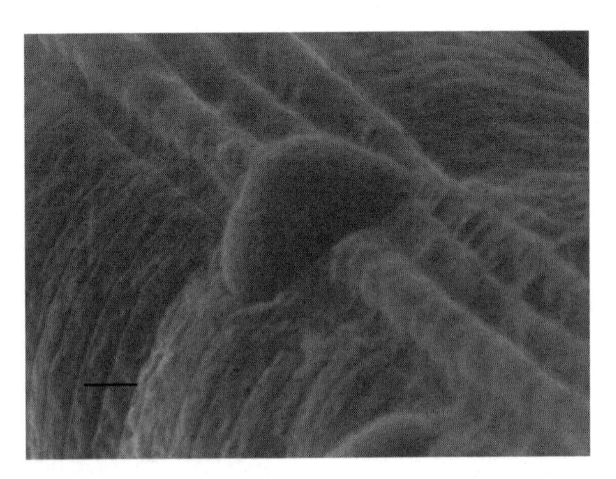

Fig. 9.3 Scanning electron micrograph of a spore of *Pasteuria penetrans* attached to the cuticle of a root knot nematode. A tight adhesion has been formed even over the undulating ridges of the lateral field that runs down the length of the body. Bar = 0.5 μm (image kindly supplied by Barbara Pembroke, University of Reading, UK)

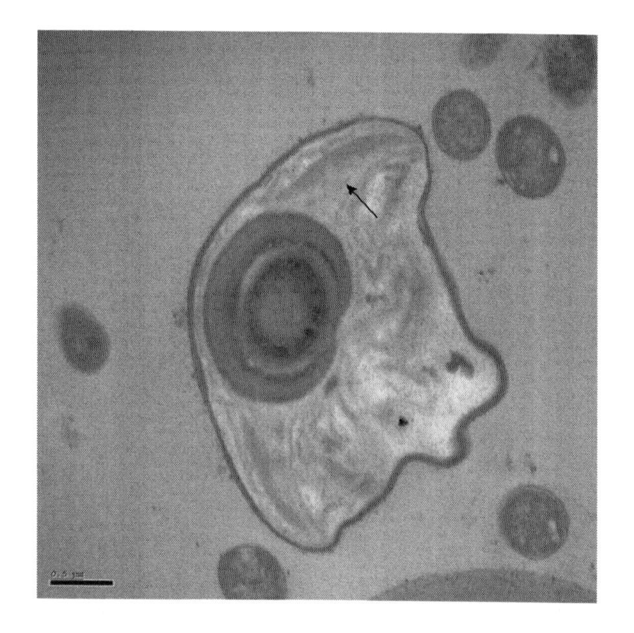

Fig. 9.4 Transmission electron micrograph of a mature spore of "*Pasteuria hartismeri*" still surrounded by the sporangial wall. The thinning of the spore coat on the underside of the spore through which the germ tube emerges is arrowed. The perisporal fibres which surround the spore are *arrowed*. Bar = 1 μm

appears to be a complicated process that may initially be reversible but which develops into a tenacious binding of the spore to the cuticle. Hydrophobic interactions may also play an initial role (Davies et al. 1996; Esnard et al. 1997). Whatever the mechanism, removal of the spore fibres results in a loss of ability to attach (Persidis et al. 1991).

Binding of a spore does not necessarily result in pathogenesis: it seems that spores may bind to nematode species that they cannot infect (Davies et al. 1990). Spores that have been killed by heating will also bind specifically to nematode cuticles (Dutky and Sayre 1978). In addition, not all spores germinate: Davies et al. (1988) estimated that a J2 had to have at least five spores attached to ensure that the nematode would eventually be infected. In the RKNs the J2s only develop into females when they establish a feeding cell in the root of a suitable plant species. It was thought that *P. penetrans*, the parasite of RKN, would only germinate once the interaction between the plant and the female nematode had begun (Fig. 9.1): it has, however, been reported that males (Hatz and Dickson 1992) and J2s (Giblin-Davis et al. 1990) of the RKN *Meloidogyne arenaria* have been observed with mature endospores of a *Pasteuria* sp. inside them.

In other groups of phytoparasitic nematodes, such as the root lesion, cyst and sting nematodes, however, germination certainly occurs without the plant interaction (Sayre et al. 1991; Giblin-Davis et al. 2003). Indeed, the *Pasteuria* isolate parasitizing the pea cyst nematode completes its life cycle in the migrating juveniles and is not found in females (Sturhan et al. 1994). A *Pasteuria* sp. reported by Davies et al. (1990) completed its life cycle in the J2s of a root cyst nematode, *Heterodera avenae*, but not in the females – while another completed its life cycle in J2s, males and females of *Heterodera glycines* (Noel and Stanger 1994;

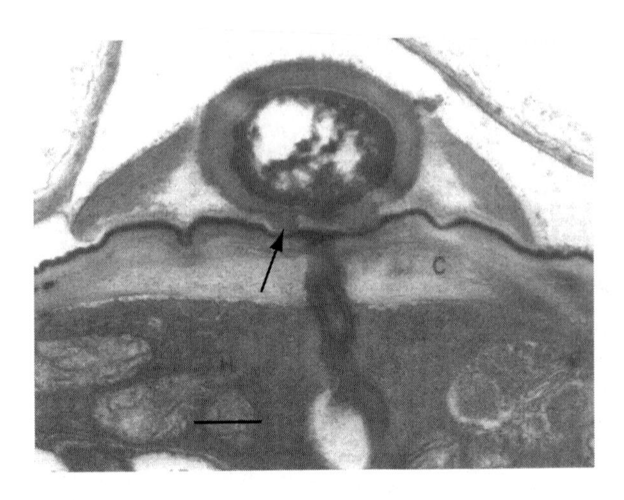

Fig. 9.5 Transmission electron micrograph of a spore of *Pasteuria penetrans* germinating into the body of a root knot nematode through the cuticle (C) and hypodermis (H). The ring or torus surrounding the germ tube is *arrowed*. Bar = 0.5 μm (reproduced with permission from Sayre and Wergin 1977)

Atibalentja and Noel 1997). This complicated picture of pathogenesis may become clearer when a definitive taxonomy of the genus *Pasteuria* is completed. The possibility that different biochemical triggers for germination are required for different *Pasteuria* spp. is an intriguing possibility, however.

Typically, 8–10 days (Giannakou et al. 1997) after attachment to an RKN the *P. penetrans* spore germinates and produces a germ tube that penetrates the body wall of the nematode (Fig. 9.5). An enzymatic degradation of the nematode tissue is assumed but no proof of this has yet been obtained. The germ tube emerges where the spore integument is at its thinnest (Figs. 9.4 and 9.5). In *P. penetrans* this region is surrounded on the outside by a ring or torus of unknown composition (Fig. 9.5). This ring also binds tightly to the cuticle of the nematode and it could be hypothesized that this seals degradative enzymes from the bacterium, facilitating penetration. Somewhat to contradict this idea, however, species such as "*P. hartismeri*" (Bishop et al. 2007) (Figs. 9.4 and 9.6) and *P. nishizawae* (Sayre et al. 1991) lack any ring on the ventral side of the spore. This ring could be a later evolutionary development because such species occur prior to *P. penetrans* in phylogenetic history (Sturhan et al. 2005; Bishop et al. 2007). Paradoxically, although the torus can bind strongly to the nematode cuticle its composition is different from the perisporal fibres because chemical treatments that dissolve the latter leave the former visibly intact (Vaid et al. 2002). This does not, however, rule out the possibility that cuticle adhesins could coat the surface of the torus.

Once the germ tube has entered the pseudocoelom of the nematode it differentiates into the vegetative form in a way that has not yet been witnessed. The vegetative form of *Pasteuria* spp. is unlike any other microorganism reported to date; in the early stages it typically resembles a ball, as if two cauliflower heads had been joined together (Figs. 9.1 and 9.7). The fine structure of the bacterium, equating to the florets of the cauliflower structure, are numerous thalli. In the

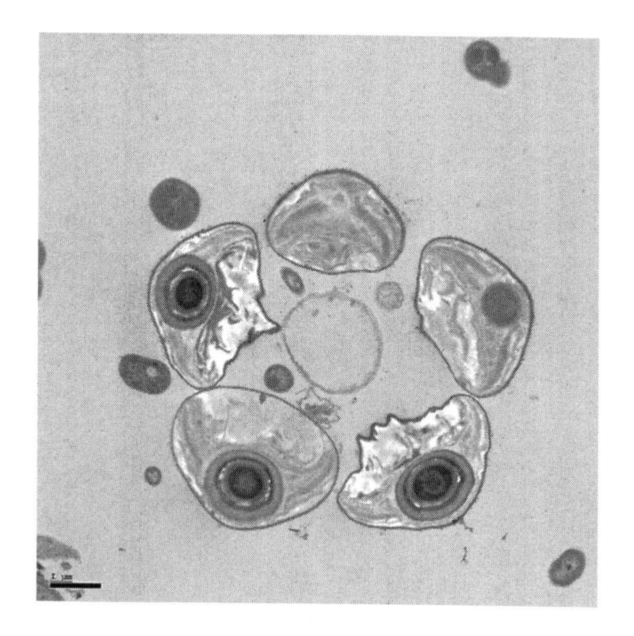

Fig. 9.6 Transmission electron micrograph of a cluster of spores of "*Pasteuria hartismeri*". The final stages in sporogenesis can be seen as a clockwise progression from the top spore. These clusters are natural, developmental stage of this species and, presumably, subsequently break up so that the spores can attach to susceptible nematodes

simplest terms, replication of the ball hyphae is by elongation and then longitudinal division (Fig. 9.7). This is dissimilar from binary fission of other eubacteria and can occur at multiple sites simultaneously in one mycelium. Any detail of the mechanism of this remarkable process is unknown. Replication of *P. penetrans* occurs predominantly in the ovaries, which normally are highly active in the production of eggs. Parasitized females produce few, if any, eggs (Bird and Brisbane 1988; Chen et al. 1997). *P. ramosa* is, similarly, described as a castrating parasite of its water flea host (Schmidt et al. 2008).

Once replication has ceased it seems that "suicide cells" (Sayre and Wergin 1977) at the centre of the septate ball lyse, allowing it to disintegrate into groups of thalli. These enlarge greatly (Figs. 9.1 and 9.7) and then disintegrate further into quartets of thalli, then doublets and so to single sporangia. The endospore develops at the tip of the sporangium and is finally released as a mature spore.

At the end of sporulation the body of the female RKN is full of spores (Fig. 9.1). When the female dies and breaks open, the spores are liberated into the rhizosphere. This is a good location to attach to juveniles that have hatched from egg masses deposited nearby. The amount of spores of *P. penetrans* liberated from one infected RKN female can approach two million (Waterman et al. 2006; Chen and Dickson 1998). Some groups of phytoparasitic nematode, though, do not develop into the sessile feeding form seen with RKN. These ectoparasitic nematodes such as cyst and sting nematodes, which do not grow to the size of female RKN nematodes, contain as few as only several hundreds of the *Pasteuria* spp. that parasitize them (Atibalentja et al. 2004).

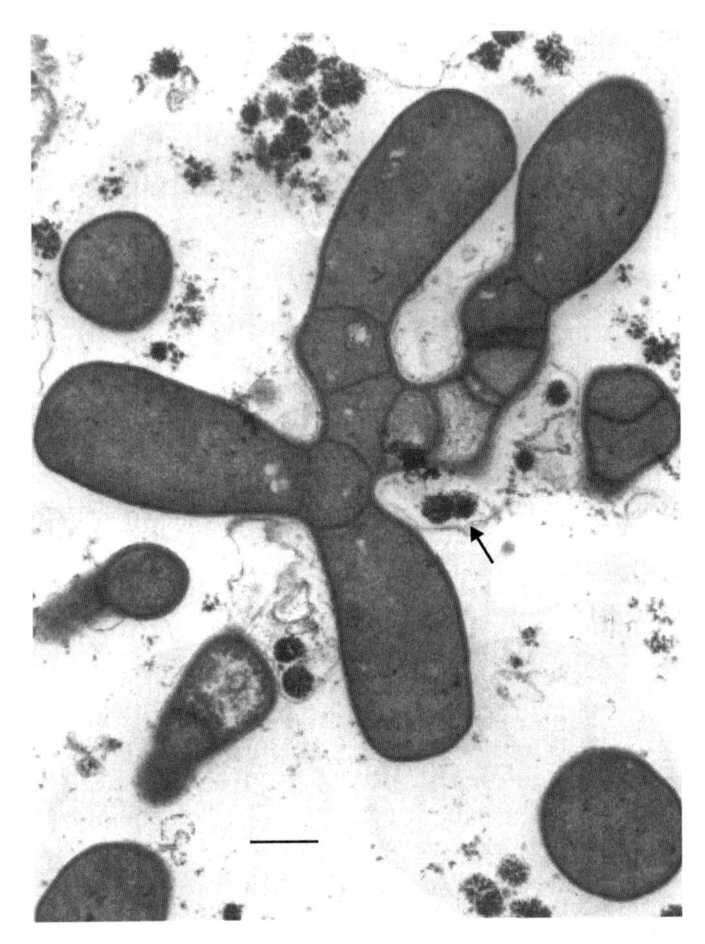

Fig. 9.7 Transmission electron micrograph showing, centrally, a quartet of thalli that have broken off from a vegetative ball mycelium and swollen immensely. Some of the intercalary, "suicide" cells joining the four thalli will then break to liberate two doublets of thalli and eventually single mature spores will form at the tips of the sporangia which develop in each of the four terminal cells of this quartet. The ball mycelia are also visible around the sporulating quartet and a dividing mycelium is *arrowed*. Reproduced with permission from Sayre and Wergin (1977)

9.3 Taxonomy and Phylogeny

As more examples of the nematode parasite were discovered, in over 51 countries (Siddiqui and Mahmood 1999), it appeared that *P. penetrans* had a very broad host range. At least 323 nematode species belonging to 116 genera of nematodes have been reported as infected (Chen and Dickson 1998; Atibalentja et al. 2000; Ciancio et al. 1994; Sturhan 1988). It is becoming increasingly clear, in fact, that an unknown number of different species or strains of *Pasteuria* exist, each with a specific host range.

From comparisons with the de facto reference species of plant parasitic nematodes, *P. penetrans* (Sayre and Starr 1985), it became noticeable that, while the parasites of other nematodes had the same vegetative and spore phases, differences were apparent in their sizes and structural detail (Sayre et al. 1988; Ciancio et al. 1994). The spore form has been used exclusively for these comparisons because the vegetative form is very prone to lysis once out of its host, and it is fragile and much less easy to isolate, while the spore has easily identifiable structural features. Thus, morphometric properties and host specificity were used as much less exacting criteria than normally accepted for bacterial classification (Sharma and Davies 1996) to define two new species for the nematode, giving three species overall: *P. penetrans* (Sayre and Starr 1985), *P. thornei* (Starr and Sayre 1988), and *P. nishizawae* (Sayre et al. 1988, 1991), these being hyper-parasites of root knot, lesion and cyst nematodes, respectively.

More definitive analysis has been made possible by the use of PCR amplification of 16S rRNA fragments. This technique was first applied to *P. ramosa* (Ebert et al. 1996). Phylogenetic analysis showed that it actually fell within the genus *Bacillus* rather than *Actinomycetales*. Strangely, from this analysis, the most closely related species were *Alicyclobacillus* spp. which are acidophilic extremophiles. A similar analysis applied to *P. penetrans* (Atibalentja et al. 2000) proved, predictably, its relatedness to the cladoceran parasite, and also their positioning within the genus *Bacillus*. Using a number of different isolates of *Pasteuria* spp., Duan et al. (2003) confirmed the positioning within the low G+C Gram-positive genera and, in particular, with the family "*Alicyclobacillaceae*". Fortunately, the differences in host range and morphometrics that had resulted in the designation of two other presumptive species, *P. thornei* and *P. nishizawae* were borne out in this analysis by genetic differences large enough to be significant.

Remarkably, comparison of fragments of the sporulation-related sigma factor genes *sigE* and *sigF* (Preston et al. 2003) with homologues from a subtype of *P. penetrans* placed its ancestor as a progenitor to both the genera *Bacillus* and *Clostridium*. Using a fragment of *sigE* Schmidt et al. (2004) aligned *P. penetrans* closely with the genera *Clostridium* and *Bacillus*, in particular, with *Paenibacillus polymyxa*. To date, the only full *Pasteuria* gene that has been cloned and expressed was that reported by Trotter and Bishop (2003). Comparison with homologues from other endospore-formers placed *P. penetrans* amongst the genus *Bacillus*. Nevertheless, expression of the cloned gene from *P. penetrans* failed to complement a *spo*0A deficient mutant of the archetypal spore former *B. subtilis*; evidence, perhaps, of the evolutionary distance between the species. Nevertheless, by virtue of the presence of these orthologues of key components of the sporulation pathway of endospores, these studies provided support for the earlier assertions based on 16S rRNA gene sequences of the taxonomical placement of *P. penetrans*. In view of the ultimate goal of the exploitation of *P. penetrans* in nematode control, an investigation into the biochemical and molecular bases of germination in *P. penetrans* would be highly desirable.

The most definitive genetic study to date was done by Charles et al. (2005). Multiple locus sequence typing was used to compare 40 orthologous sequences in

P. penetrans, identified by similarity to *B. subtilis* as house-keeping genes from a wide range of acid fast and Gram-positive and Gram-negative bacteria. This list regrettably lacked representation from the genus *Clostridium* even though previous studies had placed the genus *Pasteuria* within the *Bacillus–Clostridium* clade. Amino acid translations of the genes were analysed singly and with concatenation both with and without the removal of insertions and deletions. Results from all of these analyses placed *P. penetrans* firmly in the low G+C content *Bacillus* group. Astonishingly, as found by Preston et al. (2003) *P. penetrans* emerged as not being a recently evolved species but actually as being ancestral to the *Bacillus* species. As thought provoking a finding as this is, the use by Charles et al. (2005) of 40 genes as a basis for comparison makes this a reliable assertion. The evolutionary route by which this occurred can only be speculated upon, particularly in view of the stark differences in the vegetative forms and sporulation processes that exist between *Pasteuria* and *Bacillus* spp. Furthermore, of the organisms used in the comparison, *P. penetrans* was more closely related to the extremophile *B. halodurans* than to species with pathogenic potential such as *B. anthracis* or *B. thuringiensis* (Charles et al. 2005).

As with all other research on this bacterium, whole-genome analysis of *P. penetrans* is hampered by the inability to produce large amounts of DNA using in vitro culture. As always, DNA is extracted from spores cultured in vivo. Several genomic libraries have been constructed from the broad-host-range *P. penetrans* strain RES147 (Bird et al. 2003). The size of the *Pasteuria* genome is estimated to be less than 4.2 Mb. Preliminary analyses show that more than half of the sequences so far obtained have greater than 50% similarity to known genes on the NCBI databases.

9.4 Technical Problems

9.4.1 Obtaining the Bacteria

Taxonomic, biochemical or genetic analysis of *Pasteuria* spp. has really only been possible because it produces the robust endospore form. The vegetative stage is very fragile and prone to lysis. Of the published attempts to culture *P. penetrans* in vitro (Williams et al. 1989; Bishop and Ellar 1991) the best that has been achieved has simply been to create conditions suitable for the ball mycelia extracted from a female nematode to go through an apparently pre-programmed and limited number of divisions and differentiate into sporangia and produce spores. There is no reliable report to date of any sustained vegetative growth or triggering of germination.

To obtain sufficient supplies of *P. penetrans*, for example, J2 nematodes of an appropriate species have to be incubated with spores until attachment has occurred. The encumbered J2s are then added to the soil of pot plants of a suitable host such as tomato. After several weeks, the roots are washed free of soil to reveal the swollen and knotted roots characteristic of nematode infection. Conventionally, the infected female nematodes are then laboriously and delicately removed with the aid of fine tweezers and a magnifying lens. Less time-consuming methods involve drying the roots and then macerating them in a pestle and mortar, or partially digesting the roots with pectinase, then macerating them and sieving the spores from the root debris (Chen et al. 1996a). Such spores were found to have little contaminating material. For more stringent removal of contaminating material the spores can be further purified by ultracentrifugation in sucrose or sodium bromide density gradients (Persidis et al. 1991; Vaid et al. 2002).

For most genetic analysis the inevitable microbial contamination has to be removed from the spore preparations derived from nematode-infected roots. This is all the more important because PCR is integral to obtaining enough DNA to analyse and amplification of DNA from contaminating bacteria must be avoided at all costs. Vegetative bacteria can be destroyed, by exploiting the resistance of spores to enzymatic and chemical agents (Atibalentja et al. 2000; Ebert et al. 1996). The subsequent extraction of DNA from the spores is not without its problems, particularly because this needs to be optimized in view of the difficulty of obtaining sufficient spores in the first place; proprietary kits (Bekal et al. 1999) or laboratory protocols involving chemical agents (Schmidt et al. 2008) have been used, while bead-beating has also proven successful (Anderson et al. 1999).

It is possible to culture plant roots axenically and infect them with surface-sterilized nematode egg masses and surface-sterilized *Pasteuria* spores. The use of antifungal agents helps maintain the agar plates free of contamination, but most workers do not generally favour this approach, presumably because of the effort involved. Such a gnotobiotic approach was, however, used by Bekal et al. (1999) to clone 16S rRNA from the *Pasteuria* sp. infecting the sting nematode *Belonolaimus longicaudatus*.

9.4.2 *Genetic Heterogeneity of* Pasteuria *"Strains"*

One of the problems that brought uncertainty to the study of the interaction of this host and parasite is again due to the inability to culture the bacterium in vitro. With free-living microorganisms it is easy to obtain a clone by growth on a solid medium. On the other hand, the "populations" (Davies et al. 1988; Persidis et al. 1991; Bekal et al. 2001) of *Pasteuria* spp. that have been used by researchers are just that: undefined mixtures of unknown genetic composition. Admittedly, as a population of *Pasteuria* spp. is maintained experimentally for a long period of time a particular race of nematode selection might be expected to diminish the genetic diversity of the bacterium. Nevertheless, this population will have arisen from an initial infection of a nematode by multiple spores of different genetic constitution. This obviously makes interpretation of experiments difficult because the population of spores used cannot be expected to be homogenous.

Trotter et al. (2004) finally resolved this problem by producing a clone resulting from a single spore that had been allowed to attach to J2 of *Meloidogyne incognita*. A single spore cannot be guaranteed to result in infection of a nematode so this process had to be repeated several times and each J2 examined microscopically to ensure that only one spore was attached. Genetic proof that the resultant population of spores was a clone was obtained by cloning and sequencing the PCR amplification products of fragments of the *spo0A* gene. Due to the nature of the DNA extraction and amplification processes it could have been the case, conceivably, that the DNA amplified had resulted from very few spores. The PCR primers, however, were degenerate and by sequencing from primer binding sites within the vector the individual *spo0A* primer sequences were obtained. It could be shown that the sequences of all of the many clones were identical but each of the *spo0A* gene primers was slightly different. This proved that the single spore population was genetically identical and that these sequences had each been obtained from the DNA of different spores. Even with a true clone of *P. penetrans* spores the attachment to J2s was not complete, illustrating the complexity of the attachment process.

9.4.3 Detection of P. penetrans

Although it has limitations, the conventional method for assessing population densities of most common bacteria in soil (Ellis et al. 2003) is to plate a sample onto a solid growth medium and count the resulting colonies. Such a simple approach is obviously not possible for unculturable bacteria such as *Pasteuria* spp.

A variety of simple methods were evaluated by Chen et al. (1996a) to count the number of *P. penetrans* spores in infected tomato roots. Grinding of roots followed by microscopic enumeration gave the highest estimate (with least variance) but the method was laborious, required expertise and, should the spore density be low, was not considered sensitive enough. At best, about 1% of the spores present were retrieved. After mechanical disruption, the second most effective approach was a baiting method which is also the conventional approach for assessing spore levels in soils: a sample of the soil is taken, cultivated J2s are added and incubated for a period of time, e.g. 18 h and then recovered using sucrose density centrifugation. The nematodes effectively act like fly paper and allow the spores to be extracted from the otherwise impenetrable matrix of the soil and the average number of spores attached to each J2 is counted microscopically. Estimation of the density of spores per gram of sample is facilitated by regression analysis using a standard curve. This is prepared by adding a range of known amounts of spores to soil previously free of *Pasteuria* spp., adding J2s and, after recovery, counting the mean number of spores per nematode for each spore density (Chen et al. 1996a). Obviously, only those *Pasteuria* spp. which are able to attach to the species of J2 added will be counted. A method using enzymatic softening of roots followed by various sieving stages (Waterman et al. 2006) has been reported to be even more effective than the best method used by Chen et al. (1996a); this is claimed to give recovery rates of about 30% of the theoretical maximum.

Molecular techniques are now available for detecting and enumerating *Pasteuria* spp. in soil. These rely either on the use of antibodies or DNA amplification.

9.4.3.1 Polyclonal Antibodies

Polyclonal antibodies have been used with indirect immunofluorescence to visualize spores directly in soil suspensions (Fould et al. 2001). These authors also developed an ELISA technique for indirect quantification of *P. penetrans* spores.

A fortuitous discovery that a particular monoclonal antibody (MAb 2A41D10) recognizes an antigen common to all nematode-parasitic *Pasteuria* spp. has allowed the quantification of spores in soil. The target of MAb 2A41D10 is putatively a β-1,4-linked *N*-acetyl glucosamine carbohydrate residue that is uniformly distributed on the surface of *P. penetrans* spores and is formed in the late stages of spore maturation (Charnecki et al. 1998).

Using a chemical mixture of urea, dithiothreitol and CHES buffer, Schmidt et al. (2003) extracted this soluble spore envelope antigen which could then be assayed by ELISA using MAb 2A41D10. The lowest limit of detection was about 300 spores g^{-1} of soil. The specificity was demonstrated using extracts of *Pasteuria*-free soil or from pure cultures of other endospore-forming bacteria isolated from the soil and showing that no response was elicited from the assay. The limitation of these antibody techniques is, of course, that only mature spores will be detected; they cannot be applied to the detection and enumeration of vegetative forms within nematodes.

9.4.3.2 16S rRNA Gene Fragments

Analysis of 16S rRNA gene fragments from 30 new isolates of *Pasteuria* spp. from infected nematodes and the soil by Duan et al. (2003) allowed the design of *Pasteuria*-specific PCR primers. The resulting amplification fragments shared highly conserved restriction enzyme sites not found in any of the non-*Pasteuria* sequences that were interrogated in silico. This raised the possibility of checking presumptive *Pasteuria* amplification products prior to sequencing. Furthermore, digestion with *Hha*I showed variation between different *Pasteuria* spp. clones, suggesting that differentiation of biotypes may be possible. A sensitivity as good as 100 *Pasteuria* spores per gram of soil was obtainable with these specific primers after purification of the released DNA.

Using PCR primers to fragments of *atpF*, *sigE* and *spoIIAB* Schmidt et al. (2004) used real-time PCR to differentiate between infected and uninfected females of *M. arenaria*. Fluorescent in situ hybridization was used to show that the *sigE* probe localized within vegetative forms of *P. penetrans*. The current problem with this method is that, surprisingly, attempts to obtain amplifiable DNA from mature spores by chemical, physical and enzymatic means failed. Hence, only nematodes infected with vegetative or early sporulation stage bacteria could be detected.

One potentially significant advantage of DNA amplification techniques, however, is that sequence heterogeneity should provide a means of differentiating between different species and biotypes of *Pasteuria*. This is particularly true for the sporulation genes (Schmidt et al. 2004) and may provide a rapid method of discerning the *Pasteuria* spp. in soils that are naturally suppressive to phytoparasitic nematodes. Another method to aid in epidemiological and ecological studies is the use of repeat sequences within the genome. Variable number tandem repeats (VNTRs) are useful markers to study genetic variability and have even been used successfully to characterize highly clonal species such as *B. anthracis* (Andersen et al. 1996). A sub-set of VNTRs (Mouton et al. 2007), called short sequence repeats, have been identified in the genome of the cladoceran parasite, *P. ramosa* (Mouton and Ebert 2008). These authors were able to use this technique to demonstrate genetic differences between *P. ramosa* isolates in separate geographical locations and also to show within-population variations. Given the similarities in the genomes of the members of the genus *Pasteuria* (Schmidt et al. 2008), it seems inevitable that the same technique will be applicable to other *Pasteuria* species.

The problem for large amounts of biological material as a pre-requisite for genome analysis can be mitigated by the multiple strand-displacement assay (MDA) This technique (Dean et al. 2001) uses a viral enzyme to replicate faithfully the whole genomes of organisms from small initial samples of DNA. It is thus ideal for unculturable bacteria. Nong et al. (2007) applied this approach to females of *M. incognita* infected with vegetative cells of *P. penetrans*. They found that 2,000 vegetative *Pasteuria* cells were sufficient to give complete genome coverage of the MDA-amplified template following PCR with specific primers. While direct PCR of individual and infected females gave no amplification products, the correct-sized products were obtained for two genes in half of the females tested after an initial MDA treatment. Sequencing of the MDA-amplified DNA uncovered single nucleotide polymorphisms (SNPs) in the genes analysed from a single female; this demonstrates that infection by more than one spore had occurred and also illustrates the genetic heterogeneity of the "populations" of *P. penetrans* commonly used (Trotter et al. 2004).

9.5 Agricultural Exploitation

The unique morphology, life cycle and subsequent discovery of intriguing genetic lineages with the genus *Bacillus* and its relatives would be ample reasons to study the genus *Pasteuria* but, of course, *Pasteuria* spp. also have an agricultural and ecological values that may be exploited biotechnologically. Phytoparasitic nematodes are major pests of agriculture in temperate and particularly tropical regions. Losses of crops owing to these pests are reckoned to amount to $100 billion a year

(Oka et al. 2000). The conventional method of control has been fumigation of the soil but, because of the dense, reactive nature of soil, this could only been done for high-value crops (Aktar and Malik 2000; Siddiqui and Mahmood 1999). Some successful fumigants, such as methyl bromide, are powerful greenhouse warming gases when they reach the atmosphere and, as a group, these fumigants are unspecifically toxic agents. It is impossible, however, for a biological agent to be as effective, easy to apply and show such rapid effects as chemical nematicides (Tian et al. 2007). The need for alternative control measures is great and other agricultural methods exist such as solarization, transgenic plants and rotation of crops, but these show variable levels of success (Aktar and Malik 2000; Oka et al. 2000).

The members of the genus *Pasteuria* have obviously co-evolved with their hosts over millions of years and have managed to maintain themselves to date by this parasitic way of life, but could they be used economically to control phytoparasitic nematodes? The overriding obstacle to this, of course, is the inability to culture them in vitro. Another problem is the specificity of attachment and pathogenesis. While the narrow spectrum of activity that biological pest control agents invariably display is a big ecological advantage, it is a serious disadvantage when a particular strain or species only attacks a narrow range (even within a particular family or order) of the pest organisms that a farmer has to contend with. This certainly will be a problem in the deployment of *Pasteuria* spp. Some isolates have, however, been reported to have cross-generic host ranges (Bhattacharya and Swarup 1988; Mankau and Prasad 1972; Mankau 1975; Pan et al. 1993; Oostendorp et al. 1990; Sharma and Davies 1996). Cross-generic attachment has been proven in the laboratory (Davies et al. 1990; Hewlett and Dickson 1994; Winkelheide and Sturhan 1996). Crucially, none of the work included in these reports went on to prove whether infection and spore production took place in the secondary nematode species to which spores attached.

Other isolates of *P. penetrans* have been found to be much more host-specific even in attachment: spores which parasitized the citrus nematode *Tylenchulus semipenetrans* would not attach to *M. incognita, M. javanica or Radophilus citrophilus* (Kaplan 1994). Comparable findings were reported by Sharma and Davies (1996) and such host specificity seems very much to be normal. Even when an isolate of *P. penetrans* is capable of parasitizing two species of nematode, it will attach with higher frequency in the host that it last infected. When it has completed a life cycle in the second host its level of attachment to this host will increase while binding to the former host will be decreased. This is not just a selection of subpopulations in a heterogeneous mixture because it is apparent in single spore isolates (Bishop, unpublished data). Conversely, what must be assumed to be different *Pasteuria* spp., one with large endospores and one with smaller ones, were found parasitizing the same two species of nematode (Giblin-Davies et al. 1990).

It is not even necessary for spores of *P. penetrans* to germinate in order to exert a controlling effect on their host; the simple attachment of spores has been found to decrease plant invasion noticeably. The required number has been variously reported to be between 40 and 3 (Stirling 1984; Brown and Smart 1985; Davies et al. 1988; Kariuki et al. 2006). The movement of the juvenile is simply hindered by the attached spores and the attraction between spores attached to one juvenile and the cuticle of others causes the nematodes to become inextricably clumped together.

Notwithstanding the vagaries of the spectra of pathogenicity that *Pasteuria* spp. demonstrate, there have been numerous small-scale agricultural experiments that prove their efficacy in controlling phytonematode damage. The phenomenon of "suppressive soils" has long been reported with respect to decreased nematode damage to a particular crop in one field as opposed to another. This can be due to the presence of nematode-controlling fungi (Kerry 1988) but has also been attributed to the natural presence of *P. penetrans*. It has been shown to be possible to transfer what was, probably, a mixed population of *Pasteuria* spp. from a suppressive soil and endow a new field with an inherent protection against phytoparasitic nematodes. Kariuki and Dickson (2007) obtained endospores by collecting females of RKNs that were infesting peanut plants; these were then dried in the sun for 2 weeks to kill any nematode inoculum. This material was then incorporated into the soil in a plot that had not shown the presence of RKNs or *Pasteuria* spp. Crops of peanuts and okra were subsequently planted in the plots and the RKN *M. arenaria* added to the soil. Suppression of the phytonematode in the new plot was evident even within a year. Melki et al. (1998) followed a cumulative build up of *P. penetrans* to produce a suppressive soil but recommended that at least a year was necessary for this to occur. A similar effect, against a cyst nematode, was produced by Westphal and Becker (1999). The addition of 10,000 spores of *P. penetrans* g^{-1} soil effectively suppressed infection of peanut plants by *M. arenaria* (Chen et al. 1996b, 1997). This effect became established and magnified over the 3-year study period.

Numerous other experiments at various scales have shown that *Pasteuria* spp. can have a significant, enduring and increasing effect in controlling phytoparasitic nematodes (reviewed in Chen and Dickson 1998). Realistically, the future agricultural application of *P. penetrans* would be as part of an integrated pest management scheme. To this end, several authors, including Tzortzakakis and Gowen (1994), have shown that it is compatible with a chemical nematicide and solarization of the soil and has an additive effect in controlling nematodes when used in conjunction with these methods.

The specificity of *Pasteuria* spp. for different species of nematode will certainly be a restraining factor if it is ever used as a widespread biological control agent. A possible solution is to apply a mixture of *Pasteuria* spp. so that at least one would parasitize each of the nematode pests of a crop in a certain habitat. Such customized control measures will not add to its attraction as a biological control agent, especially for commercial exploitation. Other factors militating against this are that the RKN, which are the major agricultural pest problem, typically inhabit tropical, and hence generally poorer, parts of the world. The fact that this bacterium is obviously so fastidious and also so fragile in the vegetative stage, ensure that if any in vitro culture method is developed the product will not be cheap.

Although greatly needed it is hard to imagine that *P. penetrans* will enjoy the biotechnological exploitation that *B. thuringiensis* has (Bishop 2002). The fact that, unlike *B. thuringiensis*, once spores have been applied to a field *P. penetrans* can establish itself as a self-perpetuating control agent will also not endear it to companies wishing to exploit it commercially. Such practical considerations should not detract from the fascination that this organism holds for some nematologists, bacteriologists and molecular biologists. In many ways it could be regarded as the most remarkable bacterium yet discovered. Although progress is difficult, it is proceeding at an increasingly rapid rate, and yet there are so many areas about which we currently know very little. Even without comparison to its much simpler relatives in the supergenus *Bacillus*, the study of its unique vegetative development and sporulation and germination pathways is probably what will entice more scientists to study this most infectious bacterium.

References

Aktar M, Malik A (2000) Roles of organic amendments and soil organisms in the biological control of plant-parasitic nematodes: a review. Bioresour Technol 74:35–47

Andersen GL, Simchock JM, Wilson KH (1996) Identification of a region of genetic variability among *Bacillus anthracis* strains and related species. J Bacteriol 177:377–384

Anderson JM, Preston JF, Dickson DW, Hewlett TE, Williams NH, Maruniak JE (1999) Phylogenetic analysis of *Pasteuria penetrans* by 16S rRNA gene cloning and sequencing. J Nematol 31:319–325

Atibalentja N, Noel GR (1997) Life cycle and host specificity of *Pasteuria* sp. parasitizing *Heterodera glycines*. J Nematol 29:568 (Abstr)

Atibalentja N, Noel GR, Liao TF, Dormier LL (2000) Phylogenetic position of the American isolate of *Pasteuria* that parasitizes the soybean cyst nematode, *Heterodera glycines*, as inferred from 16S rRNA sequence analysis. Int J Syst Evol Microbiol 50:605–613

Atibalentja N, Noel GR, Ciancio A (2004) A simple method for the extraction, PCR-amplification, cloning, and sequencing of *Pasteuria* 16S rDNA from small numbers of endospores. J Nematol 36:100–105

Bekal S, Giblin-Davis RM, Becker JO (1999) Gnotobiotic culture of *Pasteuria* sp. on *Belonolaimus longicaudatus*. J Nematol 31:522

Bekal S, Borneman J, Springer MS, Giblin-Davis RM, Becker JO (2001) Phenotypic and molecular analysis of a *Pasteuria* strain parasitic to the sting nematode. J Nematol 33:110–115

Bhattacharya D, Swarup G (1988) *Pasteuria penetrans*, a pathogen of the genus *Heterodera*, its effect on nematode biology and control. Indian J Nematol 18:61–70

Bird AF, Brisbane PG (1988) The influence of *Pasteuria penetrans* in field soils on the reproduction of root-knot nematodes. Rev Nématol 11:75–81

Bird D, Opperaman CH, Davies KG (2003) Interactions between bacteria and plant parasitic nematodes: now and then. Int J Parasitol 33:1269–1276

Bishop AH (2002) The insecticidal proteins of *Bacillus thuringiensis*. In: Berkeley R, Heyndrickx M, Logan NA, De Vos P (eds) Applications and systematics of the genus *Bacillus* and related organisms. Blackwell, Oxford, pp 160–175

Bishop AH, Ellar DJ (1991) Attempts to culture *Pasteuria penetrans*. Biocontrol Sci Technol 1:101–114

Bishop AH, Gowen SR, Pembroke B, Trotter JR (2007) Morphological and molecular characteristics of a new species of *Pasteuria* parasitic on *Meloidogyne ardenensis*. J Invertebr Pathol 96:28–33

Brown SM, Smart GC (1985) Root penetration by *Meloidogyne incognita* juveniles infected with *Bacillus penetrans*. J Nematol 17:123–125

Charles L, Carbone I, Davies KG, Bird D, Burke M, Kerry BR, Opperman CH (2005) Phylogenetic analysis of *Pasteuria penetrans* by use of multiple genetic loci. J Bacteriol 187:5700–5708

Charnecki JH, Rice JD, Dickson DW, Preston JF (1998) Determinants for attachment of endospores of *Pasteuria penetrans* to phytopathogenic nematodes. In: 98th Annual Meeting of American Society of Microbiology. Abstr Q-171, p 449

Chen ZX, Dickson DW (1998) Review of *Pasteuria penetrans*: biology, ecology and biological control potential. J Nematol 30:313–340

Chen ZX, Dickson DW, Hewlett TE (1996a) Quantification of the endospore concentrations of *Pasteuria penetrans* in tomato root material. J Nematol 28:50–55

Chen ZX, Dickson DW, Freitas, LG Preston JF (1996b) Suppression of *Meloidogyne arenaria* race 1 by soil application of endospores of *Pasteuria penetrans*. J Nematol 28:159–168

Chen ZX, Dickson DW, McSorley R, Mitchell DJ, Hewlett TE (1997) Suppression of *Meloidogyne arenaria* race 1 by soil application of endospores of *Pasteuria penetrans*. J Nematol 28:159–168

Ciancio A, Bonsignore R, Vovlas N, Lamberi F (1994) Host records and spore morphometrics of *Pasteuria penetrans* group parasites of nematodes. J Invertebr Pathol 63:260–267

Cobb NA (1906) Fungus maladies of the sugar cane, with notes on associated insects and nematodes, 2nd edn. Hawai Sugar Planters Assoc Exp Stat Div Pathol Physiol Bull 19:342–348

Davies KG, Kerry BR, Flynn CA (1988) Observations on the pathogenicity of *Pasteuria penetrans*, a parasite of root-knot nematodes. Ann Appl Biol 112:491–501

Davies KG, Flynn CA, Laird V, Kerry BR (1990) The life-cycle, population dynamics, and host specificity of a parasite of *Heterodera avenae*, similar to *Pasteuria penetrans*. Rev Nématol 13:303–309

Davies KG, Afolabi P, O'Shea P (1996) Adhesion of *Pasteuria penetrans* to the cuticle of root-knot nematodes (*Meloidogyne* spp.) inhibited by fibronectin: a study of electrostatic and hydrophobic interactions. Parasitology 112:553–559

Dean FB, Nelson JR, Gielse TL, Lasken RS (2001) Rapid amplification of plasmid and phage DNA using Phi29 DNA polymerase and multiply-primed rolling circle amplification. Genome Res 11:1095–1099

Duan YP, Castro HF, Hewlett TE, White JH, Ogram AV (2003) Detection and characterization of *Pasteuria* 16S rRNA gene sequences from nematodes and soils. Int J Syst Evol Microbiol 53:105–112

Dutky EM, Sayre RM (1978) Some factors affecting infection of nematodes by the bacterial spore parasite *Bacillus penetrans*. J Nematol 10:285 (Abstr)

Ebert D, Rainey P, Embley TM, Scholz D (1996) Development, life cycle, ultrastructure, and phylogenetic position of *Pasteuria ramosa* Metchnikoff 1888: rediscovery of an obligate endoparasite of *Daphnia magna* Strauss. Philos Trans R Soc Lond Ser B 351:1689–1701

Ellis RJ, Morgan P, Weightman PAJ, Fry JC (2003) Cultivation-dependent and -independent approaches for determining bacterial diversity in heavy-metal-contaminated soil. Appl Environ Microbiol 69:3223–3230

Esnard J, McClure MA, Dickson DW, Hewlett TE, Zuckerman BM (1997) Effects of monoclonal antibodies, cationized ferritin, and other organic molecules on adhesion of *Pasteuria penetrans* endospores to *Meloidogyne incognita*. J Nematol 29:556–564

Fould S, Dieng AL, Davies KG, Normand P, Matielle T (2001) Immunological quantification of the nematode parasitic bacterium *Pasteuria penetrans* in soil. FEMS Microbiol Ecol 37:187–195

Giannakou IO, Pembroke B, Gowen SR Davies KG (1997) Effects of long-term storage and above-normal temperatures on spore adhesion of *Pasteuria penetrans* and infection of the root-knot nematode *Meloidogyne javanica*. Nematologica 43:185–192

Giblin-Davis RM, McDaniel LL, Bilz FG (1990) Isolates of the *Pasteuria penetrans* group from phytoparasitic nematodes in bermudagrass turf. J Nematol 22:750–762

Giblin-Davis RM, Williams DS, Bekal S, Dickson DW, Brito JA, Becker JO, Preston JF (2003) '*Candidatus* Pasteuria usgae' sp.nov., an obligate endoparasite of the phytoparasitic nematode, *Belonolaimus longicaudatus*. Int J Syst Evol Microbiol 53:197–200

Hatz B, Dickson DW (1992) Effect of temperature on attachment, development, and interactions of *Pasteuria penetrans* on *Meloidogyne arenaria*. J Nematol 24:512–521

Hewlett TE, Dickson DW (1994) Endospore attachment specificity of *Pasteuria penetrans* from a peanut field in Florida. J Nematol 26:103–104 (Abstr)

Kaplan DT (1994) Partial characterisation of a *Pasteuria* sp. attacking the citrus nematode, *Tylenchulus semipenetrans*, in Florida. Fundam Appl Nematol 17:509–512

Kariuki GM, Dickson DW (2007) Transfer and development of *Pasteuria penetrans*. J Nematol 39:55–61

Kariuki GM, Brito JA, Dickson DW (2006) Effects of *Pasteuria penetrans* endospores rate of attachment on root penetration and fecundity of *Meloidogyne arenaria* race 1. Nematropica 36:261–267

Kerry BR (1988) Fungal parasites of cyst nematodes. Agric Ecol Environ 24:293–305

Mankau R (1975) *Bacillus penetrans* n. comb. causing a virulent disease of plant parasitic nemtodes. J Invertebr Pathol 26:333–339

Mankau R, Prasad N (1972) Possibilities and problems in the use of a sporozoan endoparasite for biological control of plant-parasitic nematodes. Nematropica 2:7–8

Melki KC, Giannokou IO, Pembroke B, Gowen SR (1998) The cumulative build up of *Pasteuria penetrans* spores in root-knot nematodes infested soil and the effect of soil applied fungicide on its infectivity. Fundam Appl Nematol 21:679–683

Metchnikoff ME (1888) *Pasteuria ramose* un representant des bacteries a division longitudinale. Annales de L'Institut Pasteur (Paris) 2:165–170

Mohan S, Fould S, Davies KG (2001) The interaction between the gelatin-binding domain of fibronectin and the attachment of *Pasteuria penetrans* endospores to nematode cuticle. Parasitology 123:271–276

Mouton L, Ebert D (2008) Variable number of tandem repeats analysis of genetic diversity in *Pasteuria ramosa*. Curr Microbiol 56:447–452

Mouton L, Nong G, Preston JF, Ebert D (2007) Variable-number tandem repeats as molecular markers for biotypes for *Pasteuria ramosa* in *Daphnia* spp. Appl Environ Microbiol 73:3715–3718

Nannipieri C, Ascher J, Ceccherini MT, Landi L, Pietramellara G, Renella G (2003) Microbial diversity and soil functions. Eur J Soil Sci 54:655–670

Noel GR, Stanger BA (1994) First report of *Pasteuria* sp.attacking *Heterodera glycines* in North America. Suppl J Nematol 26:612–615

Nong G, Chow V, Schmidt LM, Dickson DW, Preston JF (2007) Multiple strand displacement and identification of single nucleotide polymorphisms as markers of genetic variation of *Pasteuria penetrans* biotypes infecting biotypes infecting root-knot nematodes. FEMS Microbiol Ecol 61:327–336

Oka Y, Koltai H, Bar-Eyal M, Mor M, Sharon E, Chet I, Spiegel Y (2000) New strategies for the control of plant-parasitic nematodes. Pest Manag Sci 56:983–988

Oostendorp M, Dickson DW, Mitchell DJ (1990) Host range and ecology of isolates of *Pasteuria* spp. from the southeastern United States. J Nematol 22:525–531

Pan C, Lin J, Ni Z, Wang S (1993) Study on the pathogenic bacteria parsitizing root-knot nematodes discovered in China and their application to biological control. Acta Microbiol Sin 33:313–316

Persidis A, Lay JG, Manousis T, Bishop AH, Ellar DJ (1991) Characterization of potential adhesions of the bacterium *Pasteuria penetrans*, and of putative receptors on the cuticle of *Meloidogyne incognita*, a nematode host. J Cell Sci 100:613–622

Preston JF, Dickson DW, Maruniak JE, Nong G, Brito JA, Schmidt LM, Giblin-Davis RM (2003) *Pasteuria* spp.: systematics and phylogeny of these bacterial parasites of phytopathogenic nematodes. J Nematol 35:198–207

Riesenfeld CS, Schloss PD, Handelsman J (2004) Metagenomics: genomic analysis of microbial communities. Ann Rev Genet 38:525–552

Sayre RM, Starr MP (1985) *Pasteuria penetrans* (ex Thorne 1940) nom. rev. comb. n., sp., n., a mycelial and endospores-forming bacterium parasitic in plant-parasitic nematodes. Proc Helminthol Soc Washington 52:149–165

Sayre RM, Wergin WP (1977) Bacterial parasite of a plant nematode: morphology and ultrastructure. J Bacteriol 129:1091–1101

Sayre RM, Starr MP, Golden MA, Wergin WP, Endo BY (1988) Comparison of *Pasteuria penetrans* from *Meloidogyne incognita* with a related mycelial and endospore-forming bacterial parasite from *Pratylenchus brachyurus*. Proc Helminthol Washington 55:28–49

Sayre RM, Wergin WP, Schmidt JM, Starr MP (1991) *Pasteuria nishizawae* sp. nov., a mycelial and endospore-forming bacterium parasitic on cyst nematodes of genera *Heterodera* and *Globodera*. Res Microbiol 142:551–564

Schmidt LM, Preston JF, Rice JD, Hewlett TE (2003) Environmental quantification of *Pasteuria penetrans* endospores using *in situ* antien extraction and immunodetection with monoclonal antibody. FEMS Microbiol Ecol 44:17–26

Schmidt LM, Preston JF, Nong G, Dickson DW, Aldrich HC (2004) Detection of *Pasteuria penetrans* infection in *Meloidogyne arenaria* race 1 *in planta* by polymerase chain reaction. FEMS Microbiol Ecol 48:457–464

Schmidt LM, Mouton L, Nong G, Ebert D, Preston JF (2008) Genetic and immunological comparison of the cladoceran parasite *Pasteuria ramosa* with the nematode parasite *Pasteuria penetrans*. Appl Environ Microbiol 74:259–264

Sharma SB, Davies KG (1996) Characterisation of *Pasteuria* isolated from *Heterodera cajani* using morphology, pathology and serology of endospores. Syst Appl Microbiol 19:106–112

Sharma R, Ranjan R, Kapardar RK, Grover A (2005) Unculturable bacterial diversity: an untapped resource. Curr Sci 89:72–77

Siddiqui ZA, Mahmood I (1999) Role of bacteria in the management of plant parasitic nematodes: a review. Bioresour Technol 69:167–179

Starr MP, Sayre RM (1988) *Pasteuria thornei* sp. nov. and *Pasteuria penetrans* sensu stricto emend., mycelial and endospore-forming bacteria parasitic, respectively, on plant-parasitic nematodes of the genera *Pratylenchus* and *Meloidogyne*. Ann Inst Pasteur Microbiol 139:11–31

Stirling GR (1984) Biological control of *Meloidogyne javanica* with *Bacillus penetrans*. Phytopathology 74:55–60

Stirling GR, Bird AF, Cakurs AB (1986) Attachment of *Pasteuria penetrans* spores to the cuticle of root knot nematodes. Rev Nematol 9:251–260

Sturhan D (1988) New host and geographical records of nematode-parasitic bacteria on the *Pasteuria penetrans* group. Nematologica 34:35–356

Sturhan D, Winkelheide R, Sayre RM, Wergin WP (1994) Light and electron microscopical studies of the life cycle and developmental stages of a *Pasteuria* isolate parasitizing the pea cyst nematode, *Heterodera goettingiana*. Fundam Appl Nematol 17:29–42

Sturhan D, Shutova TS, Akimov VN, Subbotin SA (2005) Occurrence, hosts, morphology, and molecular characterization of *Pasteuria* bacteria parasitic to nematodes in the family Plectidae. J Invertbr Pathol 88:17–26

Thorne G (1940) *Duboscquia penetrans* n. sp. (Sporozoa, Microsporidia Nosematidae), a parasite of the nematode *Pratylenchus pratensis* (de Man) Filipjev. Proc Helminthol Soc Washington 7:51–53

Tian B, Yang J, Zang K-Q (2007) Bacteria in the biological control of plant-parasitic nematodes: populations, mechanisms of action, and future prospects. FEMS Microbiol Ecol 61:197–213

Trotter JR, Bishop AH (2003) Phylogenetic analysis and confirmation of the endospore-forming nature of *Pasteuria penetrans* based in the *spo*0A gene. FEMS Microbiol Lett 225:249–256

Trotter JR, Darban DA, Gowen SR, Bishop AH, Pembroke B (2004) The isolation of a single spore isolate of *Pasteuria penetrans* and its pathogenicity on *Meloidogyne javanica*. Nematology 6:463–472

Tzortzakakis EA, Gowen SR (1994) Evaluation of *Pasteuria penetrans* alone and in combination with oxamyl, plant resistance and solarization from control of *Meloidogyne* spp. on vegetables grown in greenhouses in Crete. Crop Prot 13:455–462

Vaid A, Bishop AH, Davies KG (2002) The polypeptide components of the parasporal fibres of *Pasteuria penetrans*. World J Microbiol Biotechnol 18:151–157

Waterman JT, Bird DM, Opperman CH (2006) A method for isolation of *Pasteuria penetrans* endospores for bioassay and genomic studies. J Nematol 38:165–167

Westphal A, Becker JO (1999) Transfer of biological soil suppressiveness against *Heterodera schachtii*. Phytopathology 90:401–406

Williams AR, Stirling GR, Hayward AC, Perry J (1989) Properties and attempted culture of *Pasteuria penetrans*, a bacterial parasite of root-knot nematode (*Meloidogyne javanica*). J Appl Bacteriol 67:145–156

Winkelheide R, Sturhan D (1996) Studies on host specificity of *Pasteuria* isolate from *Heterodera goettingiana*. Mitteilungen der Biologischen Bundesanstalt für Land-Forstwirtschaft 317:46–53

Chapter 10
Aerobic Endospore-forming Bacteria and Soil Invertebrates

Helmut König

10.1 Introduction

The soil is not only an important natural habitat but also a constantly changing environment. Changes are caused by natural environmental factors as well as by microbial, human and animal activities. The microbial community in soil plays a major part in the degradation of organic material (Coleman and Crossley 1996; König and Varma 2005). It represents the first step of the soil food web, contributes to soil fertility, and especially, the degradation of lignin and xenobiotics is an important activity of microbes. Investigations have included both those organisms that are mainly free-living, and the particle-bound microorganisms. In addition, soil invertebrates harbour a great variety of intestinal microbes, which have a significant impact on the recycling of organic compounds. The roles of intestinal microbes occurring in the gut systems of soil animals have been studied less intensively than those of the free-living and the particle-bound microorganisms.

Important soil litter decomposers are the lumbricines (earthworms), the diplopods (millipedes), the isopods (woodlice) and dipteran larvae, as well as termites in subtropical and tropical regions. Only a small portion of the intestinal microbes of soil invertebrates has been screened axenically. Pioneering investigations into the distribution of microbes in the intestines of invertebrates were performed by Paul Buchner (1953). With the aid of molecular methods, a rough estimate of the total population of these organisms, and the as-yet uncultured microbes special to invertebrate gut systems, was obtained in recent decades. The microbiotas of collembola (springtails), earthworms, nematodes, isopods, millipedes and termites have, especially, been studied in more detail. Aerobic endospore-formers represent an important portion of the intestinal microbiota of soil invertebrates (König 2006), and this chapter summarizes our scarce knowledge about this group of bacteria.

H. König
Institut für Mikrobiologie und Weinforschung (Institute of Microbiology and Wine Research), Johannes Gutenberg Universität (Johannes Gutenberg University), Becherweg 15, 55128 Mainz, Germany
e-mail: hkoenig@uni-mainz.de

N.A. Logan and P. De Vos (eds.), *Endospore-forming Soil Bacteria*, Soil Biology 27,
DOI 10.1007/978-3-642-19577-8_10, © Springer-Verlag Berlin Heidelberg 2011

10.2 Aerobic Endospore-Formers in the Guts of Termites

Termites occur over 68% of the earth's surface. The main areas of distribution are the tropical and subtropical regions (Krishna 1970), but they can occur between the latitudes 47° north and 47° south. With an estimated population of 2.4×10^{17} individuals (Zimmerman et al. 1982) termites are among the most successful animals on earth. This fact is underlined by investigations of Fittkau and Klinge (1973) who found that termites (*Isoptera*) and ants (*Formicidae*) represented 80% of the individual insects and 30% of the total animal biomass near Manaus in Brazil. The main fermentation chamber (paunch) of an individual animal is very small (e.g., for *Mastotermes darwiniensis* it is about 10 μl; Noirot 1995; Noirot and Noirot-Timotheé 1969; Brune 1998; Berchtold et al. 1999), but because of the high number of individuals they play a major role in the degradation of organic material, especially in the recycling of lignocellulose (Wood and Sands 1978). The transit time of the food through the intestinal tract is only about 24 h (Breznak 1984).

The termite gut represents an anaerobic gradient system, which exhibits distinct spatial gradients of pH, oxygen or hydrogen (Noirot 1995; Noirot and Noirot-Timotheé 1969; Brune and Kühl 1996). Several hundred microbial clones/strains have been identified by 16S rRNA gene sequence analysis, or have been obtained in pure culture from termite guts (Fig. 10.1; Brune 1998; Breznak and Brune 1994; König et al. 2002, 2005; Ohkuma 2003; Ohkuma et al. 2005), but the most of the microbial species of more than 2,000 species of termite still remain uninvestigated.

Several bacilli from the termite gut have been described (Table 10.1; Kuhnigk 1996; Kuhnigk and König 1997; Kuhnigk et al. 1994, 1995; König et al. 2005). Enzyme activities for the degradation of polysaccharides and aromatic compounds have been detected. Because of the high volume-to-surface ratio the gut is well supplied by oxygen by the trachea; in contrast to the rumen, where strictly anaerobic cellulolytic bacteria belonging to the genera *Ruminococcus, Butyrivibrio* and *Bacteroides* are present, the cellulolytic bacteria in the termite gut are facultatively anaerobic or microaerophilic. Aerobic endospore-forming species are predominant, with populations of up to 10^7 ml^{-1} gut contents (Wenzel et al. 2002).

10.3 Aerobic Endospore-Formers in the Guts of Earthworms

While termites occur as degraders of organic soil litter mainly in tropical and subtropical regions, earthworms contribute significantly to the recycling of organic matter in soils of more temperate zones (Edwards and Bohlen 1996). Most isolates of the gut microbiota were affiliated with the class *Gammaproteobacteria* of the phylum *Proteobacteria* or with the Gram-positive phylum *Firmicutes*. The identified species of *Bacillus* and *Paenibacillus* belong to the latter group (Table 10.2). That the gut contents are habitats of these organisms were demonstrated by enhancement of spore germination from *Bacillus* species (Fischer et al. 1997) and

	Substrate	I. Stage **HydrolyticMicroorganisms**	II. Stage **Oxidative/Fermentative Microorganisms**	III. Stage **Acetogenic, Methanogenic Sulfate-reducing Bacteria**	
Wood **Mechanical degradation to µm- particles by termites**	**Cellulose** (34% - 62%)	**Cellulase** Flagellates: *Trichomitopsis termopsidis, Trichonympha sphaerica*	**β-Glucosidase**		
		Bacteria: *Alcaligenes, Azospirillum,* ***Bacillus, Brevibacillus****, Cellulomonas*-related spp., *Clavibacter, Clostridium, Corynebacterium, Klebsiella, Kocuria,* ***Paenibacillus****, Microbacterium, Micromonospora, Nocardioforme,* Rhizobia, *Ochrobactrum, Paenibacillus, Sphingomonas, Spirosoma*-related spp., *Streptomyces*	*Actinobacteria, Arthrobacter, Aureobacterium,* ***Bacillus, Brevibacillus****, Citrobacter, Enterobacter, Enterococcus, Klebsiella, Lactococcus, Nocardia, Ochrobactrum,* ***Paenibacillus****, Pseudomonas,* Rhizobia, *Rhodococcus, Serratia, Spirochaeta, Streptomyces, Treponema*		**Food source of termites:** **Acetate, Propionate, Butyrate, Microbial cells**
		Yeasts (*Cryptococcus, Filobasidium*)	Yeasts		
	Hemicellulose (14% - 32%)	**1.4-β- Xylanase** **1.3-β-Galactanase**	**β-D-Galactosidase** **α -L-Arabinosidase** **β-L-Xylosidase**	**$2 H_2 + O_2 \Rightarrow H_2O$** **Lactate ⇒ Acetate + CO_2** *Desulfovibrio*	**Excreted end products:** **CO_2, H_2, CH_4**
		Actinobacteria, Acinetobacter, ***Bacillus****, Flavobacterium*- related spp., ***Lysinibacillus****, Ochrobactrum,* ***Paenibacillus****, Pseudomonas, Streptomyces*	*Actinobacteria, Acinetobacter, Arthrobacter, Aureobacterium,* ***Bacillus****, Burkholderia, Citrobacter, Cellulomonas, Enterobacter, Escherichia, Klebsiella,* ***Lysinibacillus****, Nocardia, Ochrobactrum, Pseudomonas,* ***Paenibacillus****, Rhizobium, Serratia, Spirochaeta, Staphylococcus, Streptomyces,*	**$4 H_2 + 2CO_2 \Rightarrow$ Acetate + $2 H_2O$** *Acetonema, Clostridium, Sporomusa,* Treponema **$4 H_2 + CO_2 \Rightarrow CH_4 + 2 H_2O$** *Methanobrevibacter*	**Microbial cells**
		Yeasts (*Candida, Debaryomyces, Pichia, Sporothrix*)	Yeasts		
	Aromatic compounds (5 %)		*Acinetobacter, Arthrobacter, Alcaligenes, Aureobacterium,* ***Bacillus****, Burkholderia, Citrobacter, Comamonas, Enterobacter, Klebsiella, Listeria,* ***Lysinibacillus****, Nocardia, Ochrobactrum, Pseudomonas,* Rhizobia,		
	Lignin (18 -39%)	*Streptomyces*			**Excreted: Modified lignin**
N_2-fixation			*Enterobacter, Desulfovibrio,* Rhizobia, *Klebsiella, Treponema, Citrobacter*		

(modified from König et al. 2005); genera of bacilli in bold.

Fig. 10.1 Intestinal isolates involved in the different stages of lignocellulose degradation in the termite gut

Table 10.1 Aerobic endospore-formers isolated from the guts of termites

Species	Termite	Reference
Bacillus (now *Brevibacillus*[a]) *brevis*[b]	*Anacanthotermes ahngerianus* (h), *Zootermopsis angusticollis* (l)	Krasil'nikov and Satdykov (1969), Wenzel et al. (2002)
Bacillus cereus[c, d]	*Anacanthotermes ahngerianus* (h), *Coptotermes curvignathus* (l), *Cryptotermes brevis* (l), *Kalotermes flavicollis* (l), *Nasutitermes nigriceps* (h), *Neotermes castaneus* (l), *Reticulitermes hesperus* (l), *Reticulitermes santonensis* (l)	Krasil'nikov and Satdykov (1969), Kuhnigk and König (1997), Margulis et al. (1998), Ramin et al. (2009), Thayer (1976), Schäfer et al. (1996)
Bacillus cereus-related isolate[b]	*Zootermopsis angusticollis* (l)	Wenzel et al. (2002)
Bacillus circulans-related isolate[b]	*Zootermopsis angusticollis* (l)	Wenzel et al. (2002)
Bacillus coagulans[c]	*Mastotermes darwiniensis* (l)	Schäfer et al. (1996)
Bacillus firmus[d]	*Reticulitermes santonensis* (l)	Kuhnigk (1996)
Bacillus licheniformis[c, d]	*Reticulitermes santonensis* (l)	Kuhnigk and König (1997), Kuhnigk et al. (1994), Schäfer et al. (1996)
Bacillus megaterium[b]	*Anacanthotermes ahngerianus* (h), *Mastotermes darwiniensis* (l), *Zootermopsis angusticollis* (l)	Krasil'nikov and Satdykov (1969), Kuhnigk (1996), Mannesmann and Piechowski (1989), Wenzel et al. (2002)
Bacillus mycoides	*Anacanthotermes ahngerianus* (h)	Krasil'nikov and Satdykov (1969)
Bacillus oleronius	*Reticulitermes santonensis* (l)	Kuhnigk et al. (1995)
Bacillus sp.	*Schedorhinotermes intermedius* (l), *Coptotermes acinaciformis* (l), *Coptotermes formosanus* (l), *Heterotermes indicola* (l)	Eutick et al. (1978), Kuhnigk (1996), Mannesmann and Piechowski (1989)
Bacillus (now *Lysinibacillus*[a]) *sphaericus*[c, d]	*Odontotermes distans* (h), *Odontotermes obesus* (h), *Reticulitermes santonensis* (l), *Zootermopsis angusticollis* (l)	Kuhnigk (1996), Kuhnigk and König (1997), Schäfer et al. (1996)
Bacillus subtilis[d]	*Anacanthotermes ahngerianus* (h), *Reticulitermes santonensis* (l)	Krasil'nikov and Satdykov (1969), Kuhnigk and König (1997), Schäfer et al. (1996)
Paenibacillus macerans[c]	*Mastotermes darwiniensis* (l)	Schäfer et al. (1996)
Paenibacillus sp.[b]	*Zootermopsis angusticollis* (l)	Wenzel et al. (2002)

l lower termite; *h* higher termite
[a]Valid new genus designation
[b]Cellulolytic activity was shown
[c]Hemicellulolytic activity was shown
[d]Capability to degrade aromatic compounds was shown

Table 10.2 Aerobic endospore-formers isolated from the earthworm *Aporrectodea caliginosa*

Closest relative	Reduction of nitrate	Reduction of nitrite	% of N from nitrate or nitrite recovered in N_2O
Bacillus mycoides	+	0	<0.5
Bacillus mycoides	+	0	7
Paenibacillus sp. strain P51-3	+	+	6
Paenibacillus sp. strain P51-3	+	+	3
Paenibacillus sp. strain 61724	+	+	11
"Paenibacillus burgondia" B2	+	+	8
Paenibacillus borealis KK19	–	+	10
Paenibacillus amylolyticus NRRL B-290T	–	+	<0.5

From Ihssen et al. (2003), Gebhardt et al. (2002).

the detection of increased numbers of *Bacillus* cells compared to soil samples by fluorescence techniques (Fischer et al. 1995; Schönholzer et al. 2002). Interestingly, the emanation of nitrous oxide (N_2O) by earthworms has been attributed to their gut microbiota (Ihssen et al. 2003; Drake et al. 2005).

Although the worm gut is a habitat of several bacteria, Khomyakov et al. (2007) showed that the digestive fluid of the earthworm *Aporrectodea caliginosa* inhibited the growth or germination of certain bacteria, spores and fungal hyphae; a strain of *Bacillus megaterium* was found to be susceptible. The surface excreta of the *A. caliginosa* suppressed the dehydrogenase activity of various taxa of soil bacteria, including strains of *Bacillus subtilis*, *B. mojavensis* and *Paenibacillus* sp., as well as of *Bacillus thuringiensis*. In addition, some strains exhibited no reaction or showed stimulation of growth. These experiments demonstrated that earthworms have an effect on the soil microbiota by direct stimulation or suppression of growth or spore germination (Oleynik and Byzov 2008).

Seven aerobic endospore-forming species (*Bacillus* (now *Brevibacillus*) *brevis*, *Bacillus insolitus*, *B. megaterium*, *B.* (now *Paenibacillus*) *pabuli*, *B.* (now *Sporosarcina*) *pasteurii*, *B.* (now *Lysinibacillus*) *sphaericus* and *B. thuringensis*) have been isolated from the gut wall material of *Ochochaeta borincana*, an annelid described only from Puerto Rico (Alonso et al. 1999; Méndez et al. 2003; Valle-Molinares et al. 2007).

10.4 Aerobic Endospore-Formers in the Guts of Springtails

The springtails (*Collembola*) are small soil hexapods possessing a size of only between 0.2 mm and 1 cm. With 7,500 species and a high number of estimated individuals, springtails are not only among the most abundant arthropods on earth (Hopkins 1997; Tebbe et al. 2005), but also with an existence of about 400×10^6 years they are probably the oldest hexapods.

Collembola live mainly in soil or litter, and feed on live or dead organic material from plants as well as on nematodes or fungi (Chen et al. 1995, 1996; Thimm and Larink 1995). Their guts have volumes of only some nanolitres (Thimm et al. 1998).

In the case of *Folsomia candida*, 4×10^{11} cfu bacterial cells per g faeces were determined (Borkott and Insam 1990). Because of the high number of individuals, they have important roles in the degradation of organic material in soil; and micro-arthropods such as collembola and mites enhance the destruction of organic material.

Several collembolan gut bacteria have been identified or obtained in pure culture. Among these were the two chitin-degraders *Stenotrophomonas maltophilia* and *Curtobacterium* sp., as well as a chitin-hydrolysing *Bacillus* sp. (Hale 1967). Thimm et al. (1998) isolated 45 different bacterial strains from the gut of *Folsomia candida*; the isolates belonged to the species *Erwinia amylovora*, *Staphylococcus captitis* and *Pantoea agglomerans,* and the ascomycete *Acremonium charticola* was also found; interestingly, Tochot et al. (1982) found fungal mycelia in the gut contents of *Folsomia candida.* Additional cultured strains could be assigned to the genera *Ochrobactrum*, *Alcaligenes*, *Comomonas*, *Pseudomonas* and *Paracoccus* (Hoffmann et al. 1998). A cultivation-independent molecular approach led to the identification of clones from *Paracoccus denitrificans*, *Stenotrophomonas maltophilia* and *Bacillus weihenstephaniensis* (Jensen et al. 2003).

10.5 Aerobic Endospore-Formers in the Guts of *Isopoda*

The isopods live as cosmopolitans in diverse habitats ranging from marine environments to dry deserts. More than one-third of the described isopodan species belong to the terrestrial woodlice (Schmalfuss 2003). The guts of woodlice have been well studied with respect to their morphologies and nutritional conditions (Hames and Hopkin 1989; Zimmer 2002). *Porcellio scaber* and *Oniscus asellus* are among the most studied species with respect to the compositions of their microbiota (Hopkin 1989; Drobne 1997; Kostanjšek et al. 2002, 2004, 2005).

The common sow bug, *Porcellio scaber*, is a terrestrial arthropod, feeding on decaying leaf litter, wood and vegetable matter, substances that are rich in cellulose and other polysaccharides (Zimmer and Brune 2005). The oniscidean digestive system (Hames and Hopkin 1989; Štrus et al. 1995; Molnar et al. 1998; Zimmer 2002) contains an autochtonous flora (Kostanjšek et al. 2002, 2004, 2005). Two *Bacillus cereus* strains and another *Bacillus* strain not identified to the species level were isolated from the gut of *P. scaber* (Kostanjšek et al. 2005). Seventy percent of the investigated animals of the sow bug harboured *B. cereus* in their gut (Swiecicka and Mahillon 2006; Swiecicka 2008), and this indicates a symbiotic relationship. Margulis et al. (1998) also found *B. cereus* cells that had lost their flagella and grew filamentously in long chains (*Arthromitus* of Leidy, 1849) under certain circumstances; the microorganisms were involved in the degradation of cellulosic material (Zimmer and Topp 1998).

10.6 Aerobic Endospore-Formers in the Guts of Millipedes

Soil millipedes (*Diplopoda*) possess specific gut microbiotas that include *Gammaproteobacteria*, *Actinobacteria* and yeasts (Byzov 2005). In addition, a variety of bacterial morphotypes and yeasts are attached to the gut walls (Byzov 2005).

The foreguts of millipedes are poorly populated by microorganisms, but the cuticle-lined hindgut bears both flat cuticular surfaces and ornaments, which are colonized by microbes (Crawford et al. 1983); Bignell (1984) found actinomycete-like filaments attached to cuticular surfaces of *Cylindroiulus* sp. The inner surfaces of the intestinal walls of the millipedes *Chromatoiulus rossicus* and *Glomeris connexa* are colonized by bacteria of different morphotypes. Ineson and Anderson (1985) found 2.8×10^9 cfu/g dry weight in the whole gut of the diplopod *Glomeris marginata*. The numbers of bacteria isolated from the peritrophic membrane of *G. connexa* were similar to those in the gut tissue (ca. 10^7 cfu/g dry membrane; Byzov et al. 1996). Yeast cells colonize mainly the hindguts of *Glomeris connexa*, *Leptoiulus polonicus* and *Megaphyllum projectum* (Byzov et al. 1993b).

Most of the bacterial isolates from millipede guts and gut contents belong to the gamma subclass of Proteobacteria and the *Actinobacteria*. The dominant bacterial microbiota belong to the family of facultative anaerobes, the *Enterobacteriaceae*, including the genera *Klebsiella, Enterobacter, Plesiomonas, Salmonella, Erwinia* and *Escherichia* and *Vibrio* from the *Vibrionaceae* (Byzov 2005). Several aerobic endospore-forming isolates have also been isolated (Table 10.3). From the diplopod *Glomeris* sp. a *B. subtilis* isolate was obtained (Gebhardt et al. 2002), which produced the antifungal antibiotic bacillomycin D.

Furthermore, free methanogenic prokaryotes and ciliates (*Nyctotherus* type) with intracellular endosymbiotic methanogens have been detected microscopically by their characteristic autofluorescence in the tropical diplopods *Chicobolus* sp., *Orthoporus* sp., *Rhapidostreptus virgator* and two unidentified species (Hackstein and Stumm 1994).

Most yeast strains isolated from millipede guts and gut contents are ascomycetes. In *Pachyiulus flavipes*, the predominating species were *Debaryomyces hansenii, Torulaspora delbrueckii, Zygowilliopsis californicus* (=*Williopsis californica*) and *Pichia membranaefaciens* (Byzov et al. 1993a).

Table 10.3 Aerobic endospore-formers isolated from, or detected in, the guts and fresh faeces of millipedes

Group/Genus/Species	Host	Source
Bacillus sp.	*Schizophyllum sabulosum* var. *rubripes*	Gut contents
Bacillus spp.	*Chromatoiulus projectus, Cylindroiulus boleti*	Gut contents
Bacillus spp.	*Pachyiulus flavipes*	Intestinal walls
Bacillus spp.	*Pachyiulus flavipes*	Gut contents
Bacillus megaterium, B. cereus, B. licheniformis	*Glomeris connexa, Chromatoiulus rossicus*	Gut contents
Bacillus sp.	*Ommatoiulus sabulosus*	Gut contents
Bacillus (now *Brevibacillus*[a]) *brevis*	*Glomeris* sp.	Hindgut
Bacillus subtilis		

From Byzov (2005), Gebhardt et al. (2002)
[a]Valid new genus designation

10.7 Conclusions

Convincing evidence now exists that distinctive autochthonous communities live in the guts of soil invertebrates, and that these contribute to the degradation of recalcitrant biological materials such as chitin and lignocellulose. Intestinal aerobic endospore-formers, especially, have significant impacts in the degradation of polymeric soil material (Fig. 10.1). They possess cellulolytic and hemicellulolytic activities and some have the potential to degrade aromatic compounds if the oxygen supply is sufficient (Kuhnigk 1996; Kuhnigk and König 1997; Kuhnigk et al. 1994, 1995; Wenzel et al. 2002). The cellulolytic bacteria in the termite gut are facultatively anaerobic or microaerophilic; aerobic endospore-forming species are predominant, with counts of up to 10^7 ml^{-1} gut contents (Wenzel et al. 2002), and they play major parts in the first and second steps (Fig. 10.1) of the degradation of polymeric material in the guts of soil invertebrates.

References

Alonso A, Borges S, Betancourt C (1999) Mycotic flora of the intestinal tract and soil inhabited by *Onychocaheta borincana* (Oligochaeta: Glossoscolecidae). Pedobiologia 43:1–3

Berchtold M, Chatzinotas A, Schönhuber W, Brune A, Amann R, Hahn D, König H (1999) Differential enumeration and in situ localization of microorganisms in the hindgut of the lower termite *Mastotermes darwiniensis* by hybridization with rRNA-targeted probes. Arch Microbiol 172:407–416

Bignell DE (1984) The arthropod gut as an environment for microorganisms. In: Anderson JM (ed) Invertebrate– microbial interaction. Cambridge University Press, Cambridge, pp 205–227

Borkott H, Insam H (1990) Symbiosis with bacteria enhances the use of chitin by the springtail, *Folsomia candida* (Collembola). Biol Fertil Soils 9:126–129

Breznak JA (1984) Biochemical aspects of symbiosis between termites and their intestinal microbiota. In: Anderson JM (ed) Invertebrate – microbial interaction. Cambridge Unversity Press, Cambridge, pp 173–203

Breznak JA, Brune A (1994) Role of microorganisms in the digestion of lignocellulose by termites. Annu Rev Entomol 39:453–487

Brune A (1998) Termite guts: the world's smallest bioreactors. TIBTECH 16:16–21

Brune A, Kühl M (1996) pH profiles of extremely alkaline hindguts of soil-feeding termites (Isoptera: Termitidae) determined with microelectrodes. J Insect Physiol 42:1121–1127

Buchner P (1953) Endosymbiose der Tiere mit pflanzlichen Mikroorganismen. Verlag Birkhäuser, Stuttgart

Byzov BA (2005) Intestinal microbiota of millipedes. In: König H, Varma A (eds) Intestinal microorganisms of termites and other invertebrates. Springer, Heidelberg, pp 89–114

Byzov BA, Thanh VT, Bab'eva IP (1993a) Yeasts associated with soil invertebrates. Biol Fertil Soils 16:183–187

Byzov BA, Thanh VN, Bab'eva IP (1993b) Interrelationships between yeasts and soil diplopods. Soil Biol Biochem 25:1119–1126

Byzov BA, Chernjakovskaya TF, Zenova GM, Dobrovolskaya TG (1996) Bacterial communities associated with soil diplopods. Pedobiologia 40:67–79

Chen B, Snider RJ, Snider RM (1995) Food preference and effects of food type on the life history of some soil Collembola. Pedobiologia 39:496–505

Chen B, Snider RJ, Snider RM (1996) Food consumption by collembola from northern Michigan deciduous forest. Pedobiologia 40:149–161

Coleman DC, Crossley DA Jr (1996) Fundamentals of soil ecology. Academic, San Diego

Crawford CS, Minion GP, Bayers MD (1983) Intima morphology, bacterial morphotypes, and effects of annual molt on microflora in the hindgut of the desert millipede *Orthoporus ornatus* (Girard) (Diplopoda: Spirostreptidae). Int J Insect Morphol Embryol 12:301–312

Drake HL, Schramm A, Horn M (2005) Earthworm gut microbial biomes: their importance to soil microorganisms, denitrification, and the terrestrial production of the greenhouse gas N_2O. In: König H, Varma A (eds) Intestinal microorganisms of termites and other invertebrates. Springer, Heidelberg, pp 65–87

Drobne D (1997) Terrestrial isopods – a good choice for toxicity testing of pollutants in the terrestrial environment. Environ Toxicol Chem 16:1159–1164

Edwards CA, Bohlen PJ (1996) Biology and ecology of earthworms, 3rd edn. Chapman & Hall, London

Eutick ML, O'Brien RW, Slaytor M (1978) Bacteria from the gut of Australian termites. Appl Environ Microbiol 35:823–828

Fischer K, Hahn D, Amann RI, Daniel O, Zeyer J (1995) In-situ analysis of the bacterial community in the gut of the earthworm *Lumbricus terrestris* L. by whole-cell hybridization. Can J Microbiol 41:666–673

Fischer K, Hahn D, Hönerlage W, Zeyer J (1997) Effect of passage through the gut of the earthworm *Lumbricus terrestris* L on *Bacillus megaterium* studied by whole cell hybridization. Soil Biol Biochem 29:1149–1152

Fittkau J, Klinge H (1973) On biomass and trophic structure of the Central Amazonian rain forest ecosystem. Biotropica 5:2–14

Gebhardt K, Schimana J, Müller J, Fiedler HP, Kallenborn HG, Holzenkämpfer M, Krastel P, Zeeck A, Vater J, Höltzel A, Schmid DG, Rheinheimer J, Dettner K (2002) Screening for biologically active metabolites with endosymbiotic bacilli isolated from arthropods. FEMS Microbiol Lett 217:199–205

Hackstein JHP, Stumm CK (1994) Methane production in terrestrial arthropods. Proc Natl Acad Sci USA 91:5441–5445

Hale WG (1967) Collembola. In: Burges A, Raw R (eds) Soil biology. Academic, London, pp 397–411

Hames CAC, Hopkin SP (1989) The structure and function of the digestive system of terrestrial isopods. Zool Lond 217:599–627

Hoffmann A, Thimm T, Dröge M, Moore ERB, Munch JC, Tebbe CC (1998) Intergeneric transfer of conjugative and mobilizable plasmids harbored by *Escherichia coli* in the gut of the soil microarthropod *Folsomia candida* (Collembola). Appl Environ Microbiol 64:2652–2659

Hopkin SP (1989) Ecophysiology of metals in terrestrial invertebrates. Elsevier Applied Science, Barking

Hopkin SP (1997) Biology of springtails. Insecta: Collembola. Oxford University Press, Oxford

Ihssen J, Horn MA, Matthies C, Gößner A, Schramm A, Drake HL (2003) N_2O-producing microorganisms in the gut of the earthworm *Aporrectodea caliginosa* are indicative of ingested soil bacteria. Appl Environ Microbiol 69:1655–1661

Ineson P, Anderson JM (1985) Aerobically isolated bacteria associated with the gut and faeces of the litter feeding macroarthropods *Oniscus assellus* and *Glomeris marginata*. Soil Biol Biochem 17:843–849

Jensen GB, Hansen BM, Eilenberg J, Mahillon J (2003) The hidden lifestyles of *Bacillus cereus* and relatives. Environ Microbiol 5:631–640

Khomyakov NV, Kharin SA, Nechitailo TY, Golyshin PN, Kurakov AV, Byzov BA, Zvyagintsev DG (2007) Reaction of microorganisms to the digestive fluid of earthworms. Mikrobiologiya 76: 55–65

König H (2006) *Bacillus* species in the intestine of termites and other soil invertebrates. J Appl Microbiol 101:620–627

König H, Varma A (eds) (2005) Intestinal microorganisms of termites and other invertebrates. Springer, Heidelberg

König H, Fröhlich J, Berchtold M, Wenzel M (2002) Diversity and microhabitats of the hindgut flora of termites. Recent Res Dev Microbiol 6:125–156

König H, Fröhlich J, Hertel H (2005) Diversity and lignocellulolytic activities of cultured microorganisms. In: König H, Varma A (eds) Intestinal microorganisms of termites and other invertebrates. Springer, Heidelberg, pp 272–302

Kostanjšek R, Štrus J, Avguštin G (2002) Genetic diversity of bacteria associated with the hindgut of the terrestrial crustacean *Porcellio scaber* (Crustacea: Isopoda). FEMS Microbiol Ecol 40:171–179

Kostanjšek R, Lapanje A, Rupnik M, Štrus J, Drobne D, Avguštin G (2004) Anaerobic bacteria in the gut of terrestrial isopod crustacean *Porcellio scaber*. Folia Microbiol 49:179–182

Kostanjšek R, Štrus J, Lapanje A, Avguštin G, Rupnik M, Drobne D (2005) Intestinal microbiota of terrestrial isopods. In: König H, Varma A (eds) Intestinal microorganisms of termites and other invertebrates. Springer, Heidelberg, pp 115–131

Krasil'nikov NA, Satdykov SI (1969) Estimation of the total bacteria in the intestines of termites. Microbiologiya 38:289–292

Krishna K (1970) Taxonomy, physiology, and distribution of termites. In: Krishna K, Weesner FM (eds) Biology of termites, vol II. Academic, New York, pp 127–152

Kuhnigk T (1996) Bakterien aus dem Termitendarm. Thesis. University, Ulm

Kuhnigk T, König H (1997) Degradation of dimeric lignin model compounds by aerobic bacteria isolated from the hindgut of xylophagous termites. J Basic Microbiol 37:205–211

Kuhnigk T, Borst E, Ritter A, Kämpfer P, Graf A, Hertel H, König H (1994) Degradation of lignin monomers by the hindgut flora of termites. Syst Appl Microbiol 17:76–85

Kuhnigk T, Borst EM, Breunig A, König H, Collins MP, Hutson RA, Kämpfer P (1995) *Bacillus oleronius* sp. nov., a member of the hindgut flora of the termite *Reticulitermes santonensis*. Can J Microbiol 41:699–706

Mannesmann R, Piechowski B (1989) Verteilungsmuster von Gärkammerbakterien einiger Termitenarten. Mat Org 24:161–177

Margulis L, Jorgensen JZ, Dolan S, Kolchinsky R, Rainey FA, Lo S-Ch (1998) The Arthromitus stage of *Bacillus cereus*: intestinal symbionts of animals. Proc Natl Acad Sci USA 95:1236–1241

Méndez R, Borges S, Betancourt C (2003) A microscopical view of the intestine of *Onychochaeta borincana* (Oligochaeta: Glossoscolecidae). Pedobiologia 47:900–903

Molnar L, Pollak E, Fischer E (1998) The distribution of serotoninergic neurons in the central nervous system of *Porcellio scaber* Latr (Isopoda, Porcellionidae). Israel J Zool 44:451–452

Noirot C (1995) The gut of termites (Isoptera). Comparative anatomy, systematics, phylogeny. I. Lower termites. Ann Soc Entomol Fr 31:197–226

Noirot C, Noirot-Timotheé C (1969) The digestive system. In: Krishna K, Weesner FM (eds) Biology of termites, vol I. Academic, New York, pp 49–88

Ohkuma M (2003) Termite symbiotic systems: efficient bio-recycling of lignocellulose. Appl Microbiol Biotechnol 61:1–9

Ohkuma M, Hongoh Y, Kudo KT (2005) Diversity and molecular analyses of yet-uncultivated microorganisms. In: König H, Varma A (eds) Intestinal microorganisms of termites and other invertebrates. Springer, Heidelberg, pp 303–317

Oleynik AS, Byzov BA (2008) Response of bacteria to earthworm surface excreta. Mikrobiologiya 77:854–862

Ramin M, Alimon AR, Abdullah N (2009) Identification of cellulolytic bacteria isolated from the termite *Coptotermes curvignathus* (Holmgren). J Rapid Methods Autom Microbiol 17:103–116

Schäfer A, Konrad R, Kuhnigk T, Kämpfer P, Hertel H, König H (1996) Hemicellulose-degrading bacteria and yeasts from the termite gut. J Appl Bacteriol 80:471–478

Schmalfuss H (2003) World catalog of terrestrial isopods (Isopoda: Oniscidea). Stuttgarter Beitr Naturk, Ser A 654:341

Schönholzer F, Hahn D, Zarda B, Zeyer J (2002) Automated image analysis and in situ hybridization as tools to study bacterial populations in food resources, gut and cast of *Lumbricus terrestris* L. J Microbiol Methods 48:53–68

Štrus J, Drobne D, Ličar P (1995) Comparative anatomy and functional aspects of the digestive system in amphibious and terrestrial isopods (Isopoda: Oniscidea). In: Alikhan MA (ed) Terrestrial isopod biology. AA Baklema, Rotterdam, pp 15–23

Swiecicka I (2008) Natural occurrence of *Bacillus thuringiensis* and *Bacillus cereus* in eukaryotic organisms: a case for symbiosis. Biocontrol Sci Technol 18:221–239

Swiecicka I, Mahillon J (2006) Diversity of commensal *Bacillus cereus* sensu lato isolated from the common sow bug (*Porcellio scaber*, Isopoda). FEMS Microbiol Ecol 56:132–140

Tebbe CC, Czarnetzki AB, Thimm T (2005) Collembola as a habitat for microorganisms. In: König H, Varma A (eds) Intestinal microorganisms of termites and other invertebrates. Springer, Heidelberg, pp 133–153

Thayer DW (1976) Facultative wood-digesting bacteria from the hindgut of the termite *Reticulitermes hesperus*. J Gen Microbiol 95:287–296

Thimm T, Larink O (1995) Grazing preferences of some Collembola for endomycorrhizal fungi. Biol Fertil Soils 19:266–268

Thimm T, Hoffmann A, Borkott H, Munch JC, Tebbe CC (1998) The gut of the soil microarthropod Folsomia candida (Collembola) is a frequently changeable but selective habitat and a vector for microorganisms. Appl Environ Microbiol 64:2660–2669

Tochot F, Kilbertus G, Vannier G (1982) Rôle d'un collembole (*Folsomia candida*) au cours de la dégradation des litières de charme et de chêne, en présence ou en absence d'argile. In: Lebrun P, Andrés HM, De Mets A, Grégoire-Wibo C, Wauthy G (eds) New trends in soil biology. Proceedings of the VIII. International Colloquium of Soil Zoology. Imprimeur Dieu-Brichart, Ottignies-Louvain-laNeuve, pp 269–280

Valle-Molinares R, Borges S, Rios-Velazquez C (2007) Characterization of possible symbionts in *Onychochaeta borincana* (Annelida: Glossoscolecidae). Eur J Soil Biol 43:S14–S18

Wenzel M, Schönig M, Berchtold M, Kämpfer P, König H (2002) Aerobic and facultatively anaerobic cellulolytic bacteria from the gut of the termite *Zootermopsis angusticollis*. J Appl Microbiol 92:32–40

Wood TG, Sands WA (1978) The role of termites in ecosystems. In: Brian JV (ed) Production ecology of ants and termites. Cambridge University Press, Cambridge, pp 245–292

Zimmer M (2002) Nutrition in terrestrial isopods (Isopoda: Oniscidea): an evolutionary-ecological approach. Biol Rev 77:455–493

Zimmer M, Brune A (2005) Physiological properties of the gut lumen of terrestrial isopods (Isopoda: Oniscidea): adaptive to digesting lignocellulose? J Comp Physiol B 175:275–283

Zimmer M, Topp W (1998) Microorganisms and cellulose digestion in the gut of *Porcellio scaber* (Isopoda: Oniscidea). J Chem Ecol 24:1395–1408

Zimmerman PR, Greenberg JP, Wandiga SO, Crutzen PJ (1982) Termites: a potentially large source of atmospheric methane, carbon dioxide, and molecular hydrogen. Science 218:563–565

Chapter 11
Bacillus thuringiensis Diversity in Soil and Phylloplane

Michio Ohba

11.1 Introduction

Bacillus thuringiensis, an aerobic Gram-positive saprophyte, is allocated to the *Bacillus cereus* group together with the three other species, *Bacillus anthracis*, *Bacillus cereus* and *Bacillus mycoides*. It was first isolated early in the last century from sotto-diseased larvae of the silkworm (*Bombyx mori*), an economically important insect for the silk industry in Japan at that time (Ishiwata 1901). Later, Berliner (1915) described *B. thuringiensis*, as a new bacterial species pathogenic for insects, on the basis of a spore-forming bacterium isolated in Germany from larval cadavers of the Mediterranean flour moth, *Ephestia kühniella*. It is now well established that the organism is cosmopolitan and naturally occurring in most regions of the world (Martin and Travers 1989), even in Antarctica (Forsyth and Logan 2000).

B. thuringiensis is differentiated from other related species by the formation of noticeable crystalline parasporal inclusions during sporulation (Logan 2005). Strong and specific insecticidal activity of *B. thuringiensis* is largely due to the oral toxicity of the crystal (Cry) proteins contained in parasporal inclusions, although insecticidal activity is also associated with some other vegetative cell products: VIPs (Vegetative Insecticidal Proteins), thuringiensin and cytotoxic enzymes including phospholipases (Glare and O'Callaghan 2000). This unique property has made *B. thuringiensis* an environmentally safe and ecologically sound microbial agent that is effective for the control of insect pests of agricultural and medical importance (Charles et al. 2000; Glare and O'Callaghan 2000). For the last three decades, extensive efforts have been made to screen for *B. thuringiensis* strains that are more useful for pest control than current commercial strains, from natural populations occurring in various environments: phylloplane and plant

M. Ohba
Graduate School of Bioresource and Bioenvironmental Sciences, Kyushu University, Fukuoka 812-8581, Japan
and
Hakata-eki-mae 4-22-25-805, Hakata-ku, Fukuoka 812-0011, Japan
e-mail: ohba@brs.kyushu-u.ac.jp; michio-ohba1901@m3.gyao.ne.jp

N.A. Logan and P. De Vos (eds.), *Endospore-forming Soil Bacteria*, Soil Biology 27,
DOI 10.1007/978-3-642-19577-8_11,

material, soil of various types, insect and insect-inhabited environments, stored products, food and others (Damgaard 2000). Among the sources explored, soil and phylloplane have proved to be the good reservoirs of *B. thuringiensis* populations with great diversity in several key characteristics.

The efforts have provided novel isolates with unique insecticidal activities specific for dipteran vectors, mosquitoes and blackflies (Goldberg and Margalit 1977) and agriculturally important coleopterans (Krieg et al. 1983; Ohba et al. 1992a). It is noteworthy that numerous isolates lacking insecticidal activities have also been recovered from soils and phylloplanes, some of them producing unusual Cry proteins that preferentially kill human cancer cells (Mizuki et al. 1999b), a human-pathogenic trichononad protozoan (Kondo et al. 2002), and human- and animal-pathogenic nematode parasites (Kotze et al. 2005; Cappello et al. 2006) (see Sect. 11.4.4).

11.2 Isolation and Identification

Two steps are involved in the procedure for screening of *B. thuringiensis* from soil and phylloplane. The first step is the isolation of members of the *B. cereus* group, a complex of *B. cereus* and *B. thuringiensis*, on the basis of colony morphology characteristic of this group (Logan 2005). Two techniques are commonly used for isolation of the organism from phylloplane: the leaf-imprinting technique and the shaken-flask technique (Smith and Couche 1991). For effective isolation, several methods have been used to suppress the levels of background microbial populations or to increase the population levels of the *B. cereus* group. Examples include the screenings with pasteurization procedure (Ohba and Aizawa 1986), acetate selection (Travers et al. 1987) and selection on culture media containing antibiotics (Johnson and Bishop 1996).

The second step involves discrimination between *B. cereus* and *B. thuringiensis*. It has been well accepted that *B. thuringiensis* is phenotypically indistinguishable from *B. cereus*. Thus, selective medium specific for this organism is presently unavailable. The only key character that differentiates *B. thuringiensis* from *B. cereus* is the ability to produce parasporal inclusions in sporangia (Logan 2005). Practically, phase-contrast microscopy has been a popular choice for observation of parasporal inclusions (Ohba and Aizawa 1986). Also, parasporal inclusions can be easily identified by light microscopic observation of stained bacterial smears. It should be emphasized that the observation of fully developed sporangia, prior to cell lysis, is highly desirable for correct identification of "parasporal" inclusions. Hyakutake et al. (2001) developed a computer-based image processing technique for rapid screening of *B. thuringiensis*. The technique automatically discriminates in 5 s between *B. thuringiensis* and *B. cereus* based on the ratio of spores and parasporal inclusions.

11.3 Occurrence in Soil and Phylloplane

11.3.1 Soil

DeLucca et al. (1981) were the first to report the natural occurrence of *B. thuringiensis* in soils. A large-scale isolation experiment with many types of soils, sampled from various regions in USA, yielded 250 isolates of *B. thuringiensis*, accounting for 0.5% of *B. cereus* group colonies examined. Ohba and Aizawa (1986) showed that *B. thuringiensis* was retained in natural soils of Japan at a frequency of 2.7% among the *B. cereus* group. Thereafter, numerous studies have provided evidence that the organism is commonly distributed in soils throughout the world, from seashores to high mountains and from tropics to high latitudes.

Table 11.1 shows natural frequencies of *B. thuringiensis* in soils of the areas with no history of application of *B. thuringiensis*-based microbial insecticides. Only investigations that examined more than 3,000 isolates of the *B. cereus* group are listed. In general, *B. thuringiensis*-containing soil samples are found at the frequencies varying between 10% and 20%. Occasionally, however, the positive frequency increases up to greater than 50%. For instance, Martin and Travers (1989) recovered *B. thuringiensis* from 785 (70%) out of 1,115 soil samples collected from USA and 29 other countries (Table 11.1).

Usually, the proportion of *B. thuringiensis* to the populations of members of the *B. cereus* group ranges between 0.5% and 5%. Differences in soil types, geographic locations and population sizes of *B. cereus* group members examined, as well as technical differences, may account for variations in the frequency of *B. thuringiensis* in soil environments.

Table 11.1 Occurrence of *Bacillus thuringiensis* in soil

Locality (country)	References[a]	Frequency of *B. thuringiensis*-positive soils (%)	Frequency of *B. thuringiensis* among *B. cereus* group (%)
Greece	Aptosoglou et al. (1997)	11	2.1
Japan	Ohba and Aizawa (1986)	37	2.7
	Ohba and Aratake (1994)	8	0.6
	Sasaki et al. (1994)	18	0.8
	Ohba et al. (2000)	13	1.1
	Kikuta et al. (2001)	11	1.9
	Ohba et al. (2002b)	12	1.3
Spain	Iriarte et al. (1998)	18	4.8
	Quesada-Moraga et al. (2004)	24	7.3
USA	DeLucca et al. (1981)	17	0.5
USA[b]	Martin and Travers (1989)	70	28
Vietnam	Yasutake et al. (2006)	34	1.6

[a]The investigations examining >3,000 colonies of the *B. cereus* group are selectively listed
[b]USA and 29 other countries

11.3.2 Phylloplane

B. thuringiensis naturally occurs on phylloplanes of various plants growing in areas with no histories of the application of *B. thuringiensis*-based insecticides. This organism can be commonly found on various types of vegetation: arboreous, herbaceous, deciduous, evergreen, broad-leaved and coniferous. Also, it occurs on both abaxial (lower) and adaxial (upper) leaf surfaces. Table 11.2 is a list of several selected studies that assessed the frequency of *B. thuringiensis* among members of the *B. cereus* group occurring on phylloplanes. Most of these studies examined the leaves sampled from low-statured (≤3-m) vegetation. In an exceptional case, Noda et al. (2009) assessed the *B. thuringiensis* frequency on phylloplanes of 5- to 10-m-high canopies. The organism has also been recovered, in many other investigations not listed in Table 11.2, from a variety of species and types of plants growing in various regions of the world. In general, *B. thuringiensis* frequencies in phylloplane populations of the *B. cereus* group are greater than those in soil populations. However, this is not the case with the *B. thuringiensis* populations occurring in Japan. As shown in Table 11.1, the organism is distributed in Japanese soils at frequencies of 0.6–2.7% among the *B. cereus* group, while it constitutes 0.8–3.2% of phylloplane populations (Table 11.2). Obviously, there is no significant difference in the frequencies of *B. thuringiensis* between the two environments, soils and phylloplanes, in Japan.

Ubiquity of *B. thuringiensis* on phylloplanes has provided insights into the origin of the populations associated with certain other environments. Ohba (1996)

Table 11.2 Occurrence of *Bacillus thuringiensis* on phylloplane

Locality (country)	References[a]	Frequency of *B. thuringiensis*-positive leaf samples (%)	Frequency of *B. thuringiensis* among *B. cereus* group (%)	Plant
Colombia	Maduell et al. (2002)	74	20.2	Black pepper (13 *Piper* spp.)
Denmark	Damgaard et al. (1997)	54	11.6	Cabbage
India	Kaur and Singh (2000)	–[b]	7–20	Chickpea and three other leguminous species
Japan	Ohba (1996)	11	3.2	Mulberry
	Mizuki et al. (1999a)	47	3.4	Japanese cherry and 30 other arboreous and four herbaceous species
	Noda et al. (2009)	1.4	0.8	Camellia and eight other arboreous species
USA	Smith and Couche (1991)	50–70[c]	30–100	Brittle willow and six other arboreous species

[a]The investigations with assessment of *B. thuringiensis* frequency in *B. cereus* group are listed
[b]Unknown
[c]Positive-tree frequency

suggested that the *B. thuringiensis* population occurring on mulberry leaves is a possible source of the populations associated with silkworm-rearing insectaries of Japanese sericultural farms. Lee et al. (2002) also suggested that the high-level of *B. thuringiensis* populations in faeces of herbivorous animals is due to the intake of food plants carrying this organism. Unlike the faeces of herbivorous animals, those of carnivorous animals, in particular feline mammals, contain rather low levels of *B. thuringiensis* (Lee et al. 2002). Also, there is no reason to exclude the possibility that the *B. thuringiensis* populations occurring on phylloplanes are the major sources of the populations associated with stored product environments (Damgaard 2000).

11.4 Diversity of Soil and Phylloplane Populations

11.4.1 Vegetative Cell Surface Antigens

11.4.1.1 Flagellar (H) Antigen

Vegetative cells of the *B. cereus*/*B. thuringiensis* group are usually motile and peritrichously flagellated, having two major surface antigens, the flagellar (H) antigen and the heat-stable somatic (cell wall) antigen consisting of polysaccharides. The former antigen is not associated with *B. anthracis*, which is non-motile (Logan 2005). H-serotyping of *B. thuringiensis* was started by de Barjac and Bonnefoi (1962), for intraspecific classification of the organism. It was believed that the use of flagellar (H) antigenicity for subgrouping of *B. thuringiensis* is taxonomically sound rather than the use of insect pathogenicity, characteristics of Cry proteins and other phenotypic properties. In the last 40 years, intensive efforts to screen for *B. thuringiensis* in natural environments have provided as many as 69 H-serotypes. In several H-serotypes, there are two or more subserotypes. For instance, H-serotype 3 is divided into the following four subserotypes due to the existence of five H-antigenic subfactors (3a, 3b, 3c, 3d, and 3e): H3ac (serovar *alesti*), H3abc (serovar *kurstaki*), H3ad (serovar *sumiyoshiensis*) and H3ade (serovar *fukuokaensis*). Thus, the current H-serotyping scheme contains 82 serovars belonging to 69 H-serotypes (Lecadet et al. 1999).

Tables 11.3 and 11.4 summarize the findings on the H-serotype flora associated with soil and phylloplane environments in several countries. The observations provide evidence that the *B. thuringiensis* populations in both environments consist of highly heterogeneous multiple H-serogroups. Quesada-Moraga et al. (2004) reported that the soil populations indigenous to Spain contained as many as 24 H-serotypes, with predominance of: H7 (*aizawai*), H10ab (*darmstadiensis*), H27 (*mexicanensis*), H52 (*kim*) and H34 (*konkukian*). Ohba et al. (2000) also showed that the H-serotype flora in soils of the Ryukyus, Japan contained >21 H-serotypes, with predominant occurrence of H5ac/21 (*canadensis*/*colmeri*), H3ad (*sumiyoshiensis*), H16 (*indiana*),

Table 11.3 H-serotypes of *Bacillus thuringiensis* flora in soil[a]

Locality (country)	References	Number of isolates tested	Number of isolates			Number of H-serotypes detected	Predominant H-serotypes (serovars)
			Serotyped	UTY[b]	UTE[c]		
Indonesia	Hastowo et al. (1992)	135	63 (47%)	72 (53%)	0	11	H3ade (*fukuokaensis*), H6 (*entomocidus*), H3abc (*kurstaki*), H8ab (*morrisoni*)
Japan	Ohba and Aizawa (1986)	189	66 (35%)	84 (44%)	39 (21%)	7	H10ab (*darmstadiensis*), H11ac (*kyushuensis*), H17 (*tohokuensis*), H8ac (*ostriniae*), H3ade (*fukuokaensis*)
	Ohba and Aratake (1994)	22	14 (64%)	7 (32%)	1 (5%)	5	H17 (*tohokuensis*), H29 (*amagiensis*)
	Sasaki et al. (1994)	63	45 (71%)	0	18 (29%)	8	H16 (*indiana*), H3abc (*kurstaki*), H1 (*thuringiensis*), H7 (*aizawai*), H14 (*israelensis*)
	Ohba et al. (2000)	235	99 (42%)	12 (5%)	124 (53%)	21	H5ac/21[d] (*canadensis*/*colmeri*), H16 (*indiana*), H3ad (*sumiyoshiensis*), H10ac (*londrina*), H13 (*pakistani*)
	Ohba et al. (2002a)	72	47 (65%)	9 (13%)	16 (22%)	7	H13/29[d] (*pakistani*/*amagiensis*)
	Ohba et al. (2002b)	94	65 (69%)	13 (14%)	16 (17%)	21	H3[e]/4ac (*kenyae*), H14/19[d] (*israelensis*/*tochigiensis*), H25 (*coreanensis*), H20ab (*yunnanensis*), H49 (*muju*)
	Yasutake et al. (2007)	66	45 (68%)	1 (2%)	20 (30%)	11	H10a[e], H13 (*pakistani*), H14 (*israelensis*), H6 (*entomocidus*)
Spain	Iriarte et al. (1998)	101	87 (86%)	3 (3%)	11 (11%)	25	H34 (*konkukian*), H7 (*aizawai*), H3abc (*kurstaki*), H3abc (*kurstaki*), H10ab (*darmstadiensis*), H4ab (*sotto*)

	Quesada-Moraga et al. (2004)	–[f]	259	–[f]	–[f]	24	H7 (*aizawai*), H10ab (*darmstadiensis*), H27 (*mexicanensis*), H52 (*kim*), H34 (*konkukian*)
USA	DeLucca et al. (1981)	500	490 (98%)	10 (2%)	0	5	H3abc (*kurstaki*), H5ab (*galleriae*), H15 (*dakota*)
Vietnam	Binh et al. (2005)	479	400 (84%)	79 (16%)	0	12	H3abc (*kurstaki*), H7 (*aizawai*), H8ab (*morrisoni*), H3ac (*alesti*), H9 (*tolworthi*)
	Yasutake et al. (2006)	63	34 (54%)	5 (8%)	24 (38%)	12	H23 (*japonensis*), H29 (*amagiensis*), H21 (*colmeri*), H17 (*tohokuensis*), H36 (*malaysiensis*)

[a]Number of reference antisera used for H-serotyping varied according to the investigation owing to availability
[b]Motile but untypable by the reference H-antisera available
[c]Untestable due to non-motility and/or strong autoagglutination
[d]Reactive for two reference antisera
[e]Subfactors not identified
[f]Unknown

H10ac (*londrina*) and H13 (*pakistani*) (Table 11.3). The serological diversity on phylloplanes was comparable to those of soil. This is supported by the observations of Ohba (1996) who detected as many as 19 H-serotypes on the surface of mulberry leaves in Japan (Table 11.4).

As shown in Table 11.3, differences in geographic locations lead to the marked variations in the predominant H-serotypes. In particular, there is little similarity in the predominant serotypes between the seven local soil populations within Japan. In contrast to the populations in soil, those on phylloplanes commonly contain H3abc (*kurstaki*) as the predominant serotype, regardless of the differences in geographic locations and types of plant (Table 11.4). It should be noted that this serotype is rather rare in soil microflora (Table 11.3). Accordingly, Damgaard et al. (1997) claimed that the population of *B. thuringiensis* on phylloplane is different from that normally found in soil.

One of the most striking aspects of *B. thuringiensis* serology is the omnipresence of acrystalliferous *B. cereus* isolates in nature that have H-antigens reactive to antisera against *B. thuringiensis* reference strains. Shisa et al. (2002b) reported that the majority of the *B. cereus* isolates recovered from natural environments in Japan gave positive reactions to the reference antisera against *B. thuringiensis* H1–H55. The frequencies of seropositive isolates were 77%, 60%, and 68%, in soils, animal faeces and on phylloplanes, respectively. The observations provide a perspective that the frequency of *B. cereus* isolates with *B. thuringiensis* H-antigens may increase in future with increasing numbers of *B. thuringiensis* H-serotypes being recognized, approaching 100%. Also, it is clear from the findings that the H-serotyping is of little value in discriminating between *B. thuringiensis* and *B. cereus*; however, it is still of great value in discriminating between *B. thuringiensis* strains (Lecadet et al. 1999).

Shisa et al. (2002b) also showed that the *B. cereus* populations contained as many as 45 H-serogroups, with predominancy of five H-serotypes: H22 (*shandongiensis*), H6 (*entomocidus*), H16 (*indiana*), H13 (*pakistani*) and H24ab (*neoleonensis*). Interestingly, the phylloplane microflora, as well as the populations in soils and animal faeces, contained no *B. cereus* isolates with H3abc (*kurstaki*) antigen. This is in marked contrast to the above facts that the serotype H3abc (*kurstaki*) is among the dominant members in phylloplane populations of *B. thuringiensis* in Japan (Table 11.4).

11.4.1.2 Heat-Stable Somatic Antigens

Like many other bacteria, *B. thuringiensis* has heat-stable somatic antigens (HSSAs) in its vegetative cell walls. It produces a substantial amount of water-soluble extracellular HSSAs (polysaccharides) during vegetative cell growth, through normal cell wall turnover. Extracellular HSSAs can be easily detected by the precipitin halo formation (PHF) test on nutrient agar plates that contain specific antibodies against HSSAs. By using the PHF test, Ueda et al. (1991) classified the reference strains of 24 *B. thuringiensis* H-serotypes into 16 HSSA serogroups designated I to XVI. It is

Table 11.4 H-serotypes of *Bacillus thuringiensis* flora on phylloplane[a]

Locality (country)	References	Number of isolates tested	Number of isolates			Number of H-serotypes detected	Predominant H-serotypes (serovars)
			Serotyped	UTY[b]	UTE[c]		
Denmark	Damgaard et al. (1997)	150	111 (74%)	31 (21%)	8 (5%)	12	H3abc (*kurstaki*), H14 (*israelensis*), H21 (*colmeri*), H21/15[d] (*colmeri*/*dakota*)
	Hansen et al. (1998)	50	33 (66%)	14 (28%)	3 (6%)	9	H3abc (*kurstaki*)
Netherland	Damgaard et al. (1998)	32	30 (94%)	1 (3%)	1 (3%)	5	H14 (*israelensis*), H13 (*pakistani*), H23 (*japonensis*)
Japan	Ohba (1996)	186	137 (74%)	18 (10%)	31 (17%)	19	H13 (*pakistani*), H3abc (*kurstaki*), H6 (*entomocidus*), H16 (*indiana*), H24ab (*neoleonensis*)
	Mizuki et al. (1999a)	120	53 (44%)	20 (17%)	47 (39%)	17	H3abc (*kurstaki*), H16 (*indiana*), H22 (*shandongiensis*), H4ac (*kenyae*), H17/27[d] (*tohokuensis*/*mexicanensis*)
	Noda et al. (2009)	39	27 (69%)	11 (28%)	1 (3%)	11	H3abc (*kurstaki*), H17 (*tohokuensis*), H4ac (*kenyae*), H7 (*aizawai*), H14 (*israelensis*)

[a]Number of reference antisera for H-serotyping varied according to the investigation owing to availability
[b]Motile but untypable by the reference H-antisera available
[c]Untestable due to non-motility and/or strong autoagglutination
[d]Reactive for two reference antisera

noteworthy that the serogroup IV contained the reference strains of five H-serotypes: H3abc (*kurstaki*), H4ac (*kenyae*), H5ab (*galleriae*), H7 (*aizawai*) and H18ab (*kumamotoensis*). It is well accepted that the strains with strong toxicities against insects, in particular lepidopterans, are densely distributed in these H-serotypes. Thus, the observations may support a hypothesis that there exists a relationship, even if not direct, between insecticidal activities and HSSAs. It is clear from the findings that the degree of variation in HSSA is smaller in this organism than that of H-antigens.

Another interesting fact is that when examined with the PHF test, the isolates belonging to a single H-serotype were allocated to a single HSSA serogroup at a high frequency of 87–100%; for example, a total of 114 isolates belonging to H7 (*aizawai*), recovered from various geographic locations in Japan, were all allocated to the HSSA serogroup IV (Ohba et al. 1992b).

Relatively little is known about the serological relationships between HSSAs of *B. thuringiensis* and *B. cereus*. Earlier, Ohba and Aizawa (1978) reported that, among 22 environmental *B. cereus* isolates possessing *B. thuringiensis* H-antigens, 10 (45%) were seropositive for *B. thuringiensis* HSSA antisera, when examined by slide agglutination test. The finding suggests that common HSSAs occur between *B. thuringiensis* and *B. cereus* at high frequencies. It should be noted, however, that there was a discrepancy between the two serotypings; for instance, five *B. cereus* isolates with the H6 (*entomocidus*) antigen were not agglutinated with any of the HSSA antisera used, including the VII antiserum against the reference strain of *B. thuringiensis* H6 (Ohba and Aizawa 1978).

11.4.2 Parasporal Inclusion Morphotypes

During sporulation, *B. thuringiensis* synthesizes highly heterogeneous parasporal proteins (Cry and Cyt proteins) with hydrophobic and amphipathic domains. This leads to the formation of noticeable parasporal inclusions of several morphotypes. The shape of the inclusion is a genetically heritable character. The inclusion morphotypes commonly found in environmental isolates are: bipyramidal (rhomboidal), spherical (round), cuboidal/rectangular and irregularly pointed. Also, inclusions with atypical shapes are occasionally found. It is noteworthy that the spherical parasporal inclusions are often covered with thick, highly electron-dense envelopes (Wasano et al. 2000; Shisa et al. 2006). Little information is presently available about the origin and chemical composition of the envelopes.

It is generally accepted that the organisms producing bipyramidal inclusions are predominantly distributed in both soil and phylloplane environments, followed by those producing spherical or irregularly pointed ones (Bernhard et al. 1997; Hansen et al. 1998; Maduell et al. 2002; Armengol et al. 2007). There is a poor correlation between parasporal inclusion morphotypes and insecticidal activities. Hastowo et al. (1992) reported that (1) 29% of 41 Indonesian soil isolates with bipyramidal inclusions exhibited Lepidoptera-specific toxicity, while 71% were non-toxic to both Lepidoptera and Diptera, and (2) among 80 isolates with spherical inclusions, 27% were toxic to Lepidoptera, while 73% were non-insecticidal.

11.4.3 Cry Proteins and cry Genes

The current classification scheme for *B. thuringiensis* parasporal proteins comprises as many as 55 Cry proteins and two Cyt protein families (Crickmore et al. 2009). Furthermore, many of these families consist of multiple subfamilies. For the last three decades, intensive efforts have been made to screen for strains useful for the microbial control of insect pests.

For this purpose, Cry and Cyt proteins associated with *B. thuringiensis* natural isolates have been identified by gene typing that involves PCR tests with specific primers for the several existing major families of *cry* genes (Ibarra et al. 2003; Wang et al. 2003; Apaydin et al. 2005; Binh et al. 2005; Hernández et al. 2005; Jara et al. 2006; Armengol et al. 2007; Rosas-García et al. 2008). This method is useful for primary screening of *B. thuringiensis* strains with desirable insecticidal activities. It is conceivable, however, that the presence of *cry* genes is not always accompanied by the actual activities wanted. In fact, Shisa et al. (2002a) showed the discrepancy existing between *cry* gene-predicted and bioassay-determined insecticidal activities. In addition, previous investigators have reported the occurrence of cryptic *cry* genes in *B. thuringiensis* (Hodgman et al. 1993). However, very little is presently known about the frequency of cryptic *cry* genes in natural populations of the organism.

It is generally accepted that the genes predominantly distributed in natural environments are *cry1*-allied genes encoding Lepidoptera-specific Cry1 proteins. Also, commonly found are: *cry2* genes encoding the proteins toxic to both Lepidoptera and Diptera and *cry4*/*cry11* genes specific for Diptera. The other *cry* genes, including Coleoptera-specific *cry3* and Lepidoptera-specific *cry9*, are occasionally found, as well as *cyt* genes encoding Cyt proteins with broad-spectrum cytolytic activities. Usually, *B. thuringiensis* isolates retain heterogeneous multiple *cry* genes. For example, Maduell et al. (2002) reported that approximately 55% of the Colombian isolates, recovered from *Piper* (pepper) phylloplanes, had a gene profile consisting of five *cry1* genes (*cry1Aa*, *cry1Ab*, *cry1Ac*, *cry1Ad*, and *cry1B*), while *cry1* and *cry11* coexisted in 8%, and 2% harboured three genes (*cry1*, *cry4*, and *cry11*). It is noteworthy that the isolates with a gene profile consisting of *cry1* and *cry2* occur in natural populations of *B. thuringiensis* at high frequencies (Hernández-Rodríguez and Ferré 2009). It is very likely that the transfer of these two genes between bacteria usually occurs in pairs in natural environments. By analogy, the isolates with strong mosquitocidal activities commonly contain the combination of *cry4*/*cry11* and *cyt* genes (Delécluse et al. 2000).

It has been established that the plasmids harbouring *cry* genes can be experimentally transferred between strains of the *B. thuringiensis*/*B. cereus* group at high frequencies not only in cadavers of lepidopteran insects but also in soils (Glare and

O'Callaghan 2000). Jarrett and Stephenson (1990) suggested that the plasmid transfer between *B. thuringiensis* strains occurs in nature, leading to the production of new combinations of Cry proteins within populations of bacteria. Subsequently, in an isolation experiment, Ishii and Ohba (1993) obtained ten *B. thuringiensis* isolates, belonging to the three different H serotypes, from a single soil microhabitat. Interestingly, there was no difference between these isolates in characteristics of parasporal inclusion proteins, in mosquitocidal and hemolytic activities, SDS-PAGE profiles and immunological properties (Ishii and Ohba 1993, 1994). The observations strongly suggest that the mosquitocidal activity of these isolates is due to a single-origin-derived genetic element that is transmissible among bacterial populations inhabiting soils. Similarly, Wasano et al. (1998) reported that two Lepidoptera-specific *B. thuringiensis* isolates, derived from a single soil source, were allocated to different H-serotypes, while no significant differences were evident in other characteristics between the two isolates.

11.4.4 Cry Protein-Associated Biological Activities

11.4.4.1 Insecticidal Activity

As mentioned above (see Sect. 11.3), *B. thuringiensis* is widely distributed in both soil and phylloplane environments. It seems, however, that there is a difference between the two environments in the frequencies of insecticidal *B. thuringiensis.* Damgaard et al. (1997) reported that 68% of the isolates, recovered from cabbage phylloplanes in Denmark, exhibited larvicidal activity against Lepidoptera, suggesting that the population of *B. thuringiensis* on phylloplanes is different from that normally found in soils. Similarly, in Japan insecticidal isolates occurred on phylloplanes of various plants at relatively high frequencies: 23% (Mizuki et al. 1999a) and 49% (Noda et al. 2009). These results are in marked contrast to those obtained in Japanese soils, where the soil populations contained insecticidal *B. thuringiensis* at rather low levels of $<10\%$ (Ohba and Aratake 1994; Ohba et al. 2002a, b), and often as low as $\leq 3\%$ (Ohba et al. 2000; Yasutake et al. 2007). These findings lead to the hypothesis that the organisms with insecticidal Cry proteins have a habitat preference for the phylloplane rather than soil.

Another fact, that should be stressed, is that non-insecticidal *B. thuringiensis* strains outnumber insecticidal ones in natural environments. There is argument over the use of the term "non-insecticidal", because the insecticidal activity test usually involves only several major insect pests, representing Lepidoptera, Diptera and Coleoptera. Thus, it is likely that the non-insecticidal group contains, if not many, at least some organisms with unidentified narrow-spectrum insecticidal activities. However, a more likely possibility is that the non-insecticidal parasporal proteins may have biological activities as yet undiscovered. The examples include anti-cancer cytocidal activities as described below.

11.4.4.2 Anti-cancer Activity

B. thuringiensis populations in nature contain organisms that synthesize parasporal proteins that have unique biological activities apart from toxicity to insects. Mizuki et al. (1999b) were the first to report the occurrence of non-insecticidal *B. thuringiensis* strains, whose Cry proteins are preferentially toxic to certain human cancer cells in vitro but not to normal cells. These anti-cancer activities were found in 22 motile isolates, belonging to several H serotypes, and 20 non-motile isolates. Subsequently, a Cry protein, designated parasporin, was cloned from one of the non-motile organisms (Mizuki et al. 2000). According to the current classification scheme for parasporin (PS), a total of 12 PS proteins, belonging to the four families PS1, PS2, PS3, and PS4, have been cloned from *B. thuringiensis* isolates, mostly occurring in soils of Japan (Ohba et al. 2009). Interestingly, there are no genealogical relationships between the four PS protein families. Also, marked differences are evident in anti-cancer cytotoxicity spectra and activity levels between the four PS proteins (Ohba et al. 2009). It is very likely that each of the four PS proteins has its own cancer cell-killing mechanism through specific binding to its respective receptor.

Members of the soil microflora commonly contain aerobic and anaerobic endospore-forming bacteria that produce parasporal inclusions, belonging to various species of several genera: *Bacillus*, *Brevibacillus*, *Paenibacillus*, and *Clostridium*. It would be worthwhile to examine such bacteria, in addition to *B. thuringiensis*, for the production of novel parasporal proteins with anti-cancer activities.

11.4.4.3 Others

The diversity of *B. thuringiensis* Cry proteins has provided two more unique activities of medical and veterinary importance. One is toxicity against animal and human parasitic nematodes and the other is against protozoans causing human and animal diseases.

Wei et al. (2003) showed that the three Cry proteins (Cry5B, 14A, and 21A) were highly toxic not only to free-living and plant parasitic nematodes, but also to free-living larvae of the rodent-parasitic nematode *Nippostrongylus*. Subsequently, Kotze et al. (2005) reported that the two strains, selected from 410 *B. thuringiensis* strains, were significantly toxic to the three economically important nematode parasites of livestock in Australia. These two strains contained either Cry5A and Cry5B proteins or a Cry13 protein. It is of particular interest to note that a therapeutic activity against the human and animal hookworm parasite *Ancylostoma ceylanicum*, a blood-feeding gastrointestinal nematode, is associated with Cry5 proteins (Cappello et al. 2006).

Kondo et al. (2002) examined parasporal inclusion proteins from 816 Japanese strains of *B. thuringiensis*, isolated from soils, aquatic environments and silkworm-rearing insectaries, for anti-protozoan activity against *Trichomonas vaginalis*, the causative agent of sexually transmitted vaginal trichomoniasis. This provided two soil isolates, both belonging to H serotype 13/29 (serovar *pakistani/amagiensis*), with parasporal proteins highly toxic to the protozoan in vitro. Neither insecticidal activity nor parasporin activity against cancer cells was associated with these two isolates. The toxic proteins have not been identified as yet. In addition, it is uncertain whether the protozoan-killing factors from these strains have a therapeutic activity against vaginal trichomoniasis.

Recently, Xu et al. (2004) have claimed that the injection of parasporal proteins of seven *B. thuringiensis* soil isolates protected mice from infection with *Plasmodium berghei*, a malaria parasite of animals. It is unclear, however, from the observation that the proteins of the isolates are capable of killing *Plasmodium* cells.

11.5 Dispersal and Persistence in Nature

For years, extensive ecological studies have been made by many investigators to understand how *B. thuringiensis* is dispersed and how it persists in natural environments. Overall studies have shed some light on the life of *B. thuringiensis* strains with highly insecticidal Cry proteins (Jensen et al. 2003). However, it is still among the basic questions how non-insecticidal *B. thuringiensis*, outnumbering insecticidal ones as mentioned above (see Sect. 11.4.4.1), live in natural ecosystems.

It has been generally accepted that *B. thuringiensis* spores can survive in soil for long periods, with half-lives of 100–200 days, while the half-life on phylloplane and in water is much shorter than in soil (Glare and O'Callaghan 2000). Usually, *B. thuringiensis* hardly grows in natural soils. Akiba et al. (1980) showed that none of the 24 soils from different localities were permissive for the growth of *B. thuringiensis*, while the growth was evident in sterilized soils with pH values of 5.5–7.1. The observations strongly suggest that the failure of *B. thuringiensis* to proliferate in natural soils is largely due to the presence of competitive soil microorganisms and low pH. Later, West et al. (1985) reported that supplementing soil with additional nutrients supported a good growth of the organism. By analogy, cadavers of insects provide nutrient-rich environments supporting the proliferation of *B. thuringiensis* (Akiba 1986; Takatsuka and Kunimi 1998). These findings provide insight into the life of *B. thuringiensis*, as a saprophyte, in soil environments. Interestingly, Takatsuka and Kunimi (1998) reported that there was little evidence available for horizontal transmission of *B. thuringiensis* in larval population of *Ephestia kuehniella*, a lepidopteran pest in stored-product environments.

As described above (see Sect. 11.3.2), *B. thuringiensis* is commonly found on phylloplanes of various plants. However, its life cycle on plants has been little explored. Recently, Maduell et al. (2007) reported that *B. thuringiensis* was dispersed poorly from the soil or the seed to bean leaves, and between the leaves of the same plants. In this regard, it is noteworthy that the organism multiplies to some extent on bean phylloplanes, but much less than do the other coexisting epiphytic bacteria such as *Pseudomonas fluorescens* (Maduell et al. 2008). Apparently, *B. thuringiensis* is a poor colonist on bean phylloplanes, and the limitation of nutrients available from leaf surface is solely responsible for poor colonization of the organism.

Of particular interest are the results obtained by Bizzarri and Bishop (2007) that *B. thuringiensis* was recovered from the phylloplane of clover, *Trifolium hybridum*, over a growing season, and three simultaneous and sudden rises and declines of both spore and vegetative cell densities were observed. The findings clearly show that *B. thuringiensis* is able to complete its life cycles on the phylloplane. However, it is presently uncertain whether the organism has a plant preference for inhabiting the phylloplane.

11.6 Concluding Remarks

B. thuringiensis is a common member of the microbial flora associated with soil and phylloplane environments. It is a cosmopolitan, found in most regions of the world, even in Antarctica. Usually, the organism occurs at a frequency of $<10\%$ among the populations of the *B. cereus* group in soil and phylloplane. *B. thuringiensis* populations in nature show a great diversity of genetic and phenotypic characteristics. In fact, numerous flagellar (H) serotypes and *cry*-gene types are contained in both soil and phylloplane populations. Of particular interest is that various biological activities are associated with parasporal inclusion proteins. Examples include (1) the insecticidal activities, with different toxicity spectra and various toxicity levels, against several insect orders including Lepidoptera, Diptera and Coleoptera, (2) the toxicity specific for free-living and human and animal parasitic nematodes, (3) the protozoan toxicity against human-pathogenic *Trichomonas* and (4) the cytocidal activity preferential for human cancer cells. Another interesting fact is that the majority of *B. thuringiensis* natural isolates produce parasporal inclusion proteins with orphan activities. *B. thuringiensis* parasporal inclusion proteins have hydrophobic and amphipathic domains that easily interact with cell membranes by recognizing target proteins specifically. This encourages us to fish for orphan receptors with parasporal inclusion proteins in search of novel biological activities.

References

Akiba Y (1986) Microbial ecology of *Bacillus thuringiensis*. VII. Fate of *Bacillus thuringiensis* in larvae of the silkworm, *Bombyx mori*, and the fall webworm, *Hyphantria cunea*. Jpn J Appl Entomol Zool 30:99–105

Akiba Y, Sekijima Y, Aizawa K, Fujiyoshi N (1980) Microbial ecological studies on *Bacillus thuringiensis*. IV. The growth of *Bacillus thuringiensis* in soils of mulberry plantations. Jpn J Appl Entomol Zool 24:13–17

Apaydin Ö, Yenidünya AF, Harsa Ş, Güneş H (2005) Isolation and characterization of *Bacillus thuringiensis* strains from different grain habitats in Turkey. World J Microbiol Biotechnol 21:285–292

Aptosoglou SG, Sivropoulou A, Koliais SI (1997) Distribution and characterization of *Bacillus thuringiensis* in the environment of the olive in Greece. Microbiologica 20:69–74

Armengol G, Escobar MC, Maldonado ME, Orduz S (2007) Diversity of Colombian strains of *Bacillus thuringiensis* with insecticidal activity against dipteran and lepidopteran insects. J Appl Microbiol 102:77–88

Berliner E (1915) Über die Schlaffsucht der Mehlmottenraupe (*Ephestia kühniella* Zell.) und ihren Erreger *Bacillus thuringiensis* n. sp. Z Angew Entomol 2:29–56

Bernhard K, Jarrett P, Meadows M, Butt J, Ellis DJ, Roberts GM, Pauli S, Rodgers P, Burges HD (1997) Natural isolates of *Bacillus thuringiensis*: worldwide distribution, characterization, and activity against insect pests. J Invertebr Pathol 70:59–68

Binh ND, Chau NQ, Nguyet NA, Canh NX, Herrou J, Huong PM, Tuan ND, Asano S, Iizuka T (2005) *Bacillus thuringiensis* distribution in soil of Vietnam. In: Binh ND, Akhurst RJ, Dean DH (eds) Biotechnology of *Bacillus thuringiensis*, vol 5. Science and Technics, Hanoi, pp 141–152

Bizzarri MF, Bishop AH (2007) Recovery of *Bacillus thuringiensis* in vegetative form from the phylloplane of clover (*Trifolium hybridum*) during a growing season. J Invertebr Pathol 94:38–47

Cappello M, Bungiro RD, Harrison LB, Bischof LJ, Griffitts JS, Barrows BD, Aroian RV (2006) A purified *Bacillus thuringiensis* crystal protein with therapeutic activity against the hookworm parasite *Ancylostoma ceylanicum*. Proc Natl Acad Sci USA 103:15154–15159

Charles J-F, Delécluse A, Nielsen-LeRoux C (2000) Entomopathogenic bacteria: from laboratory to field application. Kluwer Academic, Dordrecht

Crickmore N, Zeigler D, Bravo A, Schnepf E, Lereclus D, Baum J, Van Rie J, Dean D (2009) *Bacillus thuringiensis* toxin nomenclature. http://www.lifesci.sussex.ac.uk/home/Neil_Crickmore/Bt/

Damgaard PH (2000) Natural occurrence and dispersal of *Bacillus thuringiensis* in the environment. In: Charles J-F, Delécluse A, Nielsen-LeRoux C (eds) Entomopathogenic bacteria: from laboratory to field application. Kluwer Academic, Dordrecht, pp 23–40

Damgaard PH, Hansen BM, Pedersen JC, Eilenberg J (1997) Natural occurrence of *Bacillus thuringiensis* on cabbage foliage and in insects associated with cabbage crops. J Appl Microbiol 82:253–258

Damgaard PH, Abdel-Hameed A, Eilenberg J, Smits PH (1998) Natural occurrence of *Bacillus thuringiensis* on grass foliage. World J Microbiol Biotechnol 14:239–242

de Barjac H, Bonnefoi A (1962) Essai de classification biochimique et sérologique de 24 souches de *Bacillus* du type *B. thuringiensis*. Entomophaga 7:5–31

Delécluse A, Juárez-Pérez V, Berry C (2000) Vector-active toxins: structure and diversity. In: Charles J-F, Delécluse A, Nielsen-LeRoux C (eds) Entomopathogenic bacteria: from laboratory to field application. Kluwer Academic, Dordrecht, pp 101–125

DeLucca AJ, Simonson JG, Larson AD (1981) *Bacillus thuringiensis* distribution in soils of the United States. Can J Microbiol 27:865–870

Forsyth G, Logan NA (2000) Isolation of *Bacillus thuringiensis* from Northern Victoria Land, Antarctica. Lett Appl Microbiol 30:263–266

Glare TR, O'Callaghan M (2000) *Bacillus thuringiensis*: biology, ecology and safety. Wiley, Chichester

Goldberg LJ, Margalit J (1977) A bacterial spore demonstrating rapid larvicidal activity against *Anopheles sergentii, Uranotaenia unguiculata, Culex univittatus, Aedes aegypti* and *Culex pipiens*. Mosquito News 37:355–358

Hansen BM, Damgaard PH, Eilenberg J, Pedersen JC (1998) Molecular and phenotypic characterization of *Bacillus thuringiensis* isolated from leaves and insects. J Invertebr Pathol 71:106–114

Hastowo S, Lay BW, Ohba M (1992) Naturally occurring *Bacillus thuringiensis* in Indonesia. J Appl Bacteriol 73:108–113

Hernández CS, Andrew R, Bel Y, Ferré J (2005) Isolation and toxicity of *Bacillus thuringiensis* from potato-growing areas in Bolivia. J Invertebr Pathol 88:8–16

Hernández-Rodríguez CS, Ferré J (2009) Ecological distribution and characterization of four collections of *Bacillus thuringiensis* strains. J Basic Microbiol 49:152–157

Hodgman TC, Ziniu Y, Shen J, Ellar J (1993) Identification of a cryptic gene associated with an insertion sequence not previously identified in *Bacillus thuringiensis*. FEMS Microbiol Lett 114:23–30

Hyakutake T, Ichimatsu T, Mizuki E, Ohba M (2001) A rapid screening technique for *Bacillus thuringiensis* using image processing. Jpn J Appl Entomol Zool 45:9–14

Ibarra JE, del Rincón MC, Ordúz S, Noriega D, Benintende G, Monnerat R, Regis L, de Oliveira CMF, Lanz H, Rodriguez MH, Sánchez J, Peña G, Bravo A (2003) Diversity of *Bacillus thuringiensis* strains from Latin America with insecticidal activity against different mosquito species. Appl Environ Microbiol 69:5269–5274

Iriarte J, Bel Y, Ferrandis MD, Andrew R, Murillo J, Ferré J, Caballero P (1998) Environmental distribution and diversity of *Bacillus thuringiensis* in Spain. Syst Appl Microbiol 21:97–106

Ishii T, Ohba M (1993) Characterization of mosquito-specific *Bacillus thuringiensis* strains coisolated from a soil population. Syst Appl Microbiol 16:494–499

Ishii T, Ohba M (1994) Haemolytic activity associated with parasporal inclusion proteins of mosquito-specific *Bacillus thuringiensis* soil isolates: a comparative neutralization study. FEMS Microbiol Lett 116:195–200

Ishiwata S (1901) On a kind of severe flacherie (sotto disease). Dainihon Sanshi Kaiho 114:1–5

Jara S, Maduell P, Orduz S (2006) Diversity of *Bacillus thuringiensis* strains in the maize and bean phylloplane and their respective soils in Colombia. J Appl Microbiol 101:117–124

Jarrett P, Stephenson M (1990) Plasmid transfer between strains of *Bacillus thuringiensis* infecting *Galleria mellonella* and *Spodoptera littoralis*. Appl Environ Microbiol 56:1608–1614

Jensen GB, Hansen BM, Eilenberg J, Mahillon J (2003) The hidden lifestyles of *Bacillus cereus* and relatives. Environ Microbiol 5:631–640

Johnson C, Bishop AH (1996) A technique for the effective enrichment and isolation of *Bacillus thuringiensis*. FEMS Microbiol Lett 142:173–177

Kaur S, Singh A (2000) Natural occurrence of *Bacillus thuringiensis* in leguminous phylloplanes in the New Delhi region of India. World J Microbiol Biotechnol 16:679–682

Kikuta H, Igarashi N, Tatebayashi C, Ogino S, Murata K (2001) Distribution of *Bacillus thuringiensis* in forest soil of Hokkaido. J Seric Sci Jpn 70:1–9

Kondo S, Mizuki E, Akao T, Ohba M (2002) Antitrichomonal strains of *Bacillus thuringiensis*. Parasitol Res 88:1090–1092

Kotze AC, O'Grady J, Gough JM, Pearson R, Bagnall NH, Kemp DH, Akhurst RJ (2005) Toxicity of *Bacillus thuringiensis* to parasitic and free-living life-stages of nematode parasites of livestock. Int J Parasitol 35:1013–1022

Krieg A, Huger AM, Langenbruch GA, Schnetter W (1983) *Bacillus thuringiensis* var. *tenebrionis*: ein neuer, gegenüber Larven von Coleopteren wirksamer Pathotyp. Z Angew Entomol 96:500–508

Lecadet M-M, Frachon E, Cosmao Dumanoir V, Ripouteau H, Hamon S, Laurent P, Thiéry I (1999) Updating the H-antigen classification of *Bacillus thuringiensis*. J Appl Microbiol 86:660–672

Lee D-H, Machii J, Ohba M (2002) High frequency of *Bacillus thuringiensis* in feces of herbivorous animals maintained in a zoological garden in Japan. Appl Entomol Zool 37:509–516

Logan NA (2005) *Bacillus anthracis*, *Bacillus cereus*, and other aerobic endospore-forming bacteria. In: Borriello SP, Murray PR, Funke G (eds) Topley & Wilson's microbiology and microbial infections. Bacteriology. Hodder Arnold, London, pp 922–952

Maduell P, Callejas R, Cabrera KR, Armengol G, Orduz S (2002) Distribution and characterization of *Bacillus thuringiensis* on the phylloplane of species of *Piper* (Piperaceae) in three altitudinal levels. Microb Ecol 44:144–153

Maduell P, Armengol G, Llagostera M, Lindow S, Orduz S (2007) Immigration of *Bacillus thuringiensis* to bean leaves from soil inoculum or distal plant parts. J Appl Microbiol 103:2593–2600

Maduell P, Armengol G, Llagostera M, Orduz S, Lindow S (2008) *B. thuringiensis* is a poor colonist of leaf surfaces. Microb Ecol 55:212–219

Martin PAW, Travers RS (1989) Worldwide abundance and distribution of *Bacillus thuringiensis* isolates. Appl Environ Microbiol 55:2437–2442

Mizuki E, Ichimatsu T, Hwang S-H, Park YS, Saitoh H, Higuchi K, Ohba M (1999a) Ubiquity of *Bacillus thuringiensis* on phylloplanes of arboreous and herbaceous plants in Japan. J Appl Microbiol 86:979–984

Mizuki E, Ohba M, Akao S, Yamashita S, Saitoh H, Park YS (1999b) Unique activity associated with non-insecticidal *Bacillus thuringiensis* parasporal inclusions: *in vitro* cell-killing action on human cancer cells. J Appl Microbiol 86:477–486

Mizuki E, Park YS, Saitoh H, Yamashita S, Akao T, Higuchi K, Ohba M (2000) Parasporin, a human leukemic cell-recognizing parasporal protein of *Bacillus thuringiensis*. Clin Diagn Lab Immunol 7:625–634

Noda T, Kagoshima K, Uemori A, Yasutake K, Ichikawa M, Ohba M (2009) Occurrence of *Bacillus thuringiensis* in canopies of a natural lucidophyllous forest in Japan. Curr Microbiol 58:195–200

Ohba M (1996) *Bacillus thuringiensis* populations naturally occurring on mulberry leaves: a possible source of the populations associated with silkworm-rearing insectaries. J Appl Bacteriol 80:56–64

Ohba M, Aizawa K (1978) Serological identification of *Bacillus thuringiensis* and related bacteria isolated in Japan. J Invertebr Pathol 32:303–309

Ohba M, Aizawa K (1986) Distribution of *Bacillus thuringiensis* in soils of Japan. J Invertebr Pathol 47:277–282

Ohba M, Aratake Y (1994) Comparative study of the frequency and flagellar serotype flora of *Bacillus thuringiensis* in soils and silkworm-breeding environments. J Appl Bacteriol 76:203–209

Ohba M, Iwahana H, Asano S, Suzuki N, Sato R, Hori H (1992a) A unique isolate of *Bacillus thuringiensis* serovar *japonensis* with a high larvicidal activity specific for scarabaeid beetles. Lett Appl Microbiol 14:54–57

Ohba M, Ueda K, Aizawa K (1992b) Serotyping of *Bacillus thuringiensis* environmental isolates by extracellular heat-stable somatic antigens. Can J Microbiol 38:694–695

Ohba M, Wasano N, Mizuki E (2000) *Bacillus thuringiensis* soil populations naturally occurring in the Ryukyus, a subtropical region of Japan. Microbiol Res 155:17–22

Ohba M, Shisa N, Thaithanun S, Nakashima K, Lee D-H, Ohgushi A, Wasano N (2002a) A unique feature of *Bacillus thuringiensis* H-serotype flora in soils of a volcanic island of Japan. J Gen Appl Microbiol 48:233–235

Ohba M, Tsuchiyama A, Shisa N, Nakashima K, Lee D-H, Ohgushi A, Wasano N (2002b) Naturally occurring *Bacillus thuringiensis* in oceanic islands of Japan, Daito-shoto and Ogasawara-shoto. Appl Entomol Zool 37:477–480

Ohba M, Mizuki E, Uemori A (2009) Parasporin, a new anticancer protein group from *Bacillus thuringiensis*. Anticancer Res 29:427–434

Quesada-Moraga E, García-Tóvar E, Valverde-García P, Santiago-Álvarez C (2004) Isolation, geographical diversity and insecticidal activity of *Bacillus thuringiensis* from soils in Spain. Microbiol Res 159:59–71

Rosas-García NM, Mireles-Martínez M, Hernández-Mendoza JL, Ibarra JE (2008) Screening of *cry* gene contents of *Bacillus thuringiensis* strains isolated from avocado orchards in Mexico, and their insecticidal activity towards *Argyrotaenia* sp. (Lepidoptera: Tortricidae) larvae. J Appl Microbiol 104:224–230

Sasaki J, Asano S, Bando H, Iizuka T (1994) Characteristics of *Bacillus thuringiensis* isolated from soil in Hokkaido area of Japan. J Seric Sci Jpn 63:361–366

Shisa N, Wasano N, Ohba M (2002a) Discrepancy between *cry* gene-predicted and bioassay-determined insecticidal activities in *Bacillus thuringiensis* natural isolates. J Invertebr Pathol 81:59–61

Shisa N, Wasano N, Ohgushi A, Lee D-H, Ohba M (2002b) Extremely high frequency of common flagellar antigens between *Bacillus thuringiensis* and *Bacillus cereus*. FEMS Microbiol Lett 213:93–96

Shisa N, Maeda S, Ohba M (2006) Unusual envelopes associated with parasporal inclusions of a mosquitocidal *Bacillus thuringiensis* serovar *fukuokaensis* isolate. J Basic Microbiol 46:64–67

Smith RA, Couche GA (1991) The phylloplane as a source of *Bacillus thuringiensis* variants. Appl Environ Microbiol 57:311–315

Takatsuka J, Kunimi Y (1998) Replication of *Bacillus thuringiensis* in larvae of the Mediterranean flouer moth, *Ephestia kuehniella* (Lepidoptera: Pyralidae): growth, sporulation and insecticidal activity of parasporal crystals. Appl Entomol Zool 33:479–486

Travers RS, Martin PAW, Reichelderfer CF (1987) Selective process for efficient isolation of soil *Bacillus* spp. Appl Environ Microbiol 53:1263–1266

Ueda K, Ohba M, Aizawa K (1991) Serogrouping of *Bacillus thuringiensis* by extracellular heat-stable somatic antigens. Syst Appl Microbiol 14:291–294

Wang J, Boets A, Van Rie J, Ren G (2003) Characterization of *cry1*, *cry2*, and *cry9* genes in *Bacillus thuringiensis* isolates from China. J Invertebr Pathol 82:63–71

Wasano N, Kim K-H, Ohba M (1998) Delta-endotoxin proteins associated with spherical parasporal inclusions of the four Lepidoptera-specific *Bacillus thuringiensis* strains. J Appl Microbiol 84:501–508

Wasano N, Yasunaga-Aoki C, Sato R, Ohba M, Kawarabata T, Iwahana H (2000) Spherical parasporal inclusions of the Lepidoptera-specific and Coleoptera-specific *Bacillus thuringiensis* strains: a comparative electron microscopic study. Curr Microbiol 40:128–131

Wei J-Z, Hale K, Carta L, Platzer E, Wong C, Fang S-C, Aroian RV (2003) *Bacillus thuringiensis* crystal proteins that target nematodes. Proc Natl Acad Sci USA 100:2760–2765

West AW, Burges HD, Dixon TJ, Wyborn CH (1985) Survival of *Bacillus thuringiensis* and *Bacillus cereus* spore inocula in soil: effects of pH, moisture, nutrient availability and indigenous microorganisms. Soil Biol Biochem 17:657–665

Xu Z, Yao B, Sun M, Yu Z (2004) Protection of mice infected with *Plasmodium berghei* by *Bacillus thuringiensis* crystal proteins. Parasitol Res 92:53–57

Yasutake K, Binh ND, Kagoshima K, Uemori A, Ohgushi A, Maeda M, Mizuki E, Yu YM, Ohba M (2006) Occurrence of parasporin-producing *Bacillus thuringiensis* in Vietnam. Can J Microbiol 52:365–372

Yasutake K, Uemori A, Kagoshima K, Ohba M (2007) Serological identification and insect toxicity of *Bacillus thuringiensis* isolated from the island Okinoerabu-jima, Japan. Appl Entomol Zool 42:285–290

Chapter 12
Brevibacillus, Arbuscular Mycorrhizae and Remediation of Metal Toxicity in Agricultural Soils

Juan Manuel Ruiz-Lozano and Rosario Azcón

12.1 Introduction

Increasing industrial and anthropogenic activities have raised the concentrations of toxic metals and have caused environmental pollution in agricultural soils, water and the atmosphere (Amoozegar et al. 2005). In fact, metals such as lead, cadmium, copper, zinc, nickel and mercury are continuously being added to our soils and pose a threat to food safety and have other potential health risks for humans owing to soil-to-plant transfer of metals. Toxic metals can also have considerable detrimental effects on soil ecosystems and the environment (Khan 2005). Metal contamination of soil is especially problematic because of the strong adsorption of many metals to the surfaces of soil particles. From a physiological point of view, metals fall into three main categories (1) essential and basically non-toxic (e.g. Ca and Mg), (2) essential but harmful at high concentrations (typically Fe, Mn, Zn, Cu, Co, Ni and Mo) and (3) toxic (e.g. Hg, Pb or Cd) (Valls and de Lorenzo 2002).

The remediation of soils contaminated with heavy metals is a challenging task because metals are not easily degraded; the dangers they pose are aggravated by their almost indefinite persistence in the environment. At present, methods used for the remediation of soils contaminated by heavy metals, such as physical separation, acid leaching or electrochemical processes, are not suitable for practical applications because of their high cost and low efficiency (Rajkumar et al. 2008). Bioremediation, i.e., the use of living organisms to manage or remediate heavy metal-polluted soils, is an emerging technology. It is defined as the elimination, attenuation or transformation of polluting or contaminating substances by the use of biological processes, and it is a good choice for removing industrial pollution from the environment (Wenzel 2009). In this respect, soil microorganisms play major roles in bioremediation or biotransformation processes (Kinkle et al. 1994; Amoozegar et al. 2005). Unfortunately, metals cannot be biodegraded. However, microorganisms can interact with

J.M. Ruiz-Lozano and R. Azcón (✉)
Departamento de Microbiología del Suelo y Sistemas Simbióticos, Estación Experimental del Zaidín (CSIC), Prof. Albareda, 1, 18008 Granada, Spain
e-mail: rosario.azcon@eez.csic.es

N.A. Logan and P. De Vos (eds.), *Endospore-forming Soil Bacteria*, Soil Biology 27,
DOI 10.1007/978-3-642-19577-8_12, © Springer-Verlag Berlin Heidelberg 2011

these contaminants and transform them from one chemical form to another by changing their oxidation state through the addition of (reduction) or removal of (oxidation) electrons (Tabak et al. 2005). It is well known that microorganisms already living in contaminated environments are often well-adapted to survival in the presence of existing heavy metals and to the oxidation–reduction potential of the sites. In fact, increased resistance against metal pollution comprises one of the best-documented examples of rapid microevolutionary changes in soil organisms that appear to be specifically adapted to constrained habitat conditions. Nonetheless, a cornerstone of this adaptative potential is the battery of specific stress responses that these organisms can deploy, which allow them to respond to the environmental signals by changing their pattern of gene expression (Aertsen and Michiels 2005). Certainly, much more research on bacterial heavy-metal resistance is needed if we are to understand the bacterial evolutionary adaptation mechanisms related to growth and survival in metalliferous environments.

Bacteria are usually the most numerous organisms in soil, with 10^6–10^9 viable cells cm^{-3} (Hafeburg and Kothe 2007). Due to their small sizes, bacteria have a high surface-to-volume ratio and therefore provide large contact areas for interactions with the surrounding environment. Besides their occurrence in high numbers and their high surface-to-volume ratios, it is the negative net charge of their cell envelopes that makes these organisms prone to accumulate metal cations from the environment (Hafeburg and Kothe 2007). However, although prokaryotes are usually the agents responsible for most bioremediation strategies, eukaryotes such as fungi can also transform and degrade contaminants (Tabak et al. 2005).

Microbes exist in complex biogeochemical matrices in subsurface sediments and soils. Their interactions with metals are influenced by a number of environmental factors, including solution chemistry, sorptive reactive surfaces, and the presence or absence of organic ligands and reductants. Microorganisms can interact with metals via many mechanisms, some of which may be used as the basis of potential bioremediation strategies. The major types of interaction are summarized in Fig. 12.1.

In addition to the mechanisms outlined in Fig. 12.1, accumulation of metals by plants (phytoremediation) is becoming an additional established route for the bioremediation of metal contamination (Tabak et al. 2005). Phytoremediation uses plants as a way to extract heavy metals from soil (phytoextraction) or to stabilize the metals in the soil (phytostabilization) and is receiving considerable attention due to its low cost and high efficiency (Mulligan et al. 2001). However, heavy metals at elevated levels are toxic to most plants, impairing their metabolisms and reducing plant growth. The efficiency of phytoremediation relies on the establishment of vital plants with sufficient shoot and root biomass growth, active root proliferation and/or root activities that can support a flourishing microbial consortium assisting phytoremediation in the rhizosphere (Wenzel 2009). Soil microorganisms, including plant root-associated, free-living ones, as well as symbiotic rhizobacteria and mycorrhizal fungi in particular, are an integral part of the rhizosphere biota. The overall result of plant–rhizosphere–microbe interactions is a higher microbial density and metabolic activity in the rhizosphere, even in metal-contaminated soils (Khan 2005). Thus, the interactions between plant roots and

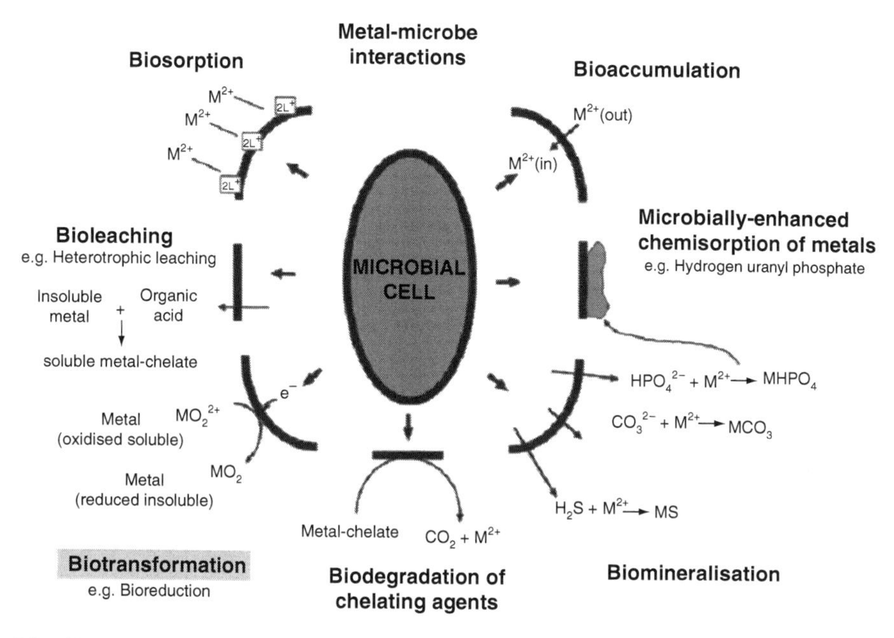

Fig. 12.1 Metal–microbe interactions impacting bioremediation. Taken from Tabak et al. (2005), with kind permission from Springer Science + Business Media

microbes in the rhizosphere may have a great influence both on the increase of nutrient uptake and on the decrease of metal toxicity (Rajkumar et al. 2008). Therefore, the potential use of alternative methods that exploit rhizosphere microbes to reduce the toxicity of metals, or to eliminate these metals, has been proposed (Zaidi et al. 2006).

Beneficial interactions between plants and rhizosphere microorganisms have been demonstrated to alleviate metal toxicity and nutrient deficiency. In fact, inoculation of heavy metal-resistant, rhizosphere bacterial strains that have been isolated from metal-polluted soils substantially improved plant growth and development (Sheng and Xia 2006; Zaidi et al. 2006). This effect is even more evident if metal-resistant bacteria are co-inoculated with indigenous arbuscular mycorrhizal (AM) fungal strains (Vivas et al. 2003a, b, c; 2005a, b; 2006a, b, c). Thus, we need a better understanding of the interactions between beneficial microbes such as AM fungi and heavy metal-resistant rhizobacteria (Whitfield et al. 2004). AM fungi improve the growth and biomass accumulation of plants and their yields mainly through the mobilization of nutrients from the soil (Smith and Read 1997), but also by protecting the host plants from heavy metals in the soil (Vivas et al. 2003a). Many rhizosphere-colonizing bacteria typically produce metabolites, such as siderophores, biosurfactants or organic acids that stimulate plant growth (Glick 1995) and also may reduce metal availability in the medium. So far, AM fungi and bacteria have been found in heavy metal-contaminated soils, which is an indication of fungal and bacterial tolerance (Azcón et al. 2009). Thus, the combination of AM fungi and soil bacteria able to remediate metal toxicity is a promising strategy that is being analyzed in several research projects. These projects allowed isolation of several AM fungal and *Brevibacillus* strains from lead-, nickel-, cadmium- and zinc-contaminated soils (Kádár 1995), showing enhanced tolerances of these metals in comparison with culture collection strains (Vivas et al. 2003a, b, c; 2005a, b; 2006a, b, c). The results obtained so far are summarized in the following section.

12.2 Remediation of Metal Toxicity in Agricultural Soils by *Brevibacillus* and Arbuscular Mycorrhizae

12.2.1 Lead

The pollution of soils with lead owing to human activities poses a major environmental and human health problem (Leyval et al. 1997). The sources of lead in the soil are diverse, including the burning of fossil fuels, mining and smelting of metalliferous ores, municipal wastes, fertilizers, pesticides, sewage sludge, pigments and spent batteries (Mercier et al. 2002). There are public health implications if plant foods accumulate high concentrations of leads; furthermore, leaf fall and dispersal may pose a hazard (Pichtel et al. 2000).

An agricultural soil from Nagyhörcsök Experimental Station (Hungary) was contaminated in 1991 with suspensions of 13 microelement salts applied separately. Each salt was applied at four levels (0, 30, 90 and 270 mg kg^{-1}) as described by Biró et al. (1998). In a later study Vivas et al. (2003a) isolated an indigenous bacterial strain (initially named strain B) from the Pb-polluted soil 10 years after contamination and tested its influence both on red clover (*Trifolium pratense*) growth and on the functioning of native mycorrhizal fungi in the face of Pb toxicity.

Results showed that bacterial strain B increased plant growth (Fig. 12.2) and nutrient accumulation, as well as legume nodule numbers and mycorrhizal infection, demonstrating its plant-growth-promoting (PGP) activity. In addition, this bacterial strain exhibited a high Pb tolerance when cultivated under increasing Pb levels in the growth medium. Therefore this bacterial strain was subjected to molecular identification. 16S rDNA sequence analysis unambiguously identified strain B as a member of the genus *Brevibacillus*, showing the highest similarity (more than 97%) with the 16S rDNA sequence of *Brevibacillus brevis*.

The mechanisms involved in the positive effect of the isolated *Brevibacillus* sp. as a plant-growth-promoting rhizobacterium (PGPR) are not totally known, but some authors have reported that PGPRs can influence plant development not only directly (production of hormones, siderophores or antibiotics, P solubilization or asymbiotic N fixation), but also indirectly through modifications to the activity of other plant–microbe interactions, such as the mycorrhizal or the *Rhizobium* symbioses (Meyer and Linderman 1986; Azcón 1993; Garbaye 1994; Barea et al. 1996), or by inducing changes in the microbial population balance, for instance by exerting

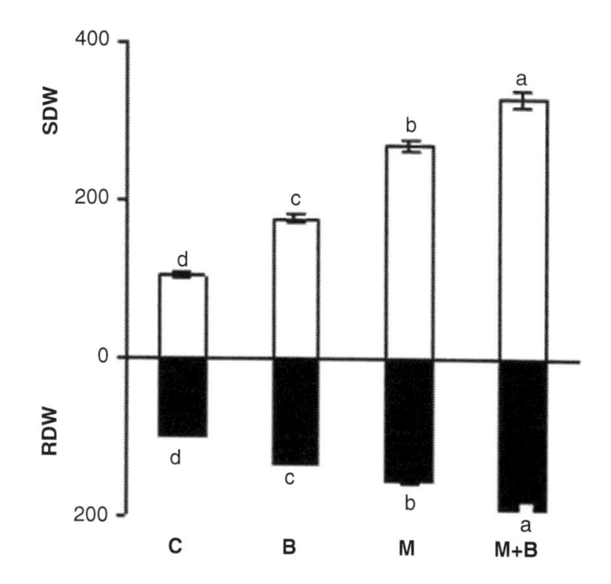

Fig. 12.2 Shoot (SDW) and root (RDW) dry weights (mg plant^{-1}) of red clover plants cultivated in soil amended with 30 mg Pb kg^{-1}. Treatments were C, Control; B, *Brevibacillus* sp.; M, mycorrhizae; M+B, mycorrhizae + *Brevibacillus* sp. a–d, values sharing the same letter are not significantly different according to Duncan's Multiple Range Test ($P < 0.05$). Taken from Vivas et al. (2003a), with kind permission from NRC Research Press

Table 12.1 Ratio of Pb concentration to root-weight-unit (mg Pb g dw root^{-1}) in red clover plants cultivated in soil amended with 30, 90 or 270 mg Pb kg^{-1}

Treatment	30 mg kg^{-1}	90 mg kg^{-1}	270 mg kg^{-1}
C	0.12ab	1.03a	6.60a
B	0.08b	0.74b	4.10b
M	0.07bc	0.35c	4.00b
M+B	0.05c	0.33c	3.30c

Taken from Vivas et al. (2003a), with kind permission from NRC Research Press
Treatments were: C, Control; B, *Brevibacillus* sp.; M, mycorrhizae; M+B, mycorrhizae + *Brevibacillus* sp.; a–c, see legend to Fig. 12.2

biological control against plant pathogens (Weller and Thomashow 1994). This *Brevibacillus* sp. was able to produce 3.8 mg L^{-1} indole-3-acetic acid (IAA) in vitro and this might have contributed to the beneficial effects observed, since the production of IAA or ethylene has been proposed as a mechanism for plant growth promotion under heavy metal stress (Pishchik et al. 2002).

The amount of Pb absorbed per root-weight-unit decreased considerably in plants inoculated with *Brevibacillus* sp. or with AM fungi plus *Brevibacillus* sp. (Table 12.1). It has been proposed that soil bacteria are associated with the clay and organic fractions of the soil microenvironment and would be expected to participate in the metal dynamics typically ascribed to these soil fractions (Mullen et al. 1989). Bacteria have a high surface-to-volume ratio (Beveridge 1988) and, as a strictly physical cellular interface, should have a high capacity for sorbing metals from solutions (Mullen et al. 1989). Our results show an important ability of *Brevibacillus* sp. for Pb biosorption (26% of the biomass weight) that may have contributed to Pb removal from soil and to alleviate Pb toxicity for plants.

Another mechanism by which *Brevibacillus* sp. could have contributed to the protection of plants against Pb toxicity is by stimulation of root exudates. Root exudates have a variety of roles, including metal chelation (Hall 2002). In this study, *Brevibacillus* sp. induced a bigger root system, probably due to IAA activity. Thus, it is likely that the amount of root exudates has also increased. In addition, it is known that the AM mycelium has a high metal sorption capacity compared to other soil microorganisms (Joner et al. 2000). The ability of soil bacteria to stimulate the development of arbuscular mycorrhizae has been well documented (Azcón 1987; Requena et al. 1997: Gryndler et al. 2000). Hence, it is also likely that the bacterial stimulation of extra-radical mycelium production by the AM fungi could have contributed to the protection of host plants from Pb toxicity.

12.2.2 Cadmium

No biological or physiological functions are attributed to cadmium, and it is toxic at certain concentrations (Biró et al. 1995). Based on the chemical properties of Cd, the possibility of conversion to a less toxic form is likely to be very low. High levels of Cd negatively affected soil biomass and had a negative effect on dehydrogenase activity, which was similar to the Cd effect on the microbial biomass (Vivas et al. 2003b). However, microbial inoculation using strains adapted to high Cd concentrations can restore the biomass values, since the stimulation of indigenous soil microbial activity is essential to enhance natural detoxification of contaminated environments (Vörös et al. 1998). In addition, the AM mycelium has been shown to have a high metal sorption capacity (Joner et al. 2000). Thus, studies were conducted with the objective of testing the effectiveness and compatibility of autochthonous *B. brevis* and *Glomus mosseae* strains isolated from a Cd-polluted soil and adapted to Cd toxicity versus a non-autochthonous *G. mosseae* strain, not adapted to Cd, in terms of plant growth, nutrition, heavy metal uptake and symbiotic development under Cd toxicity (Vivas et al. 2003b, c, 2005b).

The interaction between autochthonous *B. brevis* and *G. mosseae* isolated from Cd-polluted soil increased plant growth, AM colonization and physiological characteristics of the AM infection. Both microorganisms showed the highest functional compatibility and benefit to the plant when compared with a reference *G. mosseae* strain (Fig. 12.3).

The dually inoculated (AM fungus plus *B. brevis*) plants achieved further growth and nutrition and less Cd concentration, particularly at the highest Cd level. In fact, these plants acted in reducing Cd concentration and lowering the proportion of Cd soil/plant transfer (Table 12.2), but at the same time they were also able to increase the total Cd accumulation in the shoots, mainly in plants dually inoculated at the two lowest Cd levels (13.6 and 33.0 mg kg^{-1} Cd). Thus, treated plants can also be used to facilitate remediation since microbial ability to increase Cd content in

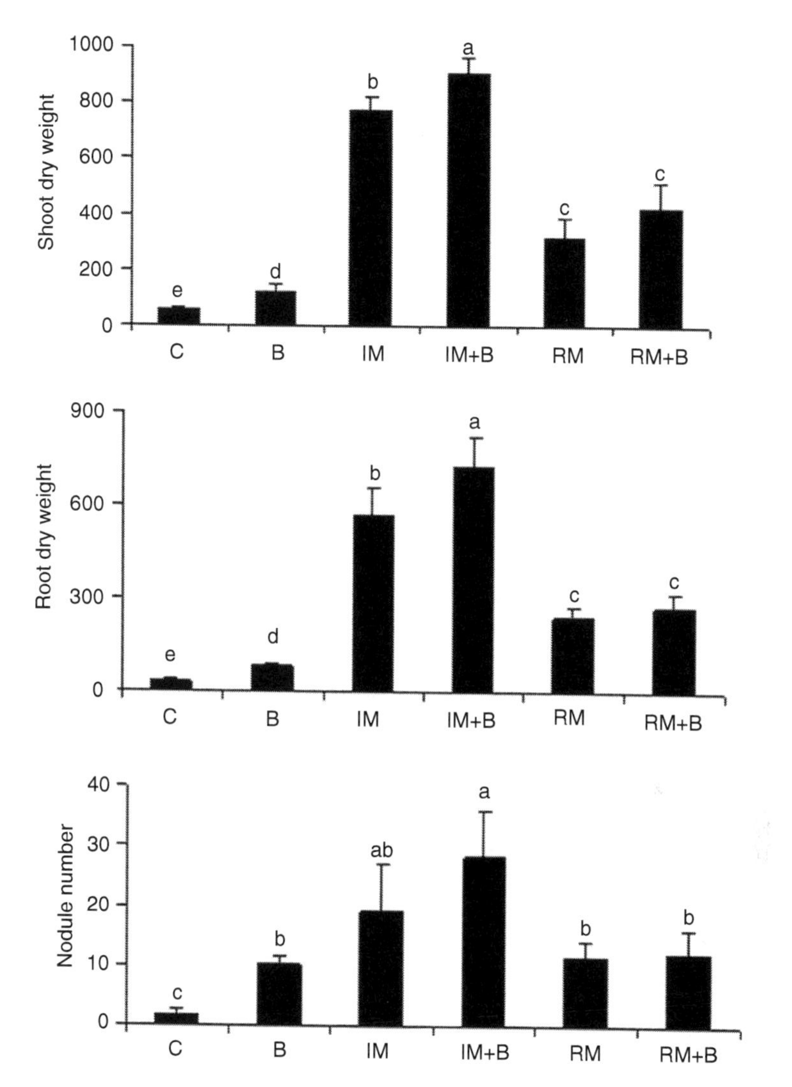

Fig. 12.3 Effect of indigenous (IM) or reference (RM) *Glomus mosseae* strains or *Brevibacillus brevis* (B) in single or in dual inoculation on shoot and root dry weight (mg) and nodule number of clover plants cultivated in soil contaminated with 8 μg Cd g^{-1}; C, control; a–e, see legend to Fig. 12.2. Taken from Vivas et al. (2005b), with kind permission from Elsevier

inoculated plants has the potential to decontaminate the soil in which they are growing. Results from this study provide evidence that the inoculation of *G. mosseae* and *B. brevis* (both adapted to Cd) decreased Cd concentration by 1.5-, 3.3- or 2.8-fold depending on the available Cd in the soil and increased Cd uptake by *Trifolium* plants by two or threefold at 13.6 and 33 mg kg^{-1} of Cd, respectively. These results are consistent with reports suggesting the beneficial use

Table 12.2 Effect of the inoculation of *Brevibacillus brevis* (B) and/or mycorrhizal fungus (M) on soil–plant Cd transfer and cadmium concentration (mg kg^{-1}) in shoots of clover plants growing in soil with increasing cadmium levels; a–d, see legend to Fig. 12.2

Microbial treatments	Available Cd (mg kg^{-1})					
	13.6		33.0		85.1	
	Cd transf.	Cd conc.	Cd transf.	Cd conc.	Cd transf.	Cd conc.
Control	1.20a	16.3a	1.78a	58.8a	2.60a	149.6a
B	1.30a	17.8a	0.82b	27.0b	1.20b	99.7b
M	0.65b	10.7b	0.45c	15.0c	0.88c	75.6c
M+B	0.83b	11.3b	0.30c	17.6c	0.64d	54.4d

Taken from Vivas et al. (2003c), with kind permission from Elsevier

of bacteria or mycorrhizal-colonization in plants growing in metal-contaminated sites (Shetty et al. 1994a, b).

In such a study, *B. brevis* behaved as a PGPR (Kloepper 1992) but its positive effect on root biomass was greater than that observed on the shoot. In a complementary study (Vivas et al. 2005b), *B. brevis* also behaved as a mycorrhizae-helper (MH) bacterium (Garbaye 1994), since it promoted mycorrhizal-colonization (quantitatively and qualitatively) mainly in interaction with autochthonous *G. mosseae*. Azcón (1987) described the stimulation of mycorrhizal root colonization by IAA-producing bacteria since growth promoting compounds were able to stimulate the plant susceptibility to AM infection and the growth of extra-radical AM mycelium. Increased amounts of IAA in the rhizosphere of *B. Brevis*-inoculated plants were also found, and they may have been correlated to the effect of this bacterium not only on root growth and nodule formation, but also on the stimulation of level and activity of mycorrhizal-colonization (Fig. 12.4).

Several mechanisms may be involved in the enhancement of plant Cd tolerance by the co-inoculation of microorganisms, e.g. the effects of increasing nutrients and of decreasing concentrations of metals (Cd, Cr, Mn, Cu, Mo, Fe and Ni). The greater intra- and extra-radical mycorrhizal development and activity by autochthonous microbial strains increased nutrient uptake. Also, the protective effect of AM colonization against Cd toxicity has been explained by the possibility that much Cd was retained in the mycorrhizal roots and thus the translocation to the shoots was inhibited (Weissenhorn and Leyval 1995). Cd biosorption by *B. brevis* seems also to contribute to the effects described in this study. In addition, in this Cd-contaminated soil, rhizospheres from non-inoculated plants showed reduced enzymatic activities, which is an indicator of the perturbations caused to the ecosystem functioning under polluted conditions (Naseby and Lynch 1997). The increase of enzymatic phosphatase, β-glucosidase and dehydrogenase activities in the rhizospheres of inoculated plants could be due to the effect of nutrient leakage from roots (quantitative and/or qualitative changes in root exudates). Microbially inoculated plants showed increased root systems that, in turn, increased carbon and nutrient leakage to the rhizosphere zone, and this could also account for the enhancement of plant Cd tolerance by the co-inoculation of these microorganisms.

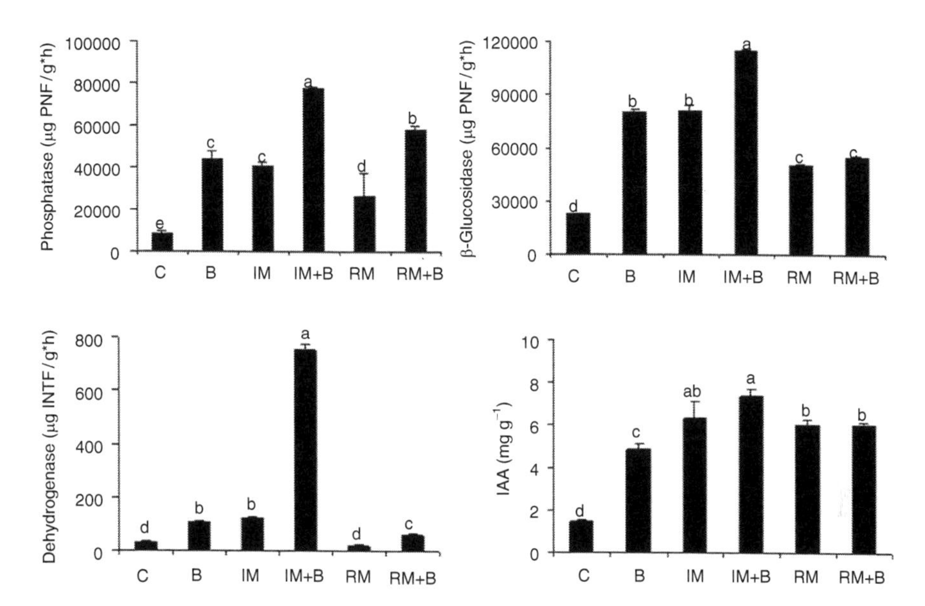

Fig. 12.4 Effect of indigenous (IM) or reference (RM) *G. mosseae* strains or *B. brevis* (B) in single or in dual inoculation with IM or RM on phosphatase, β-glucosidase, dehydrogenase activities and indole acetic acid (IAA) content in rhizospheric soil of *Trifolium* plant cultivated in soil contaminated with Cd; C, control; a–e, see legend to Fig. 12.2. Taken from Vivas et al. (2005b), with kind permission from Elsevier

12.2.3 Nickel

Although nickel is generally known as an essential microelement, excessive concentrations of this and other metals in soils are toxic to plants, bacteria and fungi. Thus, elevated Ni levels might reduce AM-root-colonization and root nodule formation even at levels which did not yet affect plant growth (Chao and Wang 1990). According to Charudhry et al. (1998), Ni was among the most toxic metals to *R. leguminosarum* bv trifoli. Reductions in nodule size and nitrogenase activity were also observed in white clover grown in metal-polluted soils (Märtensson 1992). This evidence suggests that microorganisms are far more sensitive to heavy metal stress than plants growing on the same soils.

In a study carried out with a Ni-contaminated soil 10 years after soil contamination (Kádár 1995) we could isolate two bacterial strains and a *G. mosseae* strain, both adapted to Ni-toxicity. The activities of these microorganisms against Ni toxicity were tested in a study with clover plants in soil contaminated with three different levels of Ni (Vivas et al. 2006b). The two bacterial strains showed different behaviours. Both strains exhibited higher tolerance to Ni than a reference *Brevibacillus* strain when growing axenically in culture media supplemented with Ni (Fig. 12.5), with strain B (identified as *B. brevis*) being more tolerant and effective than strain A.

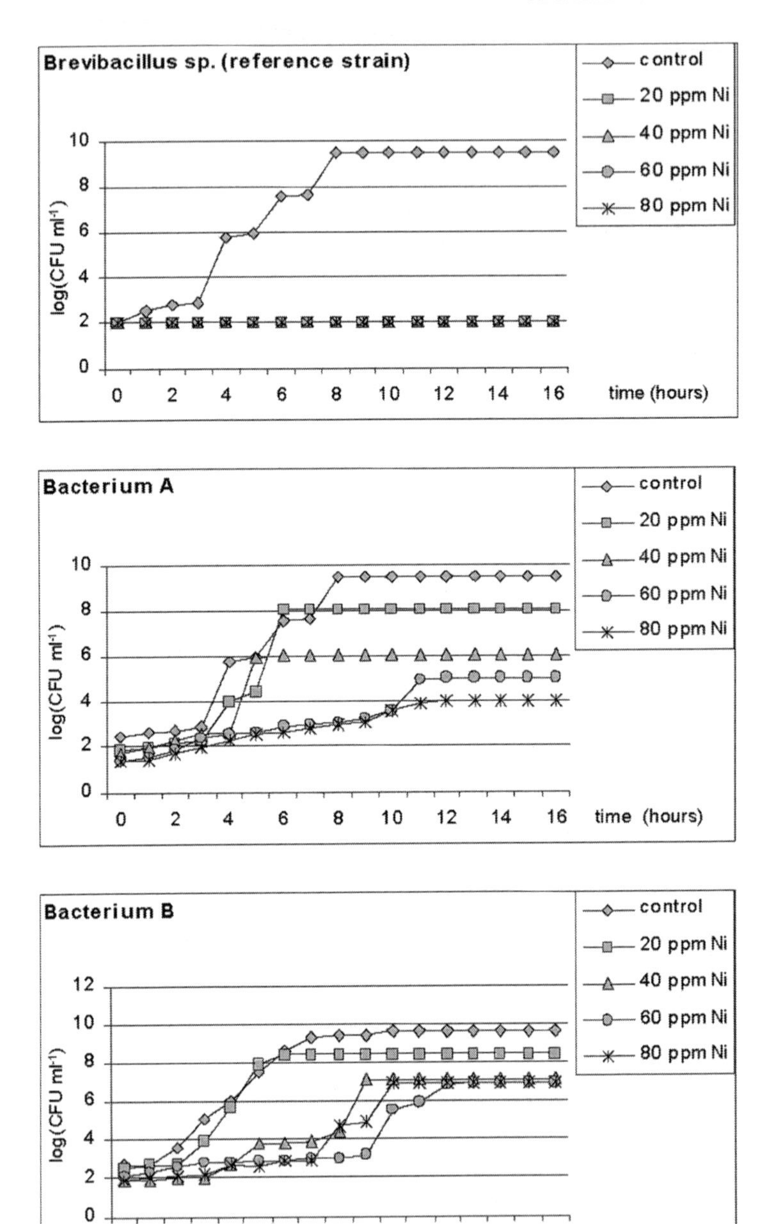

Fig. 12.5 Number of viable cells (log cfu mL^{-1}) at different time intervals of strain A, strain B (*Brevibacillus brevis*) and a reference *Brevibacillus* strain grown in nutrient broth, supplemented with 0 (control), 20, 40, 60 or 80 mg Ni L^{-1}. Taken from Vivas et al. (2006b) with kind permission from Elsevier

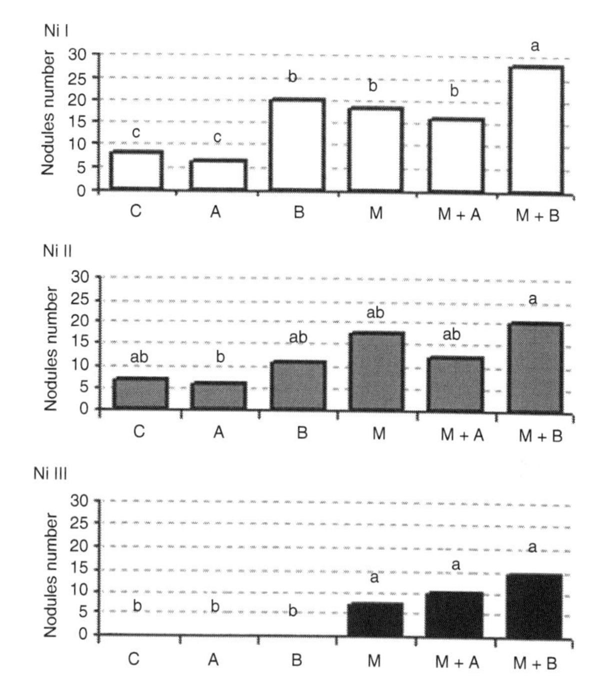

Fig. 12.6 Nodule numbers in red clover grown in soil amended with 30 (Ni I), 90 (Ni II) or 270 mg $NiSO_4$ kg^{-1} (Ni III). Treatments were: C, Control; A, bacterium A; B, *Brevibacillus brevis*; M, indigenous mycorrhizal inoculum; M+A or M+B, Mycorrhizae + bacterium A or *B. brevis*; a–c, see legend to Fig. 12.2. Taken from Vivas et al. (2006b) with kind permission from Elsevier

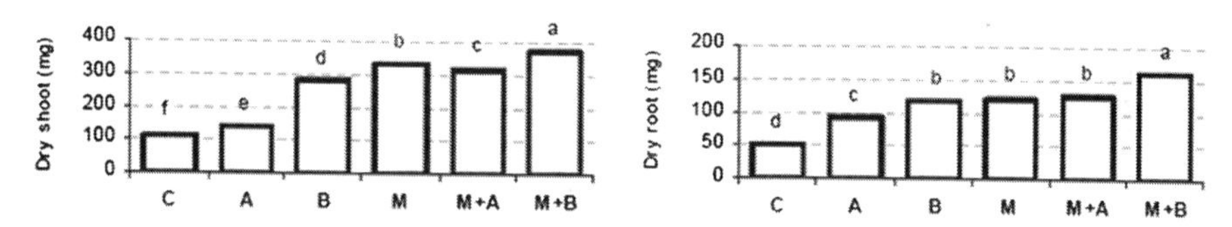

Fig. 12.7 Shoot and root dry weights (mg $plant^{-1}$) in red clover grown in soil amended with 270 mg $NiSO_4$ kg^{-1}; a–f, see legend to Fig. 12.2. Treatments were as described for Fig. 12.6. Taken from Vivas et al. (2006b) with kind permission from Elsevier

Results from this study showed that the formation of symbiotic associations (mainly in the nodules) decreased as available Ni in the soil increased (Fig. 12.6). However, under more stressed Ni conditions the combination of *B. brevis* (strain B) and AM symbiosis were very effective in enhancing growth values (Fig. 12.7) and in decreasing Ni acquisition (Fig. 12.8).

N_2-fixation by *Rhizobium* spp. is one of the most important biological processes, but it is highly sensitive to metal contamination as determined at the highest Ni level used. It is observed that an important role of microbial inoculation of Ni-polluted soils is the ability to improve symbiotic associations by AM colonization

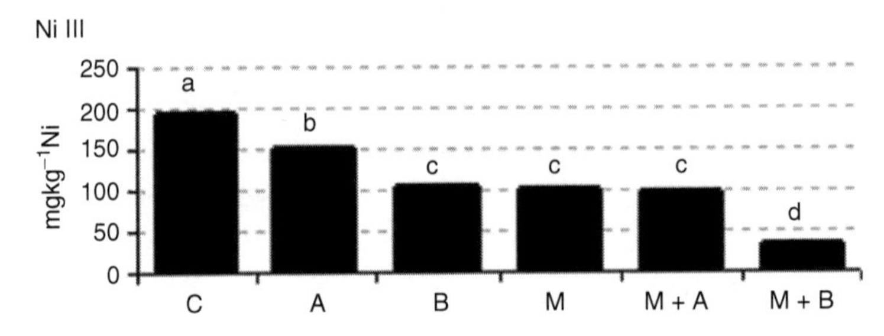

Fig. 12.8 Ni concentration in red clover plants grown in soil amended with 270 mg $NiSO_4$ kg^{-1}; a–d, see legend to Fig. 12.2. Treatments were as described for Fig. 12.6. Taken from Vivas et al. (2006b) with kind permission from Elsevier

and rhizobial nodulation of legume plants. The factors involved in improving nodule formation under such stress conditions may be a combination of effects, including the supply of available P to nodules, which is critical for legume-development in polluted habitats. The bacterial inoculation could, in addition, facilitate plant root development by altering the hormonal balance (Burd et al. 2000). Thus, mycorrhizal-colonization, particularly in *B. brevis*-inoculated plants, was very effective in increasing nodulation whatever the Ni level.

Results also showed that at whatever the level of Ni supplied to the soil, inoculated plants have decreased amounts of Ni absorbed per unit of root mass. Retention of this metal in the root zone would result in a lower amount being transported to the shoot, as found for inoculated plants. Simon et al. (2001) reported the potential importance of Ni availability on toxicity to plants. But results provided by Vivas et al. (2006b) showed that the detrimental effect of Ni could be reduced by the interactive effect of selected beneficial microbes in the soil–plant system. When roots of the plants growing in Ni-polluted soils were co-inoculated with *B. brevis* and the AM fungus *G. mosseae*, the plants' uptake of Ni per mg of root (specific absorption rate, SAR) reduced considerably, and this decreased SAR for Ni seems to be the main effect associated with the improved plant biomass of inoculated plants (Table 12.3).

In the study conducted by Vivas et al. (2006b), AM-colonized roots reduced the total Ni uptake. Both the metal tolerant mycorrhizal inoculum of *G. mosseae* and *B. brevis* protected the plants against the toxic effect of excessive concentration of Ni, by reducing Ni concentration in plant tissues. The microbial associations assayed may potentially affect the uptake of Ni in different ways depending on the combination of factors, such as via hyphal chelation or sequestration in the vacuolar membrane vesicles (Nishimura et al. 1998). Similarly, the bacterial isolates can immobilize metals by a direct mechanism as was demonstrated for Cd (Vivas et al. 2005b), and/or as mycorrhizae by helping organisms to improve extra-radical biomass of the AM fungus (Garbaye 1994).

Mycorrhizal root development when associated with *B. brevis* was less negatively affected by Ni stress conditions. This fact could be particularly relevant for

Table 12.3 Effect of native bacteria A or B (*B. brevis*) and/or native AM inoculum M (*G. mosseae*) on specific absorption rate (SAR, expressed as μg Ni g^{-1} root) values for Ni in plants growing with increasing levels of available Ni in soil; a–f, see legend to Fig. 12.2

Microbial treatments	Available Ni (mg kg^{-1})		
	11.7	27.6	65.8
C	0.070a	0.139a	0.450a
A	0.046b	0.092b	0.310b
B	0.022cd	0.067c	0.166c
M	0.022cd	0.040d	0.177c
M+A	0.032c	0.047d	0.166c
M+B	0.015f	0.018f	0.056d

Taken from Vivas et al. (2006b) with kind permission from Elsevier.

improving plant growth and nutrient acquisition. Different strategies might be involved in preventing plant toxicity damage. Changes in metal uptake and/or internal transportation storage can confer metal tolerance to the host plant (Scholeske et al. 2004). The microbial inocula used (AM fungus and/or *B. brevis*) seem to confer tolerance to Ni by affecting metal availability and uptake. Changes in root exudates, pH and physico-chemical properties of the soil (Grichko et al. 2000) may be involved and such changes could reduce metal root uptake or translocation from root to shoot tissue. The benefits that legume plants obtain from the AM symbiosis under Ni-polluted conditions are more relevant in co-inoculations with *B. brevis*.

12.2.4 Zinc

Zinc is an essential metal for normal plant growth and development since it is a constituent of many enzymes and proteins. However, excessive concentrations of this metal are well known to be toxic to most living organisms. Elevated concentrations of Zn exist in many agricultural soils owing to management practises including application of sewage sludge or animal manure, and from mining activities, and this may represent a risk to environmental quality and sustainable food production (Li and Christie 2001). Zinc only occurs as the divalent cation Zn^{2+}, which does not undergo redox changes under biological conditions. Zinc is a component in a number of enzymes and DNA-binding proteins, for example zinc-finger proteins, which exist in bacteria. In humans, zinc toxicity may be based on zinc-induced copper deficiency; however, zinc is apparently less toxic than copper. In *Escherichia coli*, the toxicity of zinc is similar to that of copper, nickel and cobalt (Crowley and Dungan 2002).

Using white clover (*Trifolium repens*), Vivas et al. (2006c) tested the effect of inoculation with an indigenous bacterial isolate and an AM fungus on Zn tolerance in terms of plant growth, nutrient uptake, Zn acquisition and symbiotic development. The microbial strain used was isolated from a long-term Zn contaminated area from a Hungarian (Nagyhörcsök) experimental field (Kádár 1995). This bacterial strain (B) stimulated biomass production by plants growing in Zn polluted soil, mainly in co-inoculation with the AM fungus (Fig. 12.9) and it was subjected to molecular identification, which unambiguously identified the strain B as *B. brevis*.

Microbially treated plants also showed reduced shoot concentrations of Zn since the proportion of soil/plant Zn transfer was decreased by the inoculants, particularly when in association (Table 12.4).

Most commonly, the mechanism of resistance in prokaryotes is an efflux of the toxic metals by the action of P-type ATPases or secondary efflux systems (Nies and Silver 1995; Paulsen and Saier 1997). An important mechanism in this respect is the synthesis of extracellular polymeric substances, a mixture of polysaccharides, mucopolysaccarides and proteins which can bind significant amounts of potentially toxic metals and entrap precipitated metal sulphides and oxides. In bacteria, peptidoglycan carboxyl groups are main cationic binding sites in Gram-positive

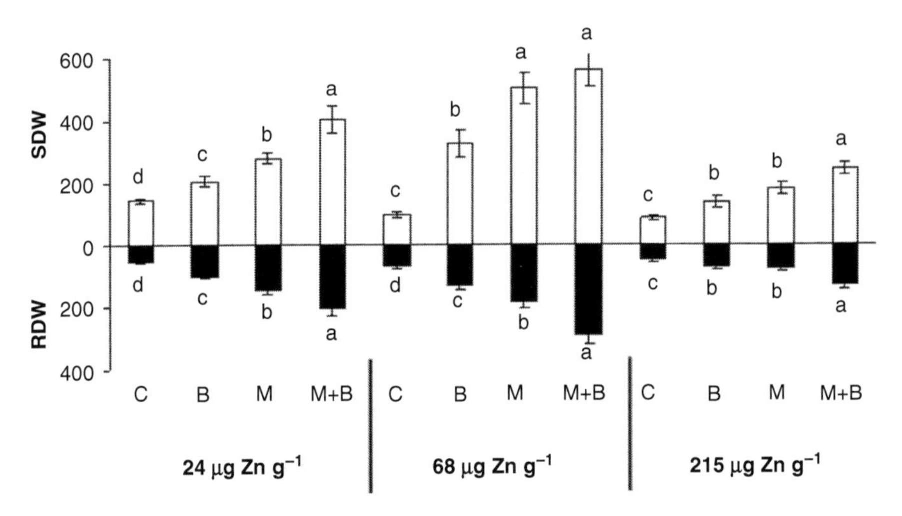

Fig. 12.9 Shoot and root dry weights (mg plant^{-1}) of white clover plants cultivated in soil amended with 24, 68 or 215 μg Zn g^{-1}. Treatments were: C, control; B, *Brevibacillus brevis*; M, mycorrhizae; M+B, mycorrhizae + *B. brevis*; a–d, see legend to Fig. 12.2. Taken from Vivas et al. (2006c), with kind permission from Elsevier

Table 12.4 Ratio Zn concentration to root-weight-unit (μg Zn g^{-1} dw root) in white clover plants cultivated in soil amended with 24, 68 or 215 μg Zn g^{-1}; a–d, see legend to Fig. 12.2

Treatment	24 μg g^{-1}	68 μg g^{-1}	215 μg g^{-1}
C	0.49a	0.45a	1.06a
B	0.13b	0.10b	0.37b
M	0.05c	0.05c	0.40c
M+B	0.03d	0.02d	0.13d

Taken from Vivas et al. (2006c), with kind permission from Elsevier.
Treatments were as described for Fig. 12.9.

species. Chitin, phenolic polymers and melanins are important structural components of fungal walls and these are also effective biosorbents for metals.

The *B. brevis* strain studied possesses cellular mechanisms (biosorption and bioaccumulation) involved in the detoxification of Zn in the growth medium. Such metabolic abilities may be related to the Zn tolerance and also to the Zn reduction in the medium (Zhou 1999). However, *Brevibacillus* cells accumulated only a 5.6% of Zn from a culture medium supplemented with Zn (Vivas et al. 2006c). Thus, this biomass sorption did not totally explain the effects found. The data obtained rather suggest that the ability of the bacteria tested to protect plants against the inhibitory effects of high concentrations of Zn is related to their capacity to stimulate plant growth by synthesis of IAA, as has been shown for many PGPRs (Burd et al. 2000). A low level of IAA produced by rhizobacteria promotes primary root elongation, whereas a high level of IAA stimulates lateral and adventitious root formation but inhibits primary root growth (Xie et al. 1996). Thus, PGP bacteria can facilitate plant growth by altering the plant hormonal balance.

In addition, in AMF-colonized plants, the expression of genes encoding plasma membrane transporters affecting element accumulation by plants has been reported (Burleigh and Bechmann 2002). The expression of a Zn transporter gene (*MtZIP2*) was decreased in roots of mycorrhizal plants at a high Zn concentration of 100 mg g^{-1}, as described by Burleigh et al. (2003). Thus, concentration of Zn in tissues of AM plants was lower than in non-mycorrhizal plants growing under Zn contamination. Also, González-Guerrero et al. (2005) suggested the role of *GintZnT1*, encoding a putative Zn transporter in Zn compartmentalization and in the protection of *Glomus intraradices* against Zn stress. Thus, if AM fungal-colonized plants accumulate Zn in the roots such as Chen et al. (2003) observed, and if on the other hand bacteria can also bind some (5.6%) of this metal, apart from the benefit to plant growth and nutrition, the additive effects can explain the alleviation of the detrimental effects caused by Zn in dually inoculated plants.

In another study aimed at investigating the effect of *B. brevis* on the axenic development of *G. mosseae* (its rate of spore germination and hyphal development) under different levels of Zn added to a water agar medium, we observed that axenic development of the AM fungus was enhanced under increasing levels of Zn when *B. brevis* was inoculated (Vivas et al. 2005a). Bacterial inoculation was a critical factor for reaching maximum growth rates of germination and mycelial development under non-polluted, and particularly, under polluted conditions. In fact, *G. mosseae* isolated from polluted sites required this biological influence from bacteria to achieve greater development. Spores of *G. mosseae* demonstrated a 56% increase in mycelial growth (without Zn) and a 133% increase (with 200 μg Zn ml^{-1}) when inoculated with the bacterium, in comparison with uninoculated spores (Table 12.5). Moreover, in a subsequent study, the same bacterium not only stimulated presymbiotic AM fungal development but also the quantity and quality (metabolic characteristics) of mycorrhizal-colonization, with the highest improvement being for arbuscular vitality and activity (Vivas et al. 2006a).

Table 12.5 Effect of *Brevibacillus brevis* on mycelial (hyphal) growth from *G. mosseae* spores on axenic medium supplemented with three different concentrations of Zn after 3 weeks of incubation; a–c, see legend to Fig. 12.2

Treatment	Zn levels (μg mL^{-1})	Hyphal development (mm $spore^{-1}$)	% of control treatment
G. mosseae	0	16.0b	100
	10	12.0b	75
	50	6.0c	38
	200	3.0c	19
G. mosseae + *B. brevis*	0	25.0a	156
	10	15.0b	94
	50	10.0bc	63
	200	7.0c	44

Taken from Vivas et al. (2005a), with kind permission from Springer Science + Business Media

12.3 Conclusions and Future Prospects

Current research in biotechnology includes investigations that use plants to facilitate remediation. Results highlight that the most adapted rhizosphere microbial strains (fungi and/or bacteria) tend to cope and to prevail in heavy metal-contaminated soils, and microbial interactions seem to be crucial for plant survival in heavy metal-polluted soils. Hyphae of metal-adapted AM fungi may have the capacity to bind metals present in roots or in the rhizosphere and this activity would decrease metal translocation from the root to the shoot, which has been proposed as a mechanism for enhancing plant tolerance. On the other hand, metal-adapted *Brevibacillus* strains have demonstrated metal biosorption abilities, and PGP and MH activities. Thus, dual inoculation of plants with native *Brevibacillus* strains and AM fungi seems to be a strategy which can be recommended for promoting plant growth in heavy metal-polluted soils. The isolation of efficient metal-adapted microorganisms may be an interesting biotechnological tool for inoculation purposes in contaminated soils. Knowledge of the mechanisms by which mycorrhizal adapted fungi and bacteria alleviate heavy metal toxicity in plants growing in contaminated environments would allow the management of microbial groups with suitable characteristics to be used in bioremediation purposes.

The future research should also take into account the combination of these microbial activities with those of transformed agrowaste residues. In fact, in order to successfully establish the plant cover in a contaminated soil, biological and physical soil characteristics need to be improved. To reach this objective, the application of transformed organic agrowaste residues has been proposed as an interesting and sustainable tool (Medina et al. 2004, 2006). Lignocellulosic materials can be used after biotransformation processes, as organic amendments. Agrowastes treated with *Aspergillus niger* and supplemented with rock phosphate (RP) have been applied in soil alone or together with PGP microorganisms such as AM fungi for revegetation purposes (Medina et al. 2006). During the agrowaste fermentation process by *A. niger*, RP is solubilized and the cellulosic product is transformed into more simple sugar compounds. These simple sugars can be used as energy sources for heterotrophic microorganisms that require such compounds for growth and metabolic activities (Bowen and Rovira 1999).

The use of agrowastes as soil amendments represents an input of organic matter that can improve soil structure and fertility (Roldán et al. 1996). Moreover, physical immobilization of heavy metals in soil could be accomplished using amendments since organic matter makes strong complexes with heavy metals (Bolan and Duraisamy 2003; Hartley et al. 2004). The influence of the amendment on AM intra- and extra-radical mycelium in contaminated soils is also important. Previous studies made by our research group have shown the positive effect of the application of *A. niger*-treated sugar beet in improving the stability of soil aggregates, soil biological characteristics, enzymatic activities, AM functionality and plant growth in single or multicontaminated soils (Medina et al. 2006; Vivas et al. 2009).

Also, *A. niger*-treated dry olive cake residue, having high lignin and phenolic contents, seems to have a direct effect on plant growth by improving plant nutrition and decreasing Cd bioavailability in soil (Medina, personal communication). Another positive effect may be the recovery of soil properties, as has been reported earlier. Microbial (*A. niger*) degradation of agrowastes of lignocellulosic composition provides compounds rich in polysaccharides and available P (from RP solubilization). The metabolic (enzymatic) activity of rhizosphere microorganisms involved in nutrient cycling increased with amendments applied to contaminated soils.

Our results showed that the combination of PGPR such as *Brevibacillus*, AM fungi and the transformed agrowaste amendments is a successful biotechnological tool for improving plant growth in contaminated soils. Thus, we conclude that this combined sustainable system can be regarded as an important strategy in order to improve bioremediation of heavy metal-contaminated soils.

12.4 Protocols

12.4.1 Molecular Identification of Bacteria

Total DNA from bacterial isolates is obtained as described by Giovannetti et al. (1990) and characterized by sequence analysis of the small ribosomal subunit (16S ribosomal DNA). PCR amplification is carried out with the eubacterial primers 27f (5′-GAGAGTTTGATCCTGGCTCAG-3′) and 1495r (5′-CTACGGCTACCTTGT-TACGA-3′) (Lane 1991) located, respectively, at the extreme 5′ and 3′ ends of the ribosomal rDNA sequence, which allows the amplification of nearly the entire gene. The amplification reactions are performed in a 20-ml volume containing 0.5 mM concentrations of each primer, 100 mM dNTPs, 1× PCR buffer (Sigma, St. Louis, MO, USA), 2.5 mM $MgCl_2$, 10 ng of genomic DNA and 0.25 U Taq DNA polymerase (Sigma). The Thermal Cycler is programmed with the following parameters: initial denaturation at 95°C for 4 min, followed by 30 cycles of denaturation at 94°C for 30 s, annealing at 56°C for 45 s, elongation at 72°C for 1 min and a final elongation at 72°C for 5 min. The amplified DNA is then purified following electrophoresis through a 1.2% agarose gel with the QIAEX II Gel Extraction Kit (Qiagen, Hilden, Germany) and cloned into a T-A-type plasmid (e.g. pGEM plasmid, Promega) for sequencing. The sequence obtained is then subjected to database searches for 16S rDNA sequence similarity using FASTA and BLAST algorithms, which will allow identification of the closest bacterial relative.

12.4.2 Bacterial Capability for Metal Biosorption

The study of the metal biosorption ability for different bacterial strains can be carried out as described by Kanazawa and Mori (1996) with some modifications. Briefly, bacteria are grown in 250 mL of nutrient broth until reaching one unit of optical density (600 nm). Then the cells are harvested by centrifugation at 7,000 $\times g$ for 30 min, and the bacterial pellet is washed twice with Ringer's solution (NaCl 0.85%, $CaCl_2$ 0.03%, KCl 0.025%, $NaHCO_3$ 0.02%). The harvested biomass is incubated for 1 h at 28°C with a solution containing known amounts of the metal under study (e.g. 267 μg Zn mL^{-1} as Zn $SO_4.7H_2O$, in the case of Zn as the metal under study).

The suspension is then centrifuged at 7,000 × g and filtered through a 0.45-μm Millipore membrane to separate the biomass from the filtrate. The biomass is dried, weighed and heavy metals are extracted by nitric acid (24 h). The metal content of both biomass and supernatants is determined by atomic absorption spectrometry. The metal biosorption capacity is determined by the ratio between the metal present in the bacterial biomass and that present in the supernatant.

12.4.3 Bacterial Production of Indole-3-Acetic Acid

The production of IAA by bacteria can be measured by the method of Wöhler (1997). The bacteria are grown overnight in nutrient broth and then collected by centrifugation at 1,000 $\times g$ for 5 min. The bacterial pellet is then incubated at 37°C for 24 h with 3 mL of phosphate buffer (pH 7.5) with glucose (1%) and 2 mL of L-tryptophan (1%). After incubation, 2 mL of 5% trichloroacetic acid and 1 mL of 0.5 M $CaCl_2$ are added to inactivate the enzymes involved in the bioassay of auxin. The resulting solution is filtered through Whatman no. 2 filter paper. Three mL of the filtrate are transferred to a test tube and mixed with 2 mL of salper solution (2 mL 0.5 M $FeCl_3$ and 98 mL 35% perchloric acid). The mixture is incubated for 30 min at 25°C in the dark. Then the absorbance of the resulting solution is measured at 535 nm. The calibration curve ranges from 0.5 to 10 μg IAA mL^{-1}.

12.4.4 Bacterial Capacity to Tolerate Increasing Amounts of Metal in the Medium

The growth of the bacterial strains isolated from heavy metal-polluted soils can be tested in comparison with reference bacterial strains from culture collections (Vivas et al. 2003a). Bacterial strains are cultivated at 28°C in nutrient broth (8 g L^{-1}) supplemented with increasing amounts of the metal under study (e.g. 0, 25, 50, 75 or 100 μg mL^{-1} Zn, supplied as $ZnSO_4$). The number of viable cells is estimated as cfu mL^{-1} at 1-h intervals from 0 to 16 h following the conventional procedure: 1 mL of suspension is spread-plated on agar nutrient broth medium (8 g L^{-1}).

12.4.5 Specific Metal Absorption Rate (SAR)

SAR for a given metal is defined as the amount of metal absorbed per unit of root biomass (Gray and Schlesinger 1983) and is calculated as follows:

SAR = Plant metal content (μg)/Root mass (g).

12.4.6 Bacterial Ability to Stimulate Germination and Development of AM Fungal Spores in Presence or Absence of Heavy Metal

The study is conducted in 9-cm diameter Petri dishes containing water-agar (1% Difco Bacto agar) adjusted to pH 7.0 with or without addition of a range of heavy metal (e.g. 0, 10, 50, or 200 μg mL^{-1} Zn levels as $ZnSO_4$). Spores of an AM fungus are obtained from the AM inoculum by the wet sieving and decanting technique (Ann et al. 1990). The spores are surface sterilized with Chloramine-T, streptomycin and Tween 80 mixture for 20 min and then washed five times in sterile water (Mosse 1962). A total of five surface-sterilized spores of the AM fungus are transferred by sterile capillary pipettes to each Petri dish and located on the vertices of a pentagon of about 3.5 cm each side. The bacterium grown in nutrient broth medium for 24–48 h at 28°C is inoculated (0.5 mL having 10^8 cfu mL^{-1}) in the centre of the dish, equidistant from the five AM spores.

The incubation is carried out at 25°C in the dark and the plates are sealed with parafilm to reduce dehydration and contamination risks. At least seven replicates per treatment must be run. Germination rate and hyphal growth of AM spores are assessed after 3 weeks of incubation. An AM spore is considered germinated if a germ tube is clearly visible. Hyphal development is determined as described by Hepper and Jakobsen (1983).

12.4.7 Soil Enzymatic Activities

In rhizosphere soil samples, acid phosphatase activity is determined using *p*-nitrophenyl phosphate disodium (PNPP, 0.115 M) as substrate. Two millilitres of 0.5 M sodium acetate buffer adjusted to pH 5.5 using acetic acid (Naseby and Lynch 1997) and 0.5 ml of substrate are added to 0.5 g of soil and incubated at 37°C for 90 min. The reaction is stopped by cooling at 2°C for 15 min. Then, 0.5 ml of 0.5 M $CaCl_2$ and 2 ml of 0.5 M NaOH are added, and the mixture is centrifuged at 15,000 $\times g$ for 5 min. The *p*-nitrophenol (PNP) formed is determined in a spectrophotometer at 398 nm (Tabatabai and Bremner 1969). Controls are made in the same way, although the substrate is added after the $CaCl_2$ and NaOH.

β-glucosidase is determined using *p*-nitrophenyl-β-D-glucopyranoside (PNG, 0.05 M; Masciandaro et al. 1994) as substrate. Two millilitres of 0.1 M maleate buffer (pH 6.5) and 0.5 mL of substrate are added to 0.5 g of sample and incubated at 37°C for 90 min. The reaction is stopped with tris-hydroxymethyl aminomethane, according to Tabatabai (1982). The amount of PNP is determined in a spectrophotometer at 398 nm (Tabatabai and Bremner 1969).

Dehydrogenase activity is determined according to García et al. (1997). For this, 1 g of soil at 60% of its water holding capacity is exposed to 0.2 mL of 0.4% INT (2-*p*-iodophenyl-3-*p*-nitrophenyl-5-phenyltetrazolium chloride) in distilled water for 20 h at 22°C in darkness. The INTF (iodo-nitrotetrazolium formazan) formed is extracted with 10 mL of methanol by shaking vigorously for 1 min and filtering through a Whatman no. 5 filter paper. INTF is measured spectrophotometrically at 490 nm.

References

Aertsen A, Michiels CV (2005) Diversify or die: generation of diversity in response to stress. Crit Rev Microbiol 31:69–78

Amoozegar MA, Hamedi J, Dadashipour M, Shariatpahahi S (2005) Effect of salinity on the tolerance to toxic metals and oxyanions in native moderately halophilic spore-forming bacilli. World J Microbiol Biotechnol 21:1237–1243

Ann ZQ, Hendrix JW, Hershman DE, Henson GT (1990) Evaluation of the "most probable number" (MPN) and wet-sieving methods for determining soil-borne populations of endogonaceous mycorrhizal fungi. Mycologia 82:516–518

Azcón R (1987) Germination and hyphal growth of *Glomus mosseae in vitro*: effects of rhizosphere bacteria and cell-free culture media. Soil Biol Biochem 19:417–419

Azcón R (1993) Growth and nutrition of nodulated mycorrhizal and non-mycorrhizal *Hedysarum coronarium* as a result of treatments with fractions from a plant growth-promoting rhizobacteria. Soil Biol Biochem 25:1037–1042

Azcón R, Medina A, Roldán A, Biró B, Vivas A (2009) Significance of treated agrowaste residue and autochthonous inoculates (arbuscular mycorrhizal fungi and *Bacillus cereus*) on bacterial community structure and phytoextraction to remediate soils contaminated with heavy metals. Chemosphere 75:327–334

Barea JM, Tobar RM, Azcón-Aguilar C (1996) Effect of a genetically modified *Rhizobium meliloti* inoculant on the development of arbuscular mycorrhizas, root morphology, nutrient uptake and biomass accumulation in *Medicago sativa*. New Phytol 134:361–369

Beveridge TJ (1988) The bacterial surface: general considerations towards design and function. Can J Microbiol 34:363–372

Biró B, Bayoumi HEAF, Balázsy S, Kecskés M (1995) Metal sensitivity of some symbiotic N_2-fixing bacteria and *Pseudomonas* strains. Acta Biol Hung 46:9–16

Biró B, Köves-Péchy K, Vörösm I, Kádár I (1998) Toxicity of some field applied heavy metal salts to the rhizobial and fungal microsymbionts of alfalfa and red clover. Agrokém Talaj 47:265–276

Bolan NS, Duraisamy VP (2003) Role of inorganic and organic soil amendments on immobilisation and phytoavailability of heavy metals: a review involving specific case studies. Aust J Soil Res 41:533–555

Bowen GD, Rovira AD (1999) The rhizosphere and its management to improve plant growth. Adv Agron 66:1–102

Burd IG, Dixon DG, Glick BR (2000) Plant growth promoting bacteria that decrease heavy metal toxicity in plants. Can J Microbiol 46:237–245

Burleigh SH, Bechmann IE (2002) Plant nutrient transporter regulation in arbuscular mycorrhizas. Plant Soil 244:247–251

Burleigh S, Kristensen BK, Bechmann IA (2003) Plasma membrane zinc transporter from *Medicago truncatula* is upregulated in roots by Zn fertilization, yet down-regulated by arbuscular mycorrhizal colonization. Plant Mol Biol 52:1077–1088

Chao CC, Wang YP (1990) Effects of heavy metals on the infection of vesicular arbuscular mycorrhizae and the growth of maize. J Agric Assoc China 153:34–45

Charudhry TM, Hayes WJ, Khan AG, Khoo CS (1998) Phytoremediation-focusing on accumulator plants that remediate metal-contaminated soils. Aust J Ecotoxicol 4:37–51

Chen BD, Li XL, Tao HQ, Christie P, Wong MH (2003) The role of arbuscular mycorrhizal in zinc uptake by red clover growing in a calcareous soil spiked with various quantities of zinc. Chemosphere 50:839–846

Crowley DE, Dungan RS (2002) Encyclopaedia of environmental microbiology, vol 4. Gabriel Bitton Wiley Interscience, New York, pp 1878–1892

Garbaye J (1994) Helper bacteria: a new dimension to the mycorrhizal symbiosis. New Phytol 128:197–210

García C, Hernández MT, Costa F (1997) Potential use of dehydrogenase activity as an index of microbial activity in degraded soils. Commun Soil Sci Plant Nutr 28:123–134

Giovannetti L, Ventura S, Bazzicalupo M, Fani R, Materassi R (1990) DNA restriction fingerprint analysis of the soil bacterium *Azospirillum*. J Gen Microbiol 136:1161–1166

Glick BR (1995) The enhancement of plant growth by free-living bacteria. Can J Microbiol 41:109–117

González-Guerrero M, Azcón-Aguilar C, Mooney M, Valderas A, MacDiarmid CW, Eide DJ, Ferrol N (2005) Characterization of a Glomus intraradices gene encoding a putative Zn transporter of the cation diffusion facilitator family. Fung Genet Biol 42:130–140

Gray JT, Schlesinger WH (1983) Nutrient use by evergreen and deciduous shrubs in Southern-California. 2. Experimental investigations of the relationship between growth, nitrogen uptake and nitrogen availability. J Ecol 71:43–56

Grichko VP, Filby B, Glick BR (2000) Increased ability of transgenic plants expressing the bacterial *enzyme ACC deaminase to accumulate Cd, Co, Cu, Ni, Pb and Zn*. J Biotechnol 81:45–53

Gryndler M, Hrselová H, Stríteská D (2000) Effect of soil bacteria on hyphal growth of the arbuscular mycorrhizal fungus *Glomus claroideum*. Folia Microbiol 45:545–551

Hafeburg G, Kothe E (2007) Microbes and metals: interactions in the environment. J Basic Microbiol 47:453–467

Hall JM (2002) Cellular mechanisms for heavy metal detoxification and tolerance. J Exp Bot 53:1–11

Hartley W, Eduards R, Lepp WN (2004) Arsenic and heavy metal mobility in iron oxide-amended contaminated soils as evaluated by short and long term leaching tests. Environ Pollut 131:495–504

Hepper CM, Jakobsen I (1983) Hyphal growth from spores of the mycorrhizal fungus *Glomus caledonius*: effect of amino acids. Soil Biol Biochem 15:55–58

Joner EJ, Briones R, Leyval C (2000) Metal-binding capacity of arbuscular mycorrhizal mycelium. Plant Soil 226:227–234

Kádár I (1995) Contamination of the soil–plant–animal–man foodchain by chemical elements in Hungary. Ministry of Environmental Protection and Land Management, Budapest (In Hungarian)

Kanazawa S, Mori K (1996) Isolation of cadmium-resistant bacteria and their resistance mechanisms. Part II. Cadmium biosorption by Cd resistant and sensitive bacteria. Soil Sci Plant Nutr 42:731–736

Khan AG (2005) Role of soil microbes in the rhizosphere of plants growing on trace metal contaminated soils in phytoremediation. J Trace Elem Med Biol 18:355–364

Kinkle BK, Sadowsky MJ, Johanstone K, Koskinen WC (1994) Tellurium and selenium resistance in Rhizobia and its potential use for direct isolation of *Rhizobium meliloti* from soil. Appl Environ Microbiol 60:1674–1677

Kloepper JW (1992) Plant growth-promoting rhizobacteria as biological control agents. In: Blaine F, Metting J Jr (eds) Applications in agriculture forestry and environmental management. Marcel Dekker, New York, pp 255–274

Lane DJ (1991) 16S/23S rRNA sequencing. In: Stackebrandt E, Goodfellow M (eds) Nucleic acid techniques in bacterial systematics. Wiley, New York, pp 115–147

Leyval C, Turnau K, Haslwandter K (1997) Effect of heavy metal pollution on mycorrhizal colonization and function: physiological, ecological and applied aspects. Mycorrhiza 7:139–153

Li XL, Christie P (2001) Changes in soil solution Zn and pH and uptake of Zn by arbuscular mycorrhizal red clover in Zn-contaminated soil. Chemosphere 42:201–207

Märtensson AM (1992) Effects of agrochemicals and heavy metals on fast-growing rhizobia and their symbiosis with small-seeded legumes. Soil Biol Biochem 24:435–445

Masciandaro G, Ceccanti B, García C (1994) Anaerobic digestion of straw and piggery wastewater: II. Optimization of the process. Agrochimica 3:195–203

Medina A, Vassilev N, Alguacil MM, Roldán A, Azcón R (2004) Increased plant growth, nutrient uptake, and soil enzymatic activities in a desertified Mediterranean soil amended with treated residues and inoculated with native mycorrhizal fungi and a plant growth-promoting yeast. Soil Sci 169:260–270

Medina A, Vassileva M, Barea JM, Azcón R (2006) The growth enhancement of clover by *Aspergillus*-treated sugar beet waste and *Glomus mosseae* inoculation in Zn contaminated soil. Appl Soil Ecol 33:87–98

Mercier G, Duchesne J, Carles-Gibergues A (2002) A simple and fast screening test to detect soils polluted by lead. Environ Pollut 118:285–296

Meyer JR, Linderman RG (1986) Response of subterranean clover to dual inoculation with vesicular-arbuscular mycorrhizal fungia and a plant growth-promoting bacterium, *Pseudomonas putida*. Soil Biol Biochem 18:185–190

Mosse B (1962) The establishment of vesicular-arbuscular mycorrhiza under aseptic conditions. J Gen Microbiol 27:509–520

Mullen MD, Wolf DC, Gerris FG, Beveridge TJ, Flemming CA, Bailey GW (1989) Bacterial sorption of heavy metals. Appl Environ Microbiol 55:3143–3149

Mulligan CN, Yong RN, Gibs BF (2001) Remediation technologies for metal-contaminated soils and groundwater: an evaluation. Eng Geol 60:193–207

Naseby DC, Lynch JM (1997) Rhizosphere soil enzymes as indicators of perturbation caused by a genetically modified strain of *Pseudomonas fluorescens* on wheat seed. Soil Biol Biochem 29:1353–1362
Nies DH, Silver S (1995) Ion efflux systems involved in bacterial metal resistances. J Ind Microbiol 14:186–199
Nishimura K, Igarashi K, Kakinuma Y (1998) Protongradient-driven nickel uptake by vacuolar membrane vesicles of *Saccharomyces cerevisiae*. J Bacteriol 180:1962–1964
Paulsen IT, Saier MH Jr (1997) A novel family of ubiquitous heavy metal ion transport proteins. J Membr Biol 156:99–103
Pichtel J, Kuroiwa K, Sawyerr HT (2000) Distribution of Pb, Cd and Ba in soils and plants of two contaminated sites. Environ Pollut 110:171–178
Pishchik VN, Vorobyev NI, Chernyaeva LL, Timofeeva SV, Kozhemyakov AP, Alexeev YV, Lukin SM (2002) Experimental and mathematical simulation of plant growth promoting rhizobacteria and plant interaction under cadmium stress. Plant Soil 243:173–186
Rajkumar M, Ma Y, Freitas H (2008) Characterization of metal-resistant plant-growth promoting *Bacillus weihenstephanensis* isolated from serpentine soil in Portugal. J Basic Microbiol 18:500–508
Requena N, Jimenez I, Toro M, Barea JM (1997) Interactions between plant-growth-promoting rhizobacteria (PGPR), arbuscular mycorrhizal fungi and *Rhizobium* spp. in the rhizosphere of *Anthyllis cytisoides*, a model legume for revegetation in mediterranean semi-arid ecosystems. New Phytol 136:667–677
Roldán A, Albaladejo J, Thornes J (1996) Aggregate stability changes in a semiarid soil after treatment with different organic amendments. Arid Soil Res Rehabil 10:139–148
Scholeske S, Maetz M, Schneider T, Hildebrandt U, Bothe H, Povh B (2004) Element distribution in mycorrhizal and non-mycorrhizal roots of the halophyte Aster tripolium determined by proton induced X-ray emission. Protoplasma 223:183–189
Sheng XF, Xia JJ (2006) Improvement of rape (*Brasica napus*) plant growth and cadmium uptake by cadmium-resistant bacteria. Chemosphere 64:1036–1042
Shetty KG, Banks MK, Hetrick BA, Schwab AP (1994a) Biological characterization of a southeast Kansas mining site. Water Air Soil Pollut 78:169–177
Shetty KG, Hetrick BAD, Figge DAH, Schwab AP (1994b) Effects of mycorrhizae and other soil microbes on revegetation of heavy metal contaminated mine spoil. Environ Pollut 86:181–188
Simon L, Szegvári I, Prokisch J (2001) Enhancement of chromium phytoextraction capacity in fodder radish with pycolinic acid. Environ Geochem Health 23:313–316
Smith SE, Read DJ (1997) Mycorrhizal symbiosis. Academic, San Diego, CA
Tabak H, Lens P, van Hullebusch ED, Dejonghe W (2005) Development in bioremediation of soils and sediments polluted with metals and radionuclides – 1. Microbial processes and mechanisms affecting bioremediation of metal contamination and influencing metal toxicity and transport. Rev Environ Sci Biotechnol 4:115–156
Tabatabai MA (1982) Soil enzymes. In: Page AL, Miller EM, Keeney DR (eds) Methods of soil analysis, part 2, 2nd edn. Agronomy Monograph 9. ASA and SSSA, Madison, WI, pp 501–538
Tabatabai MA, Bremner JM (1969) Use of p-nitrophenyl phosphate for assay of soil phosphatase activity. Soil Biol Biochem 1:301–307
Valls M, de Lorenzo V (2002) Exploiting the genetic and biochemical capacities of bacteria from the remediation of heavy metal pollutions. FEMS Microbiol Rev 26:327–338
Vivas A, Azcón R, Biró B, Barea JM, Ruíz-Lozano JM (2003a) Influence of bacterial strains isolated from lead-polluted soil and their interactions with arbuscular mycorrhizae on the growth of *Trifolium pratense* L. under lead toxicity. Can J Microbiol 49:577–588
Vivas A, Biró B, Campos E, Barea JM, Azcón R (2003b) Symbiotic efficiency of autochthonous arbuscular mycorrhizal fungus (*G. mosseae*) and *Brevibacillus* sp isolated from cadmium polluted soil under increasing cadmium levels. Environ Pollut 126:179–189
Vivas A, Vörös I, Biró B, Barea JM, Ruíz-Lozano JM, Azcón R (2003c) Beneficial effects of indigenous Cd-tolerant and Cd-sensitive *Glomus mosseae* associated with a Cd-adapted strain

of *Brevibacillus* sp. in improving plant tolerance to Cd contamination. Appl Soil Ecol 24:177–186

Vivas A, Barea JM, Azcón R (2005a) *Brevibacillus brevis* isolated from cadmium or zinc contaminated soils improves in vitro spore germination and growth of *Glomus mosseae* under high Cd or Zn concentrations. Microb Ecol 49:416–424

Vivas A, Barea JM, Azcón R (2005b) Interactive effect of *Brevibacillus brevis* and *Glomus mosseae*, both isolated from Cd contaminated soil, on plant growth, physiological mycorrhizal fungal characteristics and soil enzymatic activities in Cd-polluted soil. Environ Pollut 134:257–266

Vivas A, Barea JM, Biró B, Azcón R (2006a) Effectiveness of authochthonous bacterium and mycorrhizal fungus on *Trifolium* growth, symbiotic development and soil enzymatic activities in Zn contaminated soil. J Appl Microbiol 100:587–598

Vivas A, Biró B, Németh T, Barea JM, Azcón R (2006b) Nickel-tolerant *Brevibacillus brevis* and arbuscular mycorrhizal fungus can reduce metal acquisition and nickel toxicity effects in plant growing in nickel supplemented soil. Soil Biol Biochem 38:2694–2704

Vivas A, Biró B, Ruiz-Lozano JM, Barea JM, Azcón R (2006c) Two bacterial strains isolated from Zn-polluted soil enhance plant growth and mycorrhizal efficiency under Zn-toxicity. Chemosphere 62:1523–1533

Vivas A, Moreno B, García-Rodriguez S, Benítez E (2009) Assessing the impact of composting and vermicomposting on structural diversity of bacterial communities and enzyme activities of an olive-mill waste. Bioresour Technol 100:1319–1326

Vörös I, Biró B, Takács T, Köves-Péchy K, Bujtás K (1998) Effect of arbuscular mycorrhizal fungi on heavy metal toxicity to *Trifolium praetense* in soils contaminated with Cd, Zn and Ni salts. Agrokém Talaj 47:277–288

Weissenhorn I, Leyval C (1995) Root colonization of maize by a Cd-sensitive and a Cd-tolerant *Glomus mosseae* and cadmium uptake in sand culture. Plant Soil 175:233–237

Weller DM, Thomashow LS (1994) Current challenges in introducing beneficial microorganisms in the rhizosphere. In: O'Gara F, Dowling DN, Boesten B (eds) Molecular ecology of rhizosphere microorganisms: biotechnology and the release of GMOs. VCH Verlagsgesellchaft, Winheim, pp 1–13

Wenzel WW (2009) Rhizosphere processes and management in plant-assisted bioremediation (phytoremediation) of soils. Plant Soil 321:385–408. doi:10.1007/s11104-008-9686-1

Whitfield L, Richards AJ, Rimmer DL (2004) Effects of mycorrhizal colonisation on *Thymus polytrichus* from heavy-metal-contaminated sites in northern England. Mycorrhiza 14:47–54

Wöhler I (1997) Auxin-indole derivatives in soils determined by a colorimetric method and by high performance liquid chromatography. Microbiol Res 152:399–405

Xie H, Pasternak JJ, Glick BR (1996) Isolation and characterization of mutants of the plant growth-promoting rhizobacterium *Pseudomonas putida* GR 12–2 that overproduce indoleacetic acid. Curr Microbiol 32:67–71

Zaidi S, Usmani S, Singh BR, Musarrat J (2006) Significance of *Bacillus subtilis* strain SJ-101 as a bioinoculant for concurrent plant growth promotion and nickel accumulation in *Brassica juncea*. Chemosphere 64:991–997

Zhou JL (1999) Zn biosorption by *Rhizopus arrhizus* and other fungi. Appl Microbiol Biotechnol 51:686–693

Chapter 13
Geobacillus Activities in Soil and Oil Contamination Remediation

Ibrahim M. Banat and Roger Marchant

13.1 Introduction

The genus *Geobacillus* has only existed since the year 2001, when it was proposed by Nazina et al. (2001). That is not to say, however, that we were unaware of this group of organisms, since Nazina et al. built the new genus around previously described members of the genus *Bacillus* that had been placed in genetic group V by Ash et al. (1991), using 16S rRNA gene sequence information. Nazina et al. used *Bacillus stearothermophilus*, a well known and extensively investigated thermophilic organism, as the type species for the new genus and transferred a number of related thermophilic species of *Bacillus* into *Geobacillus*. Adding to this core of *Geobacillus* species, Nazina et al. (2001) proposed two new species that they had isolated from deep oil reservoirs. Since that original description of the genus with eight species a further nine, from a variety of sources, have been described. These include *G. toebii* from composting plant material (Sung et al. 2002), *G. debilis* from temperate soil environments (Banat et al. 2004), *Geobacillus pallidus* which was originally proposed as *Bacillus pallidus* by Scholz et al. (1987) and reassigned by Banat et al. (2004), *G. vulcani* (proposed as *Bacillus vulcani,* Caccamo et al. 2000) from marine geothermal sources and transferred to *Geobacillus* by Nazina et al. (2004) and *G. tepidamans* (Schäffer et al. 2004) from geothermal sources in Austria and Yellowstone National Park. A complete listing of current valid species for the genus *Geobacillus* can be found at http://www.bacterio.cict.fr/g/geobacillus.html. What is clear from this listing of species is that members of the genus can be isolated from a wide variety of different environmental sources. Some of these situations are those where we would expect to find obligately thermophilic bacteria, i.e., geothermally active areas, both terrestrial and marine. The observation that *Geobacillus* spp. also occur in composting plant material is logical, since this is another environment where temperatures up to 60°C are readily experienced by

I.M. Banat (✉) and R. Marchant
School of Biomedical Sciences, Faculty of Life and Health Sciences, University of Ulster, Coleraine, BT52 1SA, Northern Ireland, UK
e-mail: IM.Banat@ulster.ac.uk

N.A. Logan and P. De Vos (eds.), *Endospore-forming Soil Bacteria*, Soil Biology 27,
DOI 10.1007/978-3-642-19577-8_13,

microorganisms. On the other hand, the recovery of quite high concentrations of thermophilic *Geobacillus* spp. from temperate soil environments, and not just in the surface layers but also at considerable depth, is much harder to explain (Marchant et al. 2002a, b). Traditionally, the search for thermophilic microorganisms has focused on hot environments, but we must now examine a wider range of potential habitats and explore the means by which thermophiles can be transported to such areas if they are unable to grow and reproduce in the conditions found there. The first requirement of any such investigation is to establish the actual and potential activities of these thermophiles.

13.2 The Growth and Metabolism of *Geobacillus* Species

A starting point is to look at the description of the genus published by Nazina et al. (2001). The organisms were described as rod-shaped and producing one endospore per cell, with cells occurring either singly or in short chains and being motile by means of peritrichous flagella. Interestingly, subsequent observations of *Geobacillus* spp. have shown that extremely long cells may be produced under some conditions (Marchant et al. 2002a). These cells appear to lack cross walls and are therefore not chains of short rods, but they are subsequently capable of dividing to yield normal-sized rods through a process of progressive binary fission. The factors controlling this behaviour have not yet been identified. Cells have a Gram-positive cell wall structure but may stain Gram-variable. They are chemo-organotrophs that are aerobic or facultatively anaerobic, using oxygen as the terminal electron acceptor – replaceable by nitrate in some species. *Geobacillus* spp. are obligately thermophilic with a growth range of 37–75°C and optima of 55–65°C, and they are neutrophilic with a growth range of pH 6.0–8.5. The G+C content of DNA is 48.2–58 mol% with 16S rRNA gene sequence similarities of higher than 96.5% (Nazina et al. 2001).

From the above definition of the genus it would appear that any *Geobacillus* spp. existing in subsurface soil layers in temperate regions would be non-growing. The standard methods for evaluating growth of bacteria under laboratory conditions involve setting up cultures either on solid media or in liquid culture at appropriate temperatures and observing growth over relatively short periods of time. This does not exclude the possibility that organisms may be growing and dividing extremely slowly. To test this, Marchant et al. (2006) employed a molecular technique to examine gene expression over a temperature range in *G. thermoleovorans*. A gene not expected to be present in other thermophilic bacteria, alkane hydroxylase, was selected and the level of expression of this gene was determined at temperatures above and below the permissive growth temperature of 40°C, using reverse transcriptase PCR techniques. The system was first established in pure culture conditions and subsequently in soil microcosms. It was clear from this work that the alkane hydroxylase gene was readily expressed at or above 40°C, but that expression could not be detected below this temperature. These data seemed to eliminate the hypothesis that aerobic growth was taking place in sub-surface soil layers. There did, however, remain the possibility that *G. thermoleovorans,* which possesses a full set of denitrification genes, could grow and divide anaerobically at lower temperatures in the lower soil layers. This hypothesis was tested by examining the denitrification activity of the organism over a temperature range that spanned the permissive growth temperature (Pavlostathis et al. 2006). Once again, activity was detectable only above 40°C – seemingly eliminating the likelihood of growth and cell division taking place at ambient temperatures in soil.

One interesting observation connected with *G. toebii* was recorded by Rhee et al. (2000, 2002) who described a bacterium, "*Symbiobacterium toebii*" that has a symbiotic relationship requiring the products of cell lysis of *G. toebii* to allow growth. Further work (Kim et al. 2008) has now been carried out to characterize the cell lysis and metabolic products involved in this symbiosis, although complete identification of the molecules has not been made. This rather unusual symbiotic relationship raises the question of whether other *Geobacillus* spp. could be involved in similar interactions and whether cell lysis on a scale sufficient to provide metabolites for a symbiont is likely to occur during growth. Pavlostathis et al. (2006) have carried out an extensive study of the growth of *G. thermoleovorans* in batch and continuous flow fermentation systems. These studies have shown that under conditions of high growth rate followed by substrate exhaustion, not just of carbon but of other nutrients as well, extensive cell lysis takes place and that – beyond a certain point of commitment – reinstatement of the nutrient level fails to prevent lysis. Flow cytometry studies carried out at the same time demonstrated that at these high growth rates the death rate was correspondingly high. Since *G. toebii* and *S. toebii* have been isolated from thermophilic fermentation of plant material, it is quite possible that cell lysis sufficient to sustain the symbiotic relationship does actually take place.

13.3 Hydrocarbon Degradation by *Geobacillus* Species

Many of the species of *Geobacillus* have been isolated from oilfields and more specifically from deep oil wells. Not surprisingly, these organisms show well-developed abilities to degrade a range of hydrocarbons of different chain lengths. The ability of all *Geobacillus* spp. to degrade such molecules has not been fully investigated, and thus alkane-degrading activity has not been reported for *G. lituanicus, G. pallidus, G. tepidamans* and *G. thermoglucosidasius,* with *G. debilis* reported to have only weak ability (Banat et al. 2004). The absence of a reported alkane-degrading activity does not necessarily imply the inability of the organism in this activity, but may simply represent the fact that it has not been examined. We can say therefore that hydrocarbon degradation is a widespread capability of members of the genus *Geobacillus*, but may not be universal. There have been few reports on the range of hydrocarbons utilized by particular organisms. Marchant et al. (2002b) examined the ability of soil isolates of *G. caldoxylosilyticus*, *G. toebii* and *G. thermoleovorans,* to utilize a range of alkanes from pentane to nonadecane, including hexane, heptane, dodecane, hexadecane and octadecane and the polyaromatic hydrocarbons (PAHs) naphthalene, anthracene and kerosene. The patterns of utilization were dissimilar between the different species, and different strains of *G. thermoleovorans* also differed in their abilities to use these substrates. There was not even a clear differentiation between use of short-chain and longer-chain alkanes. The main conclusion to be drawn from this work was that the PAHs are more refractory to degradation. In a recent study (Sood and Lal 2008) it has been shown that a strain of *Geobacillus kaustophilus* has the ability to degrade paraffinic hydrocarbons (waxes) that create problems in some oilfields through deposition in pipe work. Hydrocarbons in the range C_{20}–C_{30} are broken down by this organism, which raises the possibility that the organism can be used to clean and maintain the systems in problematic oilfields.

Feng et al. (2007) have recently reported the presence of a plasmid-borne putative monooxygenase gene in *Geobacillus thermodenitrificans* NG80-2, which is responsible for the conversion of C_{15}–C_{36} alkanes into the corresponding primary alcohols. Crude oil used as the sole carbon source for growth led to a 120-fold increase in gene transcription. The gene is somewhat unusual in that it has only 33% sequence identity with DBT-5,5′-dioxide monooxygenase of *Paenibacillus* sp. A11-2 (Ishii et al. 2000) and no similarity with other alkane monooxygenases.

Feitkenhauer et al. (2003) have reported the use of *G. thermoleovorans* in a continuous-flow fermentation system to degrade phenol at 65°C, but this was a laboratory-scale set of experiments and there have been no other studies of alkane degradation either at a laboratory scale or at a larger industrial scale. One feature of other hydrocarbon-degrading bacteria is the ability to produce biosurfactant molecules that can make the hydrocarbons more amenable to degradation. There have, however, not been any reports of *Geobacillus* spp. producing biosurfactants, and we have never observed evidence of such production in any of our cultures.

13.4 The Genetics of *Geobacillus*

A considerable amount of genetic information exists for members of the genus *Geobacillus* since complete genome sequences have been published for *G. kaustophilus* HTA426 (Takami et al. 2004a) and *G. thermodenitrificans* NG80-2 (Feng et al. 2007), and draft sequences for *G. stearothermophilus* strain 10 (prepared at the University of Oklahoma) and *G. thermoleovorans* T80 (University of Ulster) are available in the public databases. Detailed examination of the genome sequences for these organisms has offered interesting insight into thermophilic adaptation in *Geobacillus* spp. Since they are closely related to a wide range of mesophilic, aerobic endospore-formers, different groups of researchers anticipated that there would be clear indications of specific adaptation to growth at high temperature in the genome sequences. These expectations have been, to a great extent, frustrated. A number of basic mechanisms for thermophily can be postulated and tested: these include the lateral transfer of specific genes from other thermophilic bacteria such as *Archaea*, the specific modification of the mesophilic gene sequences to produce more thermostable gene products and the use of chaperonins to protect gene products from damage. The only published analyses are those of Takami et al. (2004b) and Feng et al. (2007). These workers found it extremely difficult to identify specific characteristics in the genomes of *Geobacillus* spp. that could be linked to thermophily. The published genome of *G. kaustophilus* is 3.54 Mb in size and contains 37% of the genes (1,308) shared with other of mesophilic, aerobic endospore-formers. Very few of the genes that are common with mesophiles, particularly those in the core metabolic pathways, show any significant sequence variation and very few of the genes can be identified as orthologues of genes from other thermophilic organisms such as members of the *Archaea*. This situation holds also for the genome of *G. thermoleovorans* (unpublished results). An identifiable characteristic of the *Geobacillus* spp. is, however, the presence of large numbers of transposon sequences (55 in *G. thermodenitrificans*, 80 in *G. kaustophilus*), and these are in contrast with low numbers in *Bacillus subtilis, B. cereus* and *B. anthracis* (Takami et al. 2004b; Feng et al. 2007). The presence of these sequences indicates that lateral gene transfer may have been an important feature for these organisms and may also suggest that active evolution of the species is still taking place. Analysis of the draft genome sequence for *G. thermoleovorans* T80 (Matzen et al. unpublished) has led to the identification of two cassettes of genes for RAMP (Repair Associated Mysterious Proteins) that are not present in the genomes of mesophilic endospore-formers and which do not appear in the complete genomes of *G. kaustophilus* or *G. thermodenitrificans*. These groups of genes seemed to be potentially important for the mechanisms of thermophily, and therefore further wet laboratory experiments were carried out on the genomes of the previously sequenced strains of *G. kaustophilus* and *G. thermodenitrificans*. This work showed that the RAMP cassettes are indeed present in these organisms and had been either missed during the original shotgun sequencing or had been eliminated during the final assembly of the genome (Matzen et al. unpublished). The recent identification of CRISPRs (clustered regularly interspaced short palindromic repeats) in prokaryotic genomes and their association with 45 CRISPR-Associated (Cas) protein families can explain how the RAMP cassettes (which are now known to be Cas genes) could become eliminated during genome assembly (Haft et al. 2005). What role these Cas genes play, and whether they are important for thermophily remain matters for conjecture. There is little evidence from the analysis of the *Geobacillus* genomes to support the view that large numbers of key genes have been laterally transferred from other thermophilic organisms, particularly members of the *Archaea*, or that base substitution or coding changes have produced more temperature-resistant versions of pre-existing mesophile genes. An interesting question is, therefore, not so much how these organisms grow at high temperatures, but why they do not grow at the same temperatures as mesophiles?

The genetic basis of alkane degradation has been extensively reported for a number of different bacteria, including *Pseudomonas putida* (van Beilen et al. 2001), *Acinetobacter* (Ratajczak et al. 1998a, b) and *Rhodococcus erythropolis*

(Whyte et al. 2002). In these studies it has been shown that the components of the alkane degradation system are organized into two operons, with an alkane monooxygenase catalysing the first step and with up to four different gene homologues present (*alkB1–alkB4*). Marchant et al. (2006) used sequence information from the genes in *Rhodococcus* spp. and *Prauserella rugosa* to construct degenerate primers for a segment of the *alkB* gene in *G. thermoleovorans* T80. The objective of this approach was to use the expression of the alkane monooxygenase gene as a measure of the metabolic activity of the bacterium in soil conditions at various temperatures. The culture work had suggested that there was no detectable growth at temperatures below 40°C in laboratory conditions. This observation was confirmed by molecular studies using reverse transcriptase PCR methods – with no expression of the gene detectable at temperatures below 40°C, but clear evidence of expression at all higher temperatures. Since many of the species of *Geobacillus* have been isolated from high temperature environments, it is not surprising to find that they only grow at these temperatures, but what is paradoxical is that large numbers of these seemingly obligate thermophiles can be recovered from temperate soils where the temperature can never reach the growth-permissive range.

13.5 The Potential of *Geobacillus* Species for Remediation of Hydrocarbon-Contaminated Sites

The first real indication that thermophilic bacteria might have significant roles in the remediation of hydrocarbon-contaminated sites came from the application of a process for the steam stripping of volatile hydrocarbons from soil (Newmark and Aines 1998); this was developed at the Lawrence Livermore National Laboratory in America. The process involved the injection of steam through pipes inserted into the contaminated site and the collection of the volatile hydrocarbons displaced by the steam. The system was capable of removing large proportions of the contaminants but, as a corollary, elevated the temperature of the whole site; this elevated temperature persisted for extended periods (measured in weeks) following cessation of the steam injection. It was observed that levels of hydrocarbon contamination continued to decline during this period, and microbiological investigations established that thermophilic hydrocarbon-degrading bacteria were responsible. These observations support the view that elevating the temperature of a contaminated site could promote the activity of thermophiles and facilitate the process of bioremediation.

Any process of bioremediation requires a number of components to be in place, including appropriate nutritional status and the presence of active microorganisms to achieve the remediation. In many instances, specific steps are taken to ensure suitable conditions through the addition of nutrients and by bioaugmentation with selected and adapted organisms. In the case of the steam-stripping system, the temperature increase had caused the growth stimulation of existing microorganisms in the site and no bioaugmentation was undertaken.

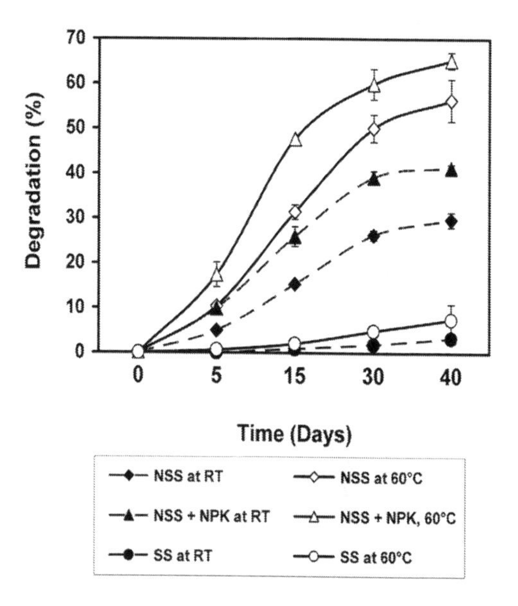

Fig. 13.1 Time course for 2% (v/v) *n*-hexadecane degradation in soil microcosms incubated at 60°C and room temperature. Abbreviation used: *SS* sterile soil; *NSS* non-sterile soil; *NPK* Augmented with nitrogen, phosphorus and potassium and *RT* room temperature

The observations from contaminated sites are supported by data on *n*-hexadecane degradation in various microcosm treatments. These were set up at room temperature and at 60°C in sterile and non-sterile soils containing 2% (v/v) *n*-hexadecane, and either with or without supplements of additional nutrient elements. The results (Fig. 13.1) showed that the controls containing sterile soil had minimal losses of hexadecane over a period of 40 days; abiotic loss of alkane in sterile soil owing to physico-chemical processes such as photo-oxidation, volatilization and evaporation were shown to be insignificant. In contrast, the microcosms containing non-sterile soil showed progressive losses of hexadecane over the incubation period, and nutrient supplementation produced enhanced degradation at both room temperature and 60°C. These results indicated that the soil, although originally uncontaminated, contained an indigenous microbial population capable of supporting bioremediation and also confirmed the ubiquity of hydrocarbon-degrading microorganisms. Interestingly, these data suggested that in this cool soil there existed a community not only of mesophilic alkane degraders but also thermophilic degraders. What was also clear is that degradation at 60°C, reaching more than 65%, was considerably greater and more rapid than that at room temperature. In parallel with the degradation there was an increase in microbial numbers, with thermophiles increasing about 10-fold while the mesophile numbers only doubled (Table 13.1; Marchant et al. 2006). It should be noted, however, that the initial numbers of mesophiles were two orders of magnitude greater than those of the thermophiles. These results demonstrated that degradation of alkanes at ambient temperature is probably carried out by mesophilic bacteria, but that thermophiles are capable of achieving higher levels of degradation at appropriate temperatures.

Table 13.1 Viable cell counts for mesophilic and thermophilic bacteria taken from soil microcosms at 0 and 40 days

Treatments		(CFU/g of soil) at room temp.	(CFU/g of soil) at 60°C
At zero time			
1	SS	ND	ND
2	NSS	9.5×10^5	1.2×10^4
3	NSS + NPK	7.5×10^5	2.0×10^4
After 40 days			
4	SS	ND	ND
5	NSS	1.3×10^6	8.5×10^4
6	NSS + NPK	1.5×10^6	2.1×10^5

Results are means of five replicates (Marchant et al. 2006).
SS sterile soil, *NSS* non-sterile soil, *NPK* nitrogen, potassium and phosphorous additive, *ND* no colonies detected.

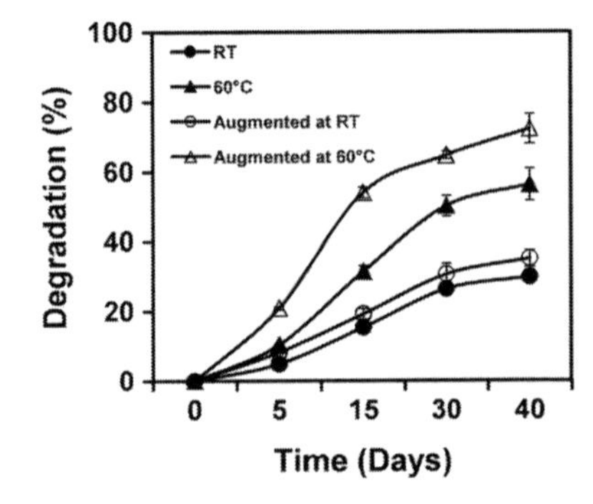

Fig. 13.2 Time course for 2% (v/v) *n*-hexadecane degradation in soil microcosms incubated at 60°C and room temperature with and without augmentation with *Geobacillus thermoleovorans* T80

Further microcosm experiments were carried out, in which non-sterile soils containing hexadecane were augmented with *G. thermoleovorans* T80, then incubated at ambient temperature and 60°C; the latter is optimum for this species (Perfumo et al. 2007). The results (Fig. 13.2) showed higher hexadecane degradation at the higher temperature, and this was significantly increased by augmentation with the bacterium (>70%); it was less than approximately half this amount for those kept at ambient temperature, where augmentation had no significant effect on the amount of degradation. *G. thermoleovorans* T80 was isolated from a similar soil (i.e. it was an autochthonous organism) and has been shown to have no apparent activity at temperatures below 40°C. Being an autochthonous soil strain was an additional advantage, as that probably helped it to avoid the survival difficulties faced by exogenous microorganisms when introduced to the soil.

In general the above results demonstrated that, irrespective of the treatment, all the microcosms incubated at 60°C had hydrocarbon degradation levels approximately twofold higher than those of microcosms kept at ambient temperature. This indicated that increased operating temperatures might significantly enhance natural attenuation in soil, by removal of contaminants from the soil particles by mobilizing them and increasing their solubilities. Moreover, the response of the indigenous microbial population at the elevated temperature showed the activation of a community of thermophilic degraders, which suggested an intrinsic potential for natural attenuation in such cool soils through thermally enhanced bioremediation techniques.

The supplementing of soils with limiting nutrients or their augmentation with thermophilic *Geobacillus* spp. produced additional degradation, resulting in a maximum of >70% removal of *n*-hexadecane within a 40-day period. Thermally enhanced bioremediation may therefore reduce in situ remediation treatment times or be used as a complementary treatment to follow other physical techniques. Further investigation into the most cost-effective methods to provide the energy needed for soil heating, and augmentation procedures to stimulate the process, remain to be carried out.

13.6 Geographical Distribution of *Geobacillus* Species

A critical question to be answered, therefore, if such remediation processes are to be more widely exploited, is whether we can be certain that hydrocarbon-degrading thermophiles are reliably present at every site. Basic studies of the distribution of these organisms, since we know that they are unable to grow at temperatures below 40°C, assume greater importance in this context and will allow us to know whether we can expect every site to contain them, without specific testing.

Since it seemed that the relatively large numbers of *Geobacillus* spp. found in sub-surface regions of temperate soils ($<10^4$/g soil Marchant et al. 2002b) could not come from growth in situ, the possible input from other sources was investigated. Surprisingly, significant numbers of thermophiles are deposited annually through rainfall, and the population structure of these organisms differs markedly from the population of purely airborne thermophiles (Marchant et al. 2008). This population structure difference between airborne and rainwater bacteria, and the fact that numbers of organisms in rainfall cannot be correlated with wind direction or with intensity of rainfall, support the hypothesis that these are not airborne bacteria washed out of the air by rain. If this is the case then we must look more widely for the sources of thermophiles. In the southern part of the Mediterranean region, and elsewhere in the world, atmospheric dust is periodically precipitated dry. This dust originates from different desert regions of the world, with the major flow being westwards across the Atlantic Ocean to the Caribbean and southern states of the United States. There is a smaller subsidiary flow northwards across Europe, coming from the Sahel-Sahara region of Africa (Griffin 2007); and preliminary examination of samples of such dust from Greece and Turkey has shown populations of thermophilic bacteria that are readily cultivable from the dust and that can be identified as organisms closely related to existing *Geobacillus* spp. (Marchant and Perfumo unpublished results). Further work is necessary to describe these organisms fully and to link them to the sources of the dust in the Sahel-Sahara region of Africa. At this time, however, it does seem likely that the thermophiles deposited by rainfall in Northern Europe have come from dust storms generated in Africa and that they have been transported in the upper atmosphere. If this is the case, we can be reasonably certain that hydrocarbon-degrading thermophilic bacteria will be ubiquitous in environments where this deposition occurs, and might be exploited directly for bioremediation without the need for bioaugmentation.

13.7 Conclusion

Members of the *Geobacillus* appear to be present in most soil environments – even those never experiencing elevated temperatures. Most have some ability to grow on hydrocarbons and may therefore be exploited in bioremediation processes. In such a process the need for bioaugmentation may only be necessary to increase the initial rate of activity. Biostimulation, on the other hand, may have an important role when depending on *Geobacillus* spp. in bioremediation processes. The hydrocarbon-degrading abilities of *Geobacillus* spp. may therefore find useful bioremediation applications in a variety of different situations in the future.

References

Ash C, Farrow JAE, Wallbanks S, Collins MD (1991) Phylogenetic heterogeneity of the genus *Bacillus* revealed by comparative analysis of small-subunit-ribosomal RNA sequences. Lett Appl Microbiol 13:202–206

Banat IM, Marchant R, Rahman TJ (2004) *Geobacillus debilis* sp. nov., a novel obligately thermophilic bacterium isolated from a cool soil environment and reassignment of *Bacillus pallidus* to *Geobacillus pallidus* comb. nov. Int J Syst Evol Microbiol 54:2197–2201

Caccamo D, Gugliandolo C, Stackebrandt E, Maugeri TL (2000) *Bacillus vulcani* sp. nov., a novel thermophilic species isolated from a shallow marine hydrothermal vent. Int J Syst Evol Microbiol 50:2009–2012

Feitkenhauer H, Schnicke S, Müller R, Märkl H (2003) Kinetic parameters of continuous cultures of *Bacillus thermoleovorans* sp. A2 degrading phenol at 65°C. J Biotechnol 103:129–135

Feng L, Wang W, Cheng J, Ren Y, Zhao G, Gao C, Tang Y, Liu X, Han W, Peng X, Liu R, Wang L (2007) Genome and proteome of long-chain alkane degrading *Geobacillus thermodenitrificans* NG80-2 isolated from a deep-subsurface oil reservoir. Proc Natl Acad Sci USA 104: 5602–5607

Griffin DW (2007) Atmospheric movement of microorganisms in clouds of desert dust and implications for human health. Clin Microbiol Rev 20:459–477

Haft DH, Selengut J, Mongodin EF, Nelson KE (2005) A guild of 45 CRISPR-Associated (Cas) protein families and multiple CRISPR/Cas subtypes exist in prokaryotic genomes. PLoS Comput Biol 1:474–483

Ishii Y, Konishi J, Okada H, Hirasawa K, Onaka T, Suzuki M (2000) Operon structure and functional analysis of the genes encoding thermophilic desulfurizing enzymes of *Paenibacillus* sp A11-2. Biochem Biophys Res Commun 270:81–88

Kim J-J, Masui R, Kuramitsu S, Seo J-H, Kim K, Sung M-H (2008) Characterization of growth-supporting factors produced by *Geobacillus toebii* for the commensal thermophile *Symbiobacterium toebii*. J Microbiol Biotechnol 18:490–496

Marchant R, Banat IM, Rahman TJ, Berzano M (2002a) What are high-temperature bacteria doing in cold environments? Trends Microbiol 10:120–121

Marchant R, Banat IM, Rahman TJ, Berzano M (2002b) The frequency and characteristics of highly thermophilic bacteria in cool soil environments. Environ Microbiol 4:595–602

Marchant R, Sharkey FH, Banat IM, Rahman TJ, Perfumo A (2006) The degradation of *n*-hexadecane in soil by thermophilic geobacilli. FEMS Microbiol Ecol 56:44–54

Marchant R, Franzetti A, Pavlostathis SG, Okutman Tas D, Erdbrűgger I, Űnyayar A, Mazmanci MA, Banat IM (2008) Thermophilic bacteria in cool temperate soil environments: are they metabolically active or continually added by global atmospheric transport? Appl Microbiol Biotechnol 5:841–852

Nazina TN, Tourova TP, Poltaraus AB, Novikova EV, Grigoryan AA, Ivanova AE, Lysenko AM, Petrunyaka VV, Osipov GA, Belyaev SS, Ivanov MV (2001) Taxonomic study of aerobic thermophilic bacilli: descriptions of *Geobacillus subterraneus* gen. nov., sp. nov. and *Geobacillus uzenensis* sp. nov. from petroleum reservoirs and transfer of *Bacillus stearothermophilus, Bacillus thermocatenulatus, Bacillus thermoleovorans, Bacillus kaustophilus, Bacillus thermoglucosidasius* and *Bacillus thermodenitrificans* to *Geobacillus* as the new combinations *G. stearothermophilus, G. thermocatenulatus, G. thermoleovorans, G. kaustophilus, G. thermoglucosidasius* and *G. thermodenitrificans*. Int J Syst Evol Microbiol 51:433–446

Nazina TN, Lebedeva EV, Poltaraus AB, Tourova TP, Grigoryan AA, Sokolova DS, Lysenko AM, Osipov GA (2004) *Geobacillus gargensis* sp. nov., a novel thermophile from a hot spring, and the reclassification of *Bacillus vulcani* as *Geobacillus vulcani* comb. nov. Int J Syst Evol Microbiol 54:2019–2024

Newmark RL, Aines RD (1998) They all like it hot: faster cleanup of contaminated soil and groundwater. Sci Technol Rev May:4–11

Pavlostathis SG, Marchant R, Banat IM, Ternan N, McMullan G (2006) High growth rate and substrate exhaustion results in rapid cell death and lysis in the thermophilic bacterium *Geobacillus thermoleovorans*. Biotechnol Bioeng 95:84–95

Perfumo A, Banat IM, Marchant R, Vezzulli L (2007) Thermally enhanced approaches for bioremediation of hydrocarbon-contaminated soils. Chemosphere 66:179–184

Ratajczak A, Geiβdörfer W, Hillen W (1998a) Alkane hydroxylase from *Acinetobacter* sp. strain ADP-1 is encoded by *alkM* and belongs to a new family of bacterial integral membrane hydrocarbon hydroxylases. Appl Environ Microbiol 64:1175–1179

Ratajczak A, Geiβdörfer W, Hillen W (1998b) Expression of the alkane hydroxlase from *Acinetobacter* sp. strain ADP-1 is induced by a broad range of n-alkanes and requires the transcriptional activator *alkR*. J Bacteriol 180:5822–5827

Rhee SK, Lee SG, Hong SP, Choi YH, Park JH, Kim CJ, Sung MH (2000) A novel microbial interaction: obligate commensalism between a new gram-negative thermophile and a thermophilic *Bacillus* strain. Extremophiles 4:131–136

Rhee SK, Jeon CO, Bae JW, Kim K, Song JJ, Kim JJ, Lee SG, Kim HI, Hong SP, Choi YH, Kim SM, Sung MH (2002) Characterization of *Symbiobacterium toebii*, an obligate commensal thermophile isolated from compost. Extremophiles 6:57–64

Schäffer C, Franck WL, Scheberl A, Kosma P, McDermott TR, Messner P (2004) Classification of isolates from locations in Austria and Yellowstone National Park as *Geobacillus tepidamans* sp. nov. Int J Syst Evol Microbiol 54:2361–2368

Scholz T, Demharter W, Hensel R, Kandler O (1987) *Bacillus pallidus* sp. nov., a new thermophilic species from sewage. Syst Appl Microbiol 9:91–96

Sood N, Lal B (2008) Isolation and characterization of a potential paraffin-wax degrading thermophilic bacterial strain *Geobacillus kaustophilus TERI NSM* for application in oil wells with paraffin deposition problems. Chemosphere 70:1445–1451

Sung MH, Kim H, Bae JW, Rhee SK, Jeon CO, Kim K, Kim JJ, Hong SP, Lee SG, Yoon JH, Park YH, Baek DH (2002) *Geobacillus toebii* sp. nov., a novel thermophilic bacterium isolated from hay compost. Int J Syst Evol Microbiol 52:2251–2255

Takami H, Nishi S, Lu J, Shinamura S, Takaki Y (2004a) Genomic characterization of thermophilic *Geobacillus* species isolated from the deepest sea mud of the Mariana Trench. Extremophiles 8: 351–356

Takami H, Takaki Y, Chee GJ, Nishi S, Shinamura S, Suzuki H, Matsui S, Uchiyama I (2004b) Thermoadaptation trait revealed by the genome sequence of thermophilic *Geobacillus kaustophilus*. Nucleic Acids Res 32:6292–6303

van Beilen JB, Panke S, Lucchini S, Franchini AG, Röthlisberger M, Witholt B (2001) Analysis of *Pseudomonas putida* alkane degradation gene clusters and flanking insertion sequences: evolution and regulation of the alk-genes. Microbiology 147:1621–1630

Whyte LG, Smits TH, Labbe D, Witholt B, Greer CW, van Beilen JB (2002) Gene cloning and characterisation of multiple alkane hydroxylase systems in Rhodococcus strains Q15 and NRR1. Appl Environ Microbiol 68:5933–5942

Chapter 14
Studying Denitrification by Aerobic Endospore-forming Bacteria in Soil

Ines Verbaendert and Paul De Vos

14.1 Introduction

14.1.1 The Genus Bacillus

Ferdinand Cohn, a German botanist, was fascinated by the heat-resistant forms of bacteria. He described the process of endospore formation and renamed the organism "*Vibrio subtilis*" (Ehrenberg 1835) as *Bacillus subtilis* (1872). This species was the first member of the very large and diverse genus that is part of the phylum *Firmicutes* and the family *Bacillaceae*. The family's salient characteristic is the production of endospores; these are formed within bacterial cells by a process called sporulation, and may be oval, round or cylindrical. These spores can remain dormant for extremely long periods, and are extremely heat-resistant and resistant to other physical agents (e.g., desiccation and UV radiation) and to various chemical agents (e.g., acids and disinfectants). Endospores are easily detected using a phase contrast microscope because of their highly refractile nature, but their presence may be discovered with a mere Gram-stain because the spore remains unstained while the vegetative cells or the vegetative parts of the cells will stain (Slepecky and Hemphill 2006).

The genus *Bacillus* belongs to the endospore-forming, low-GC Gram-positive bacteria. These organisms follow the biological cycle of spore-formers, namely, from vegetative cell to spore and from spore to vegetative cell through a very complex series of events in cellular differentiation. Although most endospore-forming bacteria seem to share a common ecological characteristic (i.e. being associated with [agricultural] soils), they form a phylogenetically heterogeneous group within the *Firmicutes*. Within this taxon, the family *Bacillaceae* encompasses the strictly aerobic or facultatively anaerobic endospore-formers that were, until the early 1990s, accommodated within the single genus *Bacillus*. Since then

I. Verbaendert (✉) and P. De Vos
Laboratory for Microbiology, University of Gent, K.L. Ledeganckstraat 35, 9000 Gent, Belgium
e-mail: Ines.Verbaendert@UGent.be, Paul.De Vos@UGent.be

N.A. Logan and P. De Vos (eds.), *Endospore-forming Soil Bacteria*, Soil Biology 27,
DOI 10.1007/978-3-642-19577-8_14, © Springer-Verlag Berlin Heidelberg 2011

a major taxonomic reshuffling has taken place, reflecting the vast variety in (1) physiology, (2) ecology, (3) genetics, (4) morphology (mainly in the size and position of the endospore within the vegetative cell), (5) nutrition and (6) growth characteristics (Madigan et al. 2003). As a consequence, the genus *Bacillus sensu stricto* is now limited to the members phylogenetically closest to *B. subtilis* (the type species) and a few others that are awaiting reclassification. We therefore focus on *Bacillus sensu stricto* members and their relatives, where it is known that they have roles in denitrification – one of the main processes in the biogeochemical nitrogen cycle.

14.1.2 Denitrification in the Genus **Bacillus** *and Related Organisms*

Denitrification refers to the dissimilatory reduction of nitrates (NO_3^-) or nitrites (NO_2^-) over nitric oxide (NO) to nitrous oxide (N_2O) or nitrogen gas (N_2) in oxygen-depleted conditions. As the oxidized nitrogen compounds are used as alternative electron acceptors, different metalloproteins encoded by specific genes catalyse the process (Fig. 14.1) (van Spanning et al. 2007). In this facultative respiratory mechanism, NO_2^- reductase (NiR) and NO reductase (NOR) are the enzymes that perform the conversion of fixed nitrogen into gaseous nitrogen, and so the involved reductases and their corresponding genes are considered to be key elements of denitrification. This part of the process is also called denitrification *sensu stricto*.

The interest in microbial denitrification exists for several reasons. First, it is known as a beneficial process in (1) the removal of nitrogen in wastewater treatment to reduce eutrophication (Park and Yoo 2009), and (2) the degradation of other organic pollutants (Liang et al. 2007; Park et al. 2007). Secondly, it is a detrimental process for agriculture since it causes nitrogen depletion in (rural) soils

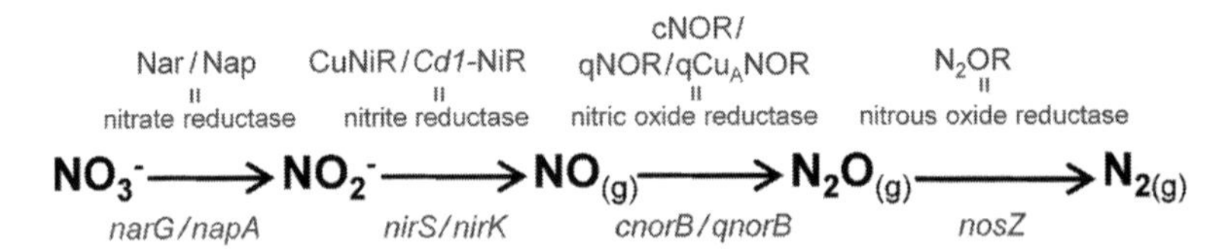

Fig. 14.1 *The denitrification process.* The sequential transformations from nitrate (NO_3^-) to nitrogen gas (N_2) are catalysed by metalloproteins which are encoded by specific genes. Nitric oxide, nitrous oxide and nitrogen gas are marked with (g), indicating the gaseous nature of these compounds. Abbreviations: Nar membrane-bound nitrate reductase encoded by *narG*; Nap periplasmic nitrate reductase encoded by *napA*; CuNiR copper-containing nitrite reductase encoded by *nirK*; *Cd1*–NiR cytochrome *cd1* containing nitrite reductase encoded by *nirS*; *cNOR* nitric oxide reductase accepting electrons from heme *c* that is encoded by *cnorB*; qNOR nitric oxide reductase accepting electrons from quinols that is encoded by *qnorB*; qCu_ANOR nitric oxide reductase from *Bacillus azotoformans* NCCB 100003 encoded by so-far-unknown genes; N_2OR nitrous oxide reductase encoded by *nosZ*

and loss of fertilizer nitrogen (Philippot et al. 2007). Thirdly, it is a mechanism that adds N_2O to the atmosphere, where N_2O is involved in the stratospheric degradation of ozone as a potent greenhouse gas (Van Cleemput and Samater 1996); in particular, the contribution of waste water systems to atmospheric N_2O is of some concern (Prendez and Lara-Gonzalez 2008). Fourthly, it is one of the main microbial processes in the global nitrogen cycle.

The ability to use nitrogen oxides as alternative electron acceptors is found in a large variety of physiological and taxonomic groups (ranging from Archaea to Gram-negative Bacteria and Fungi) and the potential for denitrification exists in a broad spectrum of habitats. However, the information about which Gram-positive organisms are jointly responsible for the denitrification activities observed in certain systems is virtually non-existent. In particular, little is known about the denitrification capacity and the distribution of the key denitrification genes in species of the genus *Bacillus* and its relatives, such as *Brevibacillus*, *Paenibacillus*, etc.

The genus *Bacillus* is considered among the groups of bacteria that are classically seen as containing true denitrifiers (Tiedje 1988). Undoubtedly, the majority of the publications on Gram-positive denitrifying taxa focus on members of the *Firmicutes* (Table 14.1). However, although the denitrifying capacity of *Bacillus* isolates has long been recognized, most published data on denitrifying *Firmicutes* are rather old, and denitrification capacities were often not accurately studied. Hence, many nitrate-respiring and ammonium-producing isolates have been misidentified as denitrifiers (Philippot et al. 2007), and others might have been overlooked. Various species are/were thought to be involved in this denitrification process (Table 14.1), but only a few denitrification genes have been sequenced from them or other denitrifying bacilli (Table 14.2). Hence, only a few denitrifying aerobic endospore-formers have so far been biochemically and genomically characterized as true denitrifiers, and as a consequence limited reliable information is available on the occurrence and the distribution of this trait and the responsible genes amongst the species. Numerous denitrifying *Bacillus sensu stricto* strains have already been isolated from soil (Kim et al. 2005; Heylen 2007) and activated sludge (Heylen 2007; Park et al. 2007). Therefore, the members of the genus *Bacillus sensu stricto* might have outstanding denitrifying capabilities in wastewater treatment systems and soil-bound ecosystems. However, their exact role in these processes has not been elucidated yet and amplification of the responsible genes in pure cultures of these isolates is currently not easy (Sect. 14.2.2).

14.2 Methods to Determine the Denitrifying Capacity of *Bacillus* Isolates

In the past, several techniques were applied to measure the denitrification process in pure cultures and in environmental samples: combinations of nitrate or nitrite removal from the medium, bubble production in Durham tubes and an increase in pH detected by a pH indicator (De Barjac and Bonnefoi 1972; Delaporte 1972;

Table 14.1 Denitrifying aerobic endospore-formers

Species and/or strain[a]	Habitat	Original literature	Remarks and/or more recent literature
Bacillus azotoformans (CIP R925) (Pichinoty et al. 1983)	Soils	Biochemical assessment of denitrification only, no analysis with N_2O gas chromatography	Purification and biochemical characterization of qCu_ANOR from *Bacillus azotoformans* (NCCB 100003); the encoding genes are not yet known (Suharti et al. 2001; Suharti and de Vries 2004, 2005)
Bacillus cereus (ATCC 8035) (Hackenthal 1966)	Foods (soddy podzolic) soil	Biochemical comparison of the enzymatic biosynthesis	The maximal denitrifying activity of strains from soddy podzolic soil was measured with gas chromatography of N_2O and the AIM[b] (Manucharova et al. 2000). *Bacillus cereus* (PK-5) was involved in aerobic denitrification with conversion of NO_3^- to N_2 (Kim et al. 2005)
Bacillus circulans (no strain number) (Manucharova et al. 2000)	Soddy podzolic soil	The maximal denitrifying activity was determined by measuring the accumulation of N_2O in the gas phase with gas chromatography and the AIM	More recent literature is not available
Bacillus firmus (NIAS 237) (Urata and Satoh 1991)	Soil	Analysis of the enzymes by cell fractionation and proton translocation measurements	More recent literature is not available
Bacillus licheniformis (multiple strains) (Pichinoty et al. 1978)	Soil	Identification and measurement of NO, N_2O and N_2 from NO_3^- was established by gas chromatography, the reductase enzymes are detected in enzymatic extracts	*Bacillus licheniformis* (PK-16) is involved in aerobic denitrification with conversion of NO_3^- to N_2 (Kim et al. 2005)
Bacillus nitritollens (multiple strains) (Delaporte 1972)	Soil (Arctic/Antarctic)	Phenotypic testing for the removal of NO_3^- with observation of gas formation (Durham tubes)	More recent literature is not available
Bacillus subtilis (I-41) (Sakai et al. 1996)	Activated sludge	N_2 and N_2O in the gaseous phase are analysed by gas chromatography	These *B. subtilis* strains require more detailed descriptions

Anoxybacillus pushchinoensis (AT-1 and AT-2) (Yamamoto et al. 2006)	Manure-amended soil	Presence of NO_3^- is tested using a modified Griess agent, and gas samples from the head space are analysed with gas chromatography	More recent literature is not available
Brevibacillus laterosporus (multiple strains) (de Barjac and Bonnefoi 1972)	Soil, water and dead honeybee larvae	Phenotypic testing for the removal of NO_3^- with observation of gas formation (Durham tubes)	More recent literature is not available
Brevibacillus ginsengisoli (Gsoil 3088^T) (Baek et al. 2006)	Soil (ginseng field)	Only reduction of NO_3^- was tested, but denitrification is claimed	More recent literature is not available
Geobacillus stearothermophilus ($TnBA_1$) (Garcia 1977)	Soil	N_2O and N_2 were studied by Warburg respirometry and gas chromatography using AIM; activity was measured for all nitrogenous reductases	Description of the membrane-bound denitrification enzymes, except N_2O reductase, from strain ATCC 12016 (Ho et al. 1993). No denitrification in *G. stearothermophilus* ATCC 12980^T and DSM 22^T(Manachini et al. 2000; Nazina et al. 2001), but *nirK* and *qnorB* genes were found in the genome of a non-denitrifying *G. stearothermophilus* strain (Heylen 2007)
Geobacillus subterraneus (strain Sam, strain K, strain 34^T) (Nazina et al. 2001)	Soil	Occurrence of denitrification (reduction of NO_3^- to N_2) was measured with gas chromatography	More recent literature not available
Geobacillus thermodenitrificans (DSM 465^T, DSM 466, strains TH6A, TH8A, TH4B, TH45A, TH33A, TH35A, TH51A, TH61A, BI5A, TU6F3) (Manachini et al. 2000)	Soil	Method of assessing denitrification was not stated	Anaerobic growth of DSM 466, reducing NO_3^- to N_2 (measurement with gas chromatography) (Nazina et al. 2001). Identification of the genes for a complete denitrification pathway in strain NG80-2 (Feng et al. 2007; Liu et al. 2008)
Geobacillus thermoleovorans (DSM 5366T) (Nazina et al. 2001)	Soil	Occurrence of denitrification (reduction of NO_3^- to N_2) was measured with gas chromatography	No denitrification in strain DSM 5366^T (Manachini et al. 2000)

(*continued*)

Table 14.1 (continued)

Species and/or strain[a]	Habitat	Original literature	Remarks and/or more recent literature
Paenibacillus anaericanus (MH21^T) (Horn et al. 2005)	Gut of the earthworm *Aporrectodea caliginosa*	Colorimetrical testing for removal of NO_3^- and NO_2^-, gas chromatographic analysis of N_2 and N_2O	Was the *Paenibacillus anaericanus* a fermentative or a denitrifying strain?
Paenibacillus larvae subsp. *pulvifaciens* (multiple strains) (de Barjac and Bonnefoi 1972)	Dead honeybee larvae	Phenotypic testing for the removal of NO_3^- with observation of gas formation (Durham tubes)	NO_3^- reduced to NO_2^-, established by API test (Heyndrickx et al. 1996), hence true denitrification capacity not tested
Paenibacillus macerans (multiple strains) (de Barjac and Bonnefoi 1972)	Plant materials	Phenotypic testing for the removal of NO_3^- with observation of gas formation (Durham tubes)	More recent literature is not available
Paenibacillus polymyxa (multiple strains) (de Barjac and Bonnefoi 1972)	Soil and decomposing vegetation	Phenotypic testing for the removal of NO_3^- with observation of gas formation (Durham tubes)	Maximal denitrifying activity determined by measuring N_2O with gas chromatography and AIM (Manucharova et al. 2000)
Paenibacillus terrae (MH72, AM141^T) (Horn et al. 2005)	Gut of the earthworm *Aporrectodea caliginosa*	Reduction of NO_3^- to NO_2^- and gas chromatographic analysis of N_2 and N_2O	Was the *Paenibacillus terrae* strain a fermentative strain or a true denitrifying strain?
Virgibacillus halodenitrificans (ATCC 49067) (Denariaz et al. 1989, 1991)	Solar saltern	Purification of a copper-containing NO_2^- reductase Gas chromatographic analysis of the products of the denitrification reaction (the end product was N_2O)	Growth under anaerobic conditions on marine agar, only in the presence of NO_3^-, which was reduced to NO_2^- (Yoon et al. 2004); no further specification of denitrification
Virgibacillus pantothenticus (multiple strains) (de Barjac and Bonnefoi 1972)	Soil	Phenotypic testing for the removal of NO_3^- with observation of gas formation (Durham tubes)	NO_3^- reduction to NO_2^- (API gallery) (Heyndrickx et al. 1998) Growth under anaerobic conditions on marine agar, only in the presence of NO_3^-, which is reduced to NO_2^- (Yoon et al. 2004); no further specification of denitrification

[a]In these columns each species is accompanied by the strain number(s) of the tested strain(s) in the corresponding literature. If no such number is present, this was not mentioned in the original article
[b]*AIM* Acetylene Inhibition Method

Table 14.2 Denitrification gene sequences in *Bacillus sensu stricto* and related organisms[a]

Species/isolate	Accession number				
	nirK	*nirS*	*cnorB*	*qnorB*	*nosZ*
Bacillus sp. R13	–	AF335924	–	–	–
Bacillus sp. R22	–	AJ626841	–	–	–
Bacillus sp. R-31770	–	–	–	AM778672	–
Bacillus sp. R-31841	–	–	–	AM778671	–
Bacillus sp. R-31856	AM404293	–	–	–	–
Bacillus sp. R-32526	–	–	–	AM403579	–
Bacillus sp. R-32546	AM404294	–	–	–	–
Bacillus sp. R-32656	–	–	–	AM778673	–
Bacillus sp. R-32694	–	–	AM403581	AM778668	–
Bacillus sp. R-32702	–	–	–	AM778670	–
Bacillus sp. R-32709	–	–	–	AM778667	–
Bacillus sp. R-32715	–	–	–	AM404295	–
Bacillus sp. R-33773	–	–	–	AM778674	–
Bacillus sp. R-33802	–	–	–	AM778669	–
Bacillus sp. SH3	–	–	–	EU374113	–
Bacillus sp. SH5	–	–	–	EU374114	–
Bacillus sp. SH8	–	–	–	EU374115	–
Bacillus sp. SH10	–	–	–	EU374116	–
Bacillus sp. SH11	–	–	–	EU374117	–
Bacillus sp. SH14	–	–	–	EU374118	–
Bacillus sp. SH19	–	–	–	EU374119	–
Bacillus sp. SH21	–	–	–	EU374120	–
Bacillus sp. SH22	–	–	–	EU374121	–
Bacillus sp. SH25	–	–	–	EU374122	–
Bacillus sp. SH27	–	–	–	EU374123	–
Bacillus sp. SH30	–	–	–	EU374124	–
Bacillus sp. SH36	–	–	–	EU374125	–
Bacillus sp. SH38	–	–	–	EU374126	–
Bacillus sp. SH41	–	–	–	EU374127	–
Bacillus sp. SH42	–	–	–	EU374128	–
Bacillus sp. SH43	–	–	–	EU374129	–
Bacillus sp. SH48	–	–	–	EU374130	–
Bacillus sp. SH51	–	–	–	EU374131	–
Bacillus sp. SH60	–	–	–	EU374132	–
Bacillus sp. SH61	–	–	–	EU374133	–
Bacillus sp. SH62	–	–	–	EU374134	–
Bacillus sp. SH63	–	–	–	EU374135	–
B. licheniformis ATCC 14580[b]	–	–	NC006270 (presence of a *norB* gene)		–
G. stearothermophilus	–	–	–	–	AB450501
G. thermodenitrificans NG80-2[b]	NC009328	–	–	–	NC009328

[a]Gene sequence accession numbers derived from http://www.ncbi.nlm.nih.gov
[b]Derived from whole-genome sequence

Pichinoty et al. 1983; Bouchard et al. 1996; Baek et al. 2006); N_2 production (Groffman et al. 2006); the acetylene inhibition method (AIM) (Manucharova et al. 2000; Yanai et al. 2008); tracer methods like the use of ^{15}N-nitrate and/or -nitrite tracers (Shoun et al. 1998; Katsuyama et al. 2008); stochiometric approaches, etc. Every procedure has its own advantages and disadvantages (Tiedje 1988;

Groffman et al. 2006; Philippot et al. 2007), but some of these techniques are more expensive and more laborious than others. Therefore, a combination of simple low-cost techniques may be an acceptable option to assess the denitrifying abilities of pure cultures of aerobic endospore-formers isolated from soil.

14.2.1 Phenotypic Determination of the Denitrifying Capacity in Bacillus *Isolates*

The first step in obtaining denitrifying bacilli from soil is to selectively isolate them and grow them in pure cultures. Spore formers can easily be isolated by subjecting the sample to 80°C for 10–30 min (pasteurization). Afterwards, the heat-resistant endospores can be plated out (Slepecky and Hemphill 2006; Madigan et al. 2003). After isolation, the true denitrifying abilities of the Gram-positive endospore-forming isolates can be evaluated. Denitrifying bacteria are usually able to grow on defined media that provide the required trace elements, vitamins and a defined carbon and nitrogen source. Mineral media (Stanier et al. 1966) complemented with succinate, ethanol or acetate as carbon sources and KNO_3 or KNO_2 as nitrogen sources have proved to be successful in the isolation of denitrifying strains, including the aerobic endospore-formers (Kim et al. 2005; Heylen 2007; Park et al. 2007).

The interpretation of the denitrification capacity in *Bacillus* may appear complex since some species only perform the first step of denitrification (nitrate reduction). This kind of reduction can point to another related dissimilatory process, namely dissimilatory nitrate reduction to ammonium (DNRA) (Fan et al. 2006). Although DNRA is carried out by some *Bacillus* species, it is not considered as denitrification *sensu stricto* but rather as "nitrate reduction" and should be ruled out in any denitrification assay, if possible. However, a combination of the low-cost Griess test and the measurement of N_2O gas by gas chromatography with the acetylene inhibition method (AIM) seems to be a good alternative approach to measure the true denitrification capacities of *Bacillus* isolates in pure culture.

14.2.1.1 The Griess Test

This is a relatively cheap and simple test method to determine whether micro-organisms are able to denitrify or not. In the first step, the strains are grown in a suitable defined broth medium supplemented with a nitrogen source (such as KNO_3 or KNO_2), a carbon source (such as succinate, ethanol or acetate), some vitamins and a pH indicator such as phenol red. This pH indicator turns red in alkaline conditions, as happens with denitrification. The most commonly used concentrations of nitrate or nitrite given in the literature vary from roughly <1 to 10 mM (Tiedje 1988; Mahne and Tiedje 1995; Goregues et al. 2005; Yamamoto et al. 2006; Liang et al. 2007). A standard amount of bacterial cells is used to inoculate the broth medium, e.g., a standard volume of a suspension with a standardized turbidity.

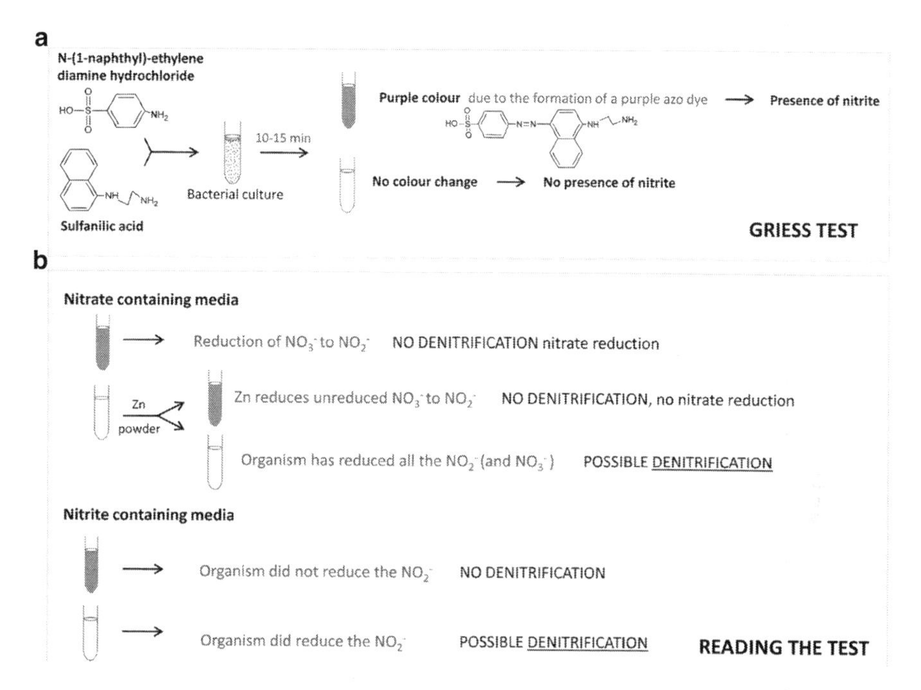

Fig. 14.2 *Phenotypic determination of the denitrifying capacity* (**a**) The Griess reaction: After addition of equal amounts of *N*-(1-naphthyl)-ethylene diamine hydrochloride and sulfanilic acid to the bacterial culture, the colour change can be read after 10–15 min. If a purple colour appears, nitrite is present in the bacterial culture; if not, nitrite is absent. (**b**) Interpretation of the Griess reaction according to the medium used. Depending on the original nitrogen source in the medium, a colour change reveals whether or not denitrification has occurred

The bacterial cultures are incubated in anaerobic conditions (e.g., in an anaerobic chamber), and at temperatures and times that suit their requirements. After incubation, the pH shift and the extent of growth of the culture are evaluated.

In the second step, the disappearance of nitrate and/or nitrite in the bacterial cultures is visualized with the Griess reaction (Fig. 14.2a). In this reaction, the reagents *N*-(1 naphthyl) ethylene diaminedihydrochloride and sulfanilic acid produce a purple-coloured end product in the presence of nitrite. After adding equal amounts of these reagents to the bacterial culture, the colour change depends on the initial nitrogen source in the medium and the denitrifying ability of the bacterial culture with which it was inoculated (Fig. 14.2b) (Wang and Skipper 2004; Slepecky and Hemphill 2006).

Since the disappearance of nitrate and/or nitrite in a *Bacillus* culture can be due to metabolic processes other than denitrification (e.g., DNRA), the results of nitrate or nitrite removal from the medium and the detection of a pH increase by an indicator dye, have to be confirmed by other methods. Given that nitrate and nitrite reduction cannot be considered as proof of denitrifying ability, the possible denitrification should be verified by measuring the production of N_2O or N_2 during anaerobic growth.

14.2.1.2 Gas Chromatography of N_2O of the "Denitrifying" Isolates

The gas produced by the so-called denitrifying aerobic endosporer-formers has not often been identified, and in most taxonomic descriptions only gas production from nitrate or nitrite is stated (Zumft and Kroneck 2007). Thus the ambiguity in the denitrification activity is caused by the lack of gas identification in species descriptions (Zumft 1992). To measure a process, the direct approach is to measure the end product of the reaction. In the case of denitrification, the end product is mainly dinitrogen gas (N_2). However, since our atmosphere is almost 80% N_2, this approach is difficult (Tiedje 1988). Therefore, the acetylene inhibition method is useful in measuring denitrification by gas chromatography (Philippot et al. 2007). Acetylene (C_2H_2) has the ability to inhibit the reduction of N_2O to N_2, thereby allowing us to use gas chromatography to quantify the N_2O produced in the denitrification process. Consequently, this method may provide evidence of denitrification activity, additional to the Griess test described above.

To measure nitrous oxide by gas chromatography, tubes or flasks are filled with approximately 10 ml of a sterile nitrate/nitrite broth medium (e.g., the same medium used for the Griess test, but without pH indicator if desired), a standard amount of bacterial cells (0.1–1 ml suspension of a standard turbidity) is added and the vessels are sealed with rubber bottle stoppers and aluminium crimp seals. The headspace of the vessels should be oxygen-free, and the oxygen is replaced by sterile argon (Manucharova et al. 2000; Nazina et al. 2001; Kumon et al. 2002) or helium (Mahne and Tiedje 1995; Shoun et al. 1998; Morley et al. 2008) by evacuating and flushing repeatedly. Acetylene gas is injected through the stoppers to give a final concentration of 10% (vol/vol) in the headspace (Tiedje 1988). Other headspace percentages of C_2H_2 have also been reported (Chèneby et al. 2000; Manucharova et al. 2000). The added acetylene will prevent N_2O from being further reduced to N_2. The bacterial cultures are incubated at temperatures and times that suit the strains' requirements. After incubation, gas samples (0.1–1 ml, depending on the gas chromatographic set-up) are removed with an air-tight syringe and injected into the gas chromatograph to detect the presence of N_2O. The syringe may additionally be flushed with the same gas that was used to flush the headspace (Mahne and Tiedje 1995). A gas chromatograph with a dual-column system (e.g., a Hayesep Q [80–100-mesh] and a Chromatopac column at 25°C, with helium as carrier gas) and equipped with a thermal conductivity detector (TCD) or an electron capture detector (ECD) should be used. Other successful column combinations and carrier gases have been used (Gamble et al. 1977; Weier and MacRae 1992; Sakai et al. 1996; Nazina et al. 2001; Dandie et al. 2007).

Although the combination of both techniques (Griess test combined with GC analysis of N_2O) provides a rather reliable, cheap and easy way to characterize a metabolic trait such as denitrification, a more efficient and refined technique is available: an automated incubation system allowing the study of gas kinetics of denitrifiers (Molstad et al. 2007). The system was primarily designed to study gas production/reduction (O_2, NO, N_2O and N_2) by denitrifying bacteria (or microbial communities), and their electron flows to the different electron acceptors during and after transition from oxic to anoxic conditions. The possibility of performing precise monitoring of N_2 gas formation is particularly interesting, and circumvents the use of more expensive ^{15}N tracer systems or the more comprehensive acetylene inhibition method. In due course, the data obtained may allow the investigation of the relationships between genotypes and phenotypes of denitrifying bacteria (Molstad et al. 2007).

14.2.2 *Genomic Determination of the Denitrifying Capacity in* Bacillus *Isolates*

Besides the phenotypic approach to assess the denitrifying ability of *Bacillus* isolates, different molecular methods exist to detect the presence of the key denitrification genes, *nir* and/or *nor*, in bacterial genomes. The principal technique for obtaining new *nir* or *nir* sequences is PCR amplification of these functional genes.

As most existing primer sets for *nirS*, *nirK* and *norB* genes are only based on gene sequences from Gram-negative strains that are closely related to each other (Braker et al. 2001; Hallin and Lindgren 1999; Casciotti and Ward 2001, 2005; Song and Ward 2003; Braker and Tiedje 2003; Goregues et al. 2005; Liang et al. 2007; Flores-Mireles et al. 2007; Nakano et al. 2008), part of the Gram-positive denitrifying diversity might be disregarded, in particular, for members of the genus *Bacillus*. Therefore, the need to re-evaluate primers for PCR amplification of denitrifying genes is essential. Moreover, new primer design based on gene sequences of Gram-positive bacteria, and in particular those of *Bacillus* isolates, is needed. Although the number of available complete and partial *nir* and *nor* gene sequences from aerobic endospore-formers has increased in the public databases, it is still limited (Table 14.2). Genome-sequencing projects and correct annotation of the genomes will probably deliver putative and new denitrifying reductases, but currently only 23 *Bacillus* genomes (covering 10 species and a limited phylogenetic area), two *Geobacillus* genomes, one *Oceanobacillus* genome, one *Brevibacillus* genome and one *Anoxybacillus* genome have been fully sequenced (Liolios et al. 2008). This still limited phylogenetic coverage in the public sequence databases may not represent the real functional denitrification capacities of *Bacillus* and related organisms.

A combination of phenotypic detection of denitrification capacity and genomic detection of one or more of the key denitrification genes might be useful to learn more about the presence and distribution of the denitrification capacity in *Bacillus* and related organisms. Surprisingly, expression of the enzymatic denitrification activity is often observed, whereas the genes are not detected, and vice versa. This shows that molecular methods are currently dependent on the state of knowledge on the functional genes in these organisms. Moreover, since these Gram-positive organisms lack a periplasmic compartment between the two membranes of the cell wall, it is possible that an undiscovered redundancy for the denitrification genes exists and reductases yet unknown catalyse denitrification in members of the genus *Bacillus* and related taxa. The membrane-bound denitrification enzyme qCu_ANOR, purified from *Bacillus azotoformans* (Suharti et al. 2001; Suharti and de Vries 2004, 2005), and the recently discovered gNOR gene (Sievert et al. 2008) serve to emphasize and support this hypothesis.

14.3 Conclusion

The denitrification process in aerobic endospore-formers is not well studied, and further characterization of their denitrification potential, primer development for their functional denitrification genes and further biochemical research will be necessary in order to clarify the genetic diversity in both pure cultures and environmental samples. Identification of the true biodiversity of denitrification in *Bacillus* and relatives will be obtained by testing for the denitrification capacity in a large number of *Bacillus* isolates through reliable cultivation approaches, phenotypic testing and whole-genome sequencing approaches. However, the methods to test for the denitrification capacity described in this chapter are simple and inexpensive starting points for assessing the denitrification capacities of aerobic endospore-formers in pure cultures. A thorough investigation of the denitrification process in these isolates could reveal novel aspects and unknown enzymes and might be an incentive for further biotechnological applications of the property of denitrification.

References

Baek S-H, Im W-T, Oh HW, Lee J-S, Oh H-M, Lee S-T (2006) *Brevibacillus ginsengisoli* sp. nov., a denitrifying bacterium isolated from soil of a ginseng field. Int J Syst Evol Microbiol 5:2665–2669

Bouchard B, Beaudet R, Villemur R, McSween G, Lépine F, Bisaillon J-G (1996) Isolation and characterization of *Desulfitobacterium frappieri* sp. nov., an anaerobic bacterium which reductively dechlorinates pentachlorophenol to 3-chlorophenol. Int J Syst Bacteriol 46:1010–1015

Braker G, Tiedje JM (2003) Nitric oxide reductase (*norB*) genes from pure cultures and environmental samples. Appl Environ Microbiol 69:3476–3483

Braker G, Ayala-del-Río HL, Devol AH, Fesefeldt A, Tiedje JM (2001) Community structure of denitrifiers, *Bacteria*, and *Archaea* along redox gradients in Pacific Northwest marine sediments by terminal restriction fragment length polymorphism analysis of amplified nitrite reductase (*nirS*) and 16S rRNA genes. Appl Environ Microbiol 67:1893–1901

Casciotti KL, Ward BB (2001) Dissimilatory nitrite reductase genes from autotrophic ammonia-oxidizing bacteria. Appl Environ Microbiol 67:2213–2221

Casciotti KL, Ward BB (2005) Phylogenetic analysis of nitric oxide reductase gene homologues from aerobic ammonia-oxidizing bacteria. FEMS Microbiol Ecol 52:197–205

Chèneby D, Philippot L, Hartmann A, Hénault C, Germon J-C (2000) 16S rDNA analysis for characterization of denitrifying bacteria isolated from three agricultural soils. FEMS Microbiol Ecol 34:121–128

Dandie CE, Burton DL, Zebarth BJ, Trevors JT, Goyer C (2007) Analysis of denitrification genes and comparison of *nosZ*, *cnorB* and 16S rDNA from culturable denitrifying bacteria in potato cropping systems. Syst Appl Microbiol 30:128–138

De Barjac H, Bonnefoi A (1972) Essai de classification biochimique de 64 *Bacillus* des groups II et III représentant 11 espèces différentes. Ann Inst Pasteur (Paris) 122:463–473

Delaporte B (1972) Trois nouvelles éspèces de *Bacillus*: *Bacillus similibadius* n. sp., *Bacillus longisporus* n. sp. and *Bacillus nitritollens* n. sp. Ann Inst Pasteur (Paris) 123:821–834

Denariaz G, Payne WJ, LeGall J (1989) A halophilic denitrifier, *Bacillus halodenitrificans* sp. nov. Int J Syst Bacteriol 39:145–151

Denariaz G, Payne WJ, LeGall J (1991) The denitrifying nitrite reductase of *Bacillus halodenitrificans*. Biochim Biophys Acta 1056:225–232

Ehrenberg CG (1835) Physikalische Abhandlungen der Koeniglichen Akademie der Wissenschaften zu Berlin aus den Jahren 1833–1835, p. 145–336

Fan LF, Shieh WY, Wu WF, Chen CP (2006) Distribution of nitrogenous nutrients and denitrifier strains in estuarine sediment profiles of the Tanshui river, northern Taiwan. Est Coast Shelf Sci 69:543–553

Feng L, Wang W, Cheng J, Ren Y, Zhao G, Gao C, Tang Y, Liu X, Han W, Peng X, Liu R, Wang L (2007) Genome and proteome of long-chain alkane degrading *Geobacillus thermodenitrificans* NG80-2 isolated from a deep-subsurface oil reservoir. Proc Natl Acad Sci USA 104:5602–5607

Flores-Mireles AL, Winans SC, Holguin G (2007) Molecular characterization of diazotrophic and denitrifying bacteria associated with mangrove roots. Appl Environ Microbiol 73:7308–7321

Gamble T, Betlach MR, Tiedje JM (1977) Numerically dominant denitrifying bacteria from world soils. Appl Environ Microbiol 33:926–939

Garcia JL (1977) Étude de la dénitrification chez une bactérie thermophile sporulée. Ann Microbiol (Inst Past) 128A:447–458

Goregues CM, Michotey VD, Bonin PC (2005) Molecular, biochemical and physiological approaches for understanding the ecology of denitrification. Microb Ecol 49:198–208

Groffman PM, Altabet MA, Böhlke JK, Butterbach-Bahl K, David MB, Firestone MK, Giblin AE, Kana TM, Nielsen LP, Voytek MA (2006) Methods for measuring denitrification: diverse approaches to a difficult problem. Ecol Appl 16:2091–2122

Hackenthal E (1966) Die parallele induktion von nitratreduktase und nitritreduktase bei *Bacillus cereus* durch verschiedene anionen. Biochem Pharmacol 15:1119–1126

Hallin S, Lindgren P-E (1999) PCR detection of genes encoding nitrite reductase in denitrifying bacteria. Appl Environ Microbiol 65:1652–1657

Heylen K (2007) Study of the genetic basis of denitrification in pure culture denitrifiers isolated from activated sludge and soil. PhD Thesis, Laboratory of Microbiology (LM-UGent), Ghent University, Belgium

Heyndrickx M, Vandemeulebroecke K, Hoste B, Janssen P, Kersters K, De Vos P, Logan NA, Ali N, Berkeley RCW (1996) Reclassification of *Paenibacillus* (formerly *Bacillus*) *pulvifaciens* (Nakamura 1984) Ash et al. 1994, a later subjective synonym of *Paenibacillus* (formerly *Bacillus*) *larvae* (White 1906) Ash et al. 1994, as a subspecies of *P. larvae*, with emended descriptions of *P. larvae* as *P. larvae* subsp. *larvae* and *P. larvae* subsp. *pulvifaciens*. Int J Syst Bacteriol 46:270–279

Heyndrickx M, Lebbe L, Kersters K, De Vos P, Forsyth G, Logan NA (1998) *Virgibacillus*: a new genus to accommodate *Bacillus pantothenticus* (Proom and Knight 1950). Emended description of *Virgibacillus pantothenticus*. Int J Syst Bacteriol 48:99–106

Ho TP, Jones AM, Hollocher TC (1993) Denitrification enzymes of *Bacillus stearothermophilus*. FEMS Microbiol Lett 114:135–138

Horn MA, Ihssen J, Matthies C, Schramm A, Acker G, Drake HL (2005) *Dechloromonas denitrificans* sp.nov., *Flavobacterium denitrificans* sp. nov., *Paenibacillus anaericanus* sp. nov. and *Paenibacillus terrae* strain MH72, N_2O-producing bacteria isolated from the gut of the earthworm *Aporrectodea caliginosa*. Int J Syst Evol Microbiol 55:1255–1265

Katsuyama C, Kondo N, Suwa Y, Yamagishi T, Itoh M, Ohte N, Kimura H, Nagasoa K, Kato K (2008) Denitrification activity and relevant bacteria revealed by nitrite reductase gene fragments in soil of temperate mixed forests. Microb Environ 23:337–345

Kim JK, Park KJ, Cho KS, Nam S-W, Park T-J, Bajpad R (2005) Aerobic nitrification-denitrification by heterotrophic *Bacillus* strains. Bioresour Technol 96:1897–1906

Kumon Y, Sasaki Y, Kato I, Takaya N, Shoun H, Beppu T (2002) Codenitrification and denitrification are dual metabolic pathways through which dinitrogen evolves from nitrate in *Streptomyces antibioticus*. J Bacteriol 184:2963–2968

Liang D-W, Zhang T, Fang HHP (2007) Denitrifying degradation of dimethyl phthalate. Appl Microbiol Biotechnol 74:221–229

Liolios K, Mavrommatis K, Tavernakarakis N, Kyrpides NC (2008) The Genomes On Line Database (GOLD) in 2007: status of genomic and metagenomic projects and their associated metadata. NAR 36:D475–D479

Liu X, Gao C, Zhang A, Jin P, Wang L, Feng L (2008) The *nos* gene cluster from gram-positive bacterium *Geobacillus thermodenitrificans* NG80-2 and functional characterization of the recombinant NosZ. FEMS Microbiol Lett 289:46–52

Madigan MT, Martinko JM, Parker J (2003) Brock biology of microorganisms, 10th edn. Pearson Education, London

Mahne I, Tiedje JM (1995) Criteria and methodology for identifying respiratory denitrifiers. Appl Environ Microbiol 61:1110–1115

Manachini PL, Mora D, Nicastro G, Parini C, Stackebrandt E, Pukall R, Fortina MG (2000) *Bacillus thermodenitrificans* sp. nov., nom. rev. Int J Syst Evol Microbiol 50:1331–1337

Manucharova NA, Dobrovol'skaya TG, Stepanov AL (2000) Taxonomic composition of denitrifying bacteria in soddy podzolic soil. Mikrobiologiia 69:234–237

Molstad L, Dörsch P, Bakken LR (2007) Robotized incubation system for monitoring gases (O_2, NO, N_2O, N_2) in denitrifying cultures. J Microbiol Methods 71:202–211

Morley N, Baggs EM, Dörsch P, Bakken L (2008) Production of NO, N_2O and N_2 by extracted soil bacteria, regulation by NO_2^- and O_2 concentrations. FEMS Microbiol Ecol 65:102–112

Nakano M, Shimizu Y, Okumura H, Sugahara I, Maeda H (2008) Construction of a consortium comprising ammonia-oxidizing bacteria and denitrifying bacteria isolated from marine sediment. Biocontrol Sci 13:73–89

Nazina TN, Tourova TP, Poltaraus AB et al (2001) Taxonomic study of aerobic thermophilic bacilli: descriptions of *Geobacillus subterraneus* gen. nov., sp. nov. and *Geobacillus uzenensis* sp. nov. from petroleum reservoirs and transfer of *Bacillus stearothermophilus*, *Bacillus thermocatenulatus*, *Bacillus thermoleovorans*, *Bacillus kaustophilus*, *Bacillus thermoglucosidasius* and *Bacillus thermodenitrificans* to *Geobacillus* as the new combinations *G. stearothermophilus*, *G. thermocatenulatus*, *G. thermoleovorans*, *G.* kaustophilus, G. *thermoglucosidasius* and *G. thermodenitrificans*. Int J Syst Evol Microbiol 51:433–446

Park JY, Yoo YJ (2009) Biological nitrate removal from in industrial wastewater treatment: which electron donor we can choose. Appl Microbiol Biotechnol 82:415–429

Park SJ, Yoon JC, Shin KS, Kim EH, Yim S, Cho YJ, Sung GM, Lee DG, Kim SP, Lee DU, Woo SH, Koopman B (2007) Dominance of endospore forming bacteria on a Rotating Activated Bacillus Contactor biofilm for advanced wastewater treatment. J Microbiol 45:113–121

Philippot L, Hallin S, Schloter M (2007) Ecology of denitrifying prokaryotes in agricultural soil. Adv Agron 96:135–191

Pichinoty F, Garcia JL, Job C, Durand M (1978) La dénitrification chez *Bacillus licheniformis*. Can J Microbiol 24(1):45–49

Pichinoty F, de Barjac H, Mandel M, Asselineau J (1983) Description of *Bacillus azotoformans* sp. nov. Int J Syst Bacteriol 33:660–662

Prendez M, Lara-Gonzalez S (2008) Application of strategies for sanitation management in wastewater treatment plants in order to control/reduce greenhouse gas emissions. J Environ Manage 88:658–664

Sakai K, Ikehata Y, Ikenaga Y, Wakayama M, Moriguchi M (1996) Nitrite oxidation by heterotrophic bacteria under various nutritional and aerobic conditions. J Ferment Bioeng 82:613–617

Shoun H, Mitsuyoshi K, Bada I, takaya N, Matsuo M (1998) Denitrification by actinomycetes and purification of dissimilatory nitrite reductase and azurin from *Streptomyces thioluteus*. J Bacteriol 180:4413–4415

Sievert SM, Scott KM, Klotz MG, Chain PS, Hauser LJ, Hemp J, Hügler M, Land M, lapidus A, Larimer FW, Lucas S, Malfatti SA, Meyer F, Paulsen IT, Ren Q, Simon J (2008) Genome of the epsilonproteobacterial chemolithoautotroph *Sulfurimonas denitrificans*. Appl Environ Microbiol 74:1145–1156

Slepecky RA, Hemphill HE (2006) The genus *Bacillus* – Non-medical. In: Dworkin M, Falkow S, Rosenberg E, Schleifer K-H, Stackebrandt E (eds) The prokaryotes. A handbook on the biology of bacteria. Bacteria: Firmicutes, Cyanobacteria, vol 4. Springer, Berlin, pp 530–562

Song B, Ward BB (2003) Nitrite reductase genes in halobenzoate degrading denitrifying bacteria. FEMS Microbiol Ecol 43:349–357

Stanier RY, Palleroni NJ, Doudoroff M (1966) The aerobic Pseudomonads: a taxonomic study. J Gen Microbiol 43:159–271

Suharti HHA, de Vries S (2004) NO Reductase from *Bacillus azotoformans* is a bifunctional enzyme accepting electrons from menaquinol and a specific endogenous membrane-bound cytochrome c_{551}. Biochemistry 43:13487–13495

Suharti, de Vries S (2005) Membrane-bound denitrification in the Gram-positive bacterium *Bacillus azotoformans*. Biochem Soc Trans 33:130–133

Suharti SMJF, Schröder I, de Vries S (2001) A novel copper A containing menaquinol NO reductase from *Bacillus azotoformans*. Biochemistry 40:2632–2639

Tiedje JM (1988) Ecology of denitrification and dissimilatory nitrate reduction to ammonium. In: Zehnder AJB (ed) Environmental microbiology of anaerobes. Wiley, New York, pp 179–244

Urata K, Satoh T (1991) Enzyme localization and orientation of the active site of dissimilatory nitrite reductase from *Bacillus firmus*. Arch Microbiol 156:24–27

Van Cleemput O, Samater AH (1996) Nitrite in soils: accumulation and role in the formation of gaseous N compounds. Fertil Res 45:81–89

van Spanning RJM, Richardson DJ, Ferguson SJ (2007) Introduction to the biochemistry and molecular biology of denitrification. In: Bothe H, Ferguson SJ, Newton WE (eds) Biology of the nitrogen cycle. Elsevier, Amsterdam, pp 3–20

Wang G, Skipper HD (2004) Identification of denitrifying rhizobacteria from bentgrass and bermudagrass golf greens. J Appl Microbiol 97:827–837

Weier KL, MacRae IC (1992) Denitrifying bacteria in the profile of a brigalow clay soil beneath a permanent pasture and cultivated crop. Soil Biol Biochem 24:919–923

Yamamoto M, Ishii A, Nogi Y, Inoue A, Ito M (2006) Isolation and characterization of novel denitrifying alkalithermophiles AT-1 and AT-2. Extremophiles 10:421–426

Yanai Y, Hatano R, Okazaki M, Toyota K (2008) Analysis of the C_2H_2 inhibition-based N_2O production curve to characterize the N_2O-reducing activity of denitrifying communities in soil. Geoderma 146:269–276

Yoon J-H, Oh T-K, Park Y-H (2004) Transfer of *Bacillus halodenitrificans* Denariaz *et al.* 1989 to the genus *Virgibacillus* as *Virgibacillus halodenitrificans* comb. nov. Int J Syst Evol Microbiol 54:2163–2167

Zumft WG (1992) The denitrifying prokaryotes. In: Balows A, Trüper HG, Dworkin M, Harder W, Schleifer K-H (eds) The prokaryotes. A handbook on the biology of bacteria: ecophysiology, isolation, identification, applications, vol 1. Springer, New York, pp 554–582

Zumft WG, Kroneck PMH (2007) Respiratory transformation of nitrous oxide (N_2O) to dinitrogen by *Bacteria* and *Archaea*. Adv Microb Physiol 52:107–227

Chapter 15
Paenibacillus, Nitrogen Fixation and Soil Fertility

Lucy Seldin

15.1 The Genus *Paenibacillus* and the Nitrogen-Fixing Species

The genus *Paenibacillus* was defined in 1993 after an extensive comparative analysis of 16S rRNA sequences of 51 species of the genus *Bacillus* (Ash et al. 1991, 1993). At that time, the genus comprised 11 species, with *P. polymyxa* as the type species. Currently, the genus comprises more than 100 species and two subspecies (http://www.ncbi.nlm.nih.gov) and harbours strains of industrial and agricultural importance. Different enzymes (such as cyclodextrin glucanotransferase, chitinase, amylase, cellobiohydrolase, agarase and proteases) and antimicrobial substances (antibiotics, bacteriocins and/or small peptides) are produced by different *Paenibacillus* spp. (Priest 1993; Rosado and Seldin 1993; Mavingui and Heulin 1994; Walker et al. 1998; Aguilera et al. 2001; Sakiyama et al. 2001; von der Weid et al. 2003; Reynaldi et al. 2004; Alvarez et al. 2006; Aktuganov et al. 2008; Fortes et al. 2008; Tupinambá et al. 2008). Furthermore, different strains of *Paenibacillus* spp. can degrade polyaromatic hydrocarbons (Daane et al. 2002), be toxic for insects (Pettersson et al. 1999), produce phytohormones (Lebuhn et al. 1997; Nielsen and Sorensen 1997; Bent et al. 2001; Çakmakçi et al. 2007; da Mota et al. 2008), solubilize phosphate (Seldin et al. 1998), suppress phytopathogens through antagonistic functions (Piuri et al. 1998; Budi et al. 1999; Beatty and Jensen 2002; Selim et al. 2005; von der Weid et al. 2005; Aktuganov et al. 2008) and/or furnish nutrients to the plants by nitrogen fixation (Seldin et al. 1984; Chanway et al. 1988; Holl et al. 1988; Elo et al. 2001; Berge et al. 2002; von der Weid et al. 2002, among others). Some of these *Paenibacillus* species (or strains) that can influence plant growth and health by presenting one or more of the characteristics mentioned above are considered to be plant growth-promoting rhizobacteria (PGPR). They are usually free-living bacteria, having been found in

L. Seldin
Laboratório de Genética Microbiana, Instituto de Microbiologia Prof. Paulo de Góes, CCS – Centro de Ciências da Saúde – Bloco I, Avenida Carlos Chagas Filho, 373, Cidade Universitária, Ilha do Fundão, CEP. 21941-902 Rio de Janeiro, RJ, Brazil
e-mail: lseldin@micro.ufrj.br, lucy@seldin.com.br

N.A. Logan and P. De Vos (eds.), *Endospore-forming Soil Bacteria*, Soil Biology 27,
DOI 10.1007/978-3-642-19577-8_15, © Springer-Verlag Berlin Heidelberg 2011

different kinds of soils (Axelrood et al. 2002; Garbeva et al. 2003; Lee et al. 2007; Kim et al. 2008; Lee and Yoon 2008, and many others), in a variety of plant rhizospheres (Seldin et al. 1983; Berge et al. 2002; von der Weid et al. 2002; Coelho et al. 2007; Beneduzi et al. 2008, and many others), and inside plant tissues (Shishido et al. 1999; Garbeva et al. 2001). Some species were even found in association with arbuscular mycorrhizal hyphae (Mansfeld-Giese et al. 2002). Within the genus *Paenibacillus*, 16 species are considered to harbour nitrogen-fixing strains: *P. polymyxa, P. macerans, P. peoriae, P. durus, P. brasilensis, P. graminis, P. odorifer, P. borealis, P. wynnii, P. massiliensis, P. sabinae, "P. donghaensis"*, *P. zanthoxyli*, *P. forsythiae*, *P. riograndensis* and *P. sonchi*. Some of these species are well studied, while our knowledge of others is largely restricted to their taxonomic descriptions. In the following sections, the contribution of these species to soil fertility is considered.

It is well established that biological nitrogen fixation (BNF) is particularly important in agricultural systems, where nitrogen is usually the limiting nutrient for crop growth; the reduction of atmospheric nitrogen into bioavailable ammonium by BNF is an important source of nitrogen input (Demba Diallo et al. 2004; Wartiainen et al. 2008). Although nitrogen fixation by free-living soil microorganisms is sometimes considered a minor source of nitrogen input in soil when compared to systems such as the *Rhizobium*–legume symbiosis (Peoples and Craswell 1992), it has been shown to be the dominant source of fixed nitrogen in different soils (Widmer et al. 1999; Lovell et al. 2000; Poly et al. 2001). Therefore, due to the relevant ecological characteristics of the nitrogen-fixing *Paenibacillus* species, different studies have been performed to help the correct identification of their members. Coelho et al. (2003) examined the eight nitrogen-fixing *Paenibacillus* species recognized at that time by restriction fragment length polymorphism (RFLP) analysis of part of 16S and 23S rRNA genes amplified by polymerase chain reaction (PCR) and by multilocus enzyme electrophoresis (MLEE) assay. Both methods provided rapid tools for the characterization and the establishment of the taxonomic position of isolates belonging to the nitrogen-fixing group. Furthermore, da Mota et al. (2004) investigated the usefulness of the RNA polymerase beta-subunit encoding gene (*rpoB*) as an alternative to the 16S rRNA gene for taxonomic studies. For amplification of *Paenibacillus rpoB* gene sequences, a primer set consisting of a forward primer (*rpoB*1698f) and a reverse primer (*rpoB*2041r) was used as described by Dahllöf et al. (2000). *rpoB* gene fragments (375 bp) were amplified from the eight reference strains of the nitrogen-fixing *Paenibacillus* species, and the nucleotide sequences were determined and compared. An *rpoB* database for nitrogen-fixing *Paenibacillus* species was then formed (accession numbers AY493861 to AY493868). Phylogenetic relationships among the nitrogen-fixing *Paenibacillus* species based on comparisons of partial *rpoB* sequences and complete 16S rRNA gene sequences indicated that the partial *rpoB* sequences were as informative as the full 16S rRNA gene sequence. Therefore, *rpoB* was recommended for the identification of nitrogen-fixing *Paenibacillus* strains (da Mota et al. 2004). Later, the same group developed and applied a specific PCR-Denaturing Gradient Gel Electrophoresis (DGGE) system based on *rpoB* as a molecular marker for amplification and fingerprinting of *Paenibacillus* populations in environmental samples (da Mota et al. 2005). More recently, a specific PCR system for the amplification of the nitrogen-fixing species of *Paenibacillus* was developed based on *nifH* gene sequences (Coelho et al. 2009). The PCR system consists of an 18-mer forward primer – nifHPAENf: 5′TKATYCTGAACACGA-AAG3′ – and a 19-mer reverse primer – nifHPAENr: 5′CTCRCGGATTGGC-ATTGCG3′. The expected band of 280 bp was observed when the genomic DNA from the ten different species of *Paenibacillus* was used as template, whereas no PCR products were obtained with DNA from any of the other bacterial strains tested. The specificity of these primers was also confirmed in rhizosphere soil samples by sequence analyses of randomly chosen soil-derived clones obtained from the rhizosphere of a sorghum cultivar that had been treated with large and small amounts of nitrogen fertilizer. In this case, a nested PCR was necessary to increase the sensitivity of the detection method. All of the 187 clones that were sequenced were identified as clones affiliated with the genus *Paenibacillus* (Coelho et al. 2009). Therefore, this molecular approach was effective for assessing the presence of *nifH* gene-containing *Paenibacillus* in the environment, and may be used in the future to determine the ecological roles of this group of microorganisms for supplying nitrogen to plants.

15.2 *Paenibacillus polymyxa*

P. polymyxa is one of the best studied PGPR within the genus *Paenibacillus*. Strains belonging to this species have been isolated from the rhizospheres of wheat, barley, white clover, perennial ryegrass, crested wheatgrass, lodgepole pine, green bean, soybean, garlic, banana tree, sugarcane, lemon grass, sorghum and maize, among many other plants. *P. polymyxa* can be also found in association with the external mycelium of the arbuscular mycorrhizal fungus *Glomus intraradices* (Mansfeld-Giese et al. 2002). Due to its broad host range and its ability to produce different kinds of antimicrobial substances (Rosado and Seldin 1993; Piuri et al. 1998; Dijksterhuis et al. 1999; Seldin et al. 1999; Cho et al. 2007; He et al. 2007; Li et al. 2007; Timmusk et al. 2009), *P. polymyxa* is potentially a commercially useful biocontrol agent. For example, *P. polymyxa* strain PKB1 has been identified as a potential agent for biocontrol of blackleg disease of canola (rapeseed) caused by the pathogenic fungus *Leptosphaeria maculans*. Factors presumed to contribute to disease suppression by strain PKB1 include the production of fusaricidin-type antifungal metabolites (Li et al. 2007). In addition, Zhou et al. (2008) showed that an antagonistic protein produced by *P. polymyxa* HT16 had a strong antifungal activity against the phytopathogenic fungus *Penicillum expansum*. The same antagonistic effect was previously shown (Timmusk and Wagner 1999) in *Arabidopsis thaliana* (thale cress). Yao et al. (2004) demonstrated that *P. polymyxa* WY110, a strain isolated from rice rhizosphere, could suppress the growth of various plant pathogens, including the fungus *Pyricularia oryzae*, while the strain GS01 showed antifungal activity in ginseng (Cho et al. 2007). Also, lytic enzymes of *P. polymyxa*, such as proteases and chitinases, could have roles in interactions between antagonistic microorganisms and soil-borne plant pathogens, as shown by Nielsen and Sorensen (1997).

P. polymyxa is also known to have important roles in the rhizospheres of different crops, with the ability of many strains to secrete plant growth-enhancing substances such as cytokinins and auxins (Lebuhn et al. 1997; Timmusk et al. 1999; da Mota et al. 2008). Auxin production contributes to colonization efficiency and to the growth and survival of bacteria on their host plants (Vandeputte et al. 2005). Indole-3-acetic acid (IAA) is the main naturally occurring auxin excreted by *Paenibacillus* species. IAA production has been described in different *P. polymyxa* strains, such as L6 and Pw-2 (Bent et al. 2001), in E681 colonizing *A. thaliana* (Jeong et al. 2006) and in 68 strains prevalent in the rhizospheres of maize and sorghum sown in Brazil (da Mota et al. 2008). In lodgepole pine (*Pinus contorta*), the strain Pw-2 significantly stimulated root-biomass accumulation and total root elongation at 12 weeks after inoculation (Bent et al. 2001). Crested wheatgrass and white clover also responded positively when they were inoculated with *P. polymyxa* (Chanway et al. 1988; Holl et al. 1988). Moreover, auxin production and plant growth increment (foliar fresh weight and total leaf area) have also been reported in *P. polymyxa* E681 (Jeong et al. 2006). Other studies have also demonstrated the colonization of *P. polymyxa* in *A. thaliana* and in barley (*Hordeum vulgare*) (Timmusk et al. 2005). *P. polymyxa* colonized not only the root tip but also the intercellular spaces outside the vascular cylinder (Timmusk et al. 2005). Recently, the Auxin Efflux Carrier protein from the type strain of *P. polymyxa* was sequenced, and its presence was demonstrated in different strains of *P. polymyxa* (da Mota et al. 2008).

P. polymyxa is also known to enhance plant drought–stress tolerance and change plant gene expression after being inoculated into plant roots (Timmusk and Wagner 1999), to stimulate increased *Rhizobium etli* populations and nodulation when co-resident in the rhizosphere of *Phaseolus vulgaris* (Petersen et al. 1996), to improve nitrogen fixation by the production of degradation and fermentation products of pectin which can be used by *Azospirillum* species (Khammas and Kaiser 1992), to aggregate soil in the wheat rhizosphere (Bezzate et al. 2000), and to promote nodulation in the symbiosis *Bradyrhizobium*–cowpea (*Vigna unguiculata*) (Silva et al. 2007).

Concerning the ability to fix nitrogen, about 20–50% of the *P. polymyxa* strains tested showed neither the presence of structural *nif* gene sequences (coding for the enzyme nitrogenase) nor the ability to reduce acetylene to ethylene (Grau and Wilson 1962; Seldin et al. 1983; Oliveira et al. 1993). Furthermore, no evidence of reiteration or rearrangement of *nif* genes was found in Nif$^+$ *P. polymyxa* strains (Oliveira et al. 1993).

Finally, different studies were performed in order to assess the diversity of *P. polymyxa* strains. Santos et al. (2002) evaluated the diversity of 102 strains by using the DNA of a bacteriophage as a probe in hybridization studies. Fifty-three genotypic groups were formed, demonstrating the diversity of the species. In the same context, von der Weid et al. (2000) also demonstrated that 67 *P. polymyxa* strains isolated from the rhizosphere of maize planted in a tropical soil were very heterogeneous, when compared using phenotypic and genetic characteristics. Furthermore, the resulting data showed that strains isolated during the different stages of maize growth were significantly related to the times of sampling. Da Mota et al. (2002) analysed the genetic diversity of *P. polymyxa* populations associated with the rhizospheres of different maize cultivars, 90 days after plant sowing. Again, the *P. polymyxa* isolates were significantly different among the cultivars studied. These findings emphasize that understanding the genetic structure of *P. polymyxa* and the selection for specific groups of strains (at least in maize) can be of great agricultural interest.

15.3 *Paenibacillus macerans*

Strains of *P. macerans* are often isolated from soil and from the rhizosphere of different grasses and they show many of the important characteristics found among other species of *Paenibacillus*. *P. macerans* can be considered a PGPR as some strains show nitrogen-fixing capability (Witz et al. 1967), and also are able to produce antimicrobial substances active against phytopathogenic bacteria and fungi (Fogarty 1983). Halsall and Gibson (1985) have already shown the applicability of *P. macerans* in the decomposition of cellulose in wheat, and Mansfeld-Giese et al. (2002) showed that this species together with *P. polymyxa* was found in association with the external mycelium of the arbuscular mycorrhizal fungus *Glomus intraradices*.

One important characteristic of *P. macerans* is the production of crystalline dextrins or cyclodextrins (CDs) which are cyclic (α-1,4)-linked oligosaccharides consisting of six, seven or eight glycosyl units produced from starch (Gordon et al. 1973). The enzyme cyclodextrin glucanotransferase (CGTase) is responsible not only for the degradation of starch to smaller oligosaccharides, but also for the additional cyclization reaction yielding mixtures of cyclic oligosaccharides (Qi and Zimmermann 2005). The CGTase from *P. macerans* was the first one to be studied at the genetic level (Takano et al. 1986). As exploring the cyclodextrin-producing species of bacteria can be very important for the discovery of novel bioactive compounds, Vollú et al. (2003) described a PCR detection technique based on the 16S rRNA gene for the identification of *P. macerans* strains. Two 20-mer primers were presented – MAC 1: 5′ATCAAGTCTTCCGCATGGGA3′ and MAC 2: 5′ACTCTAGAGTGCCCAMCWTT3′ and the amplification conditions were described (Vollú et al. 2003). An agarose gel showing the PCR products obtained after the amplification, using a colony and different amounts of DNA from *P. macerans* strain LMD 24.10^{T} as template, and primers MAC 1 and 2 is presented in Fig. 15.1. This method not only detects *P. macerans* introduced to environmental samples, at least in soil, but also can be used in taxonomic studies for the quick identification of strains suspected to be *P. macerans*.

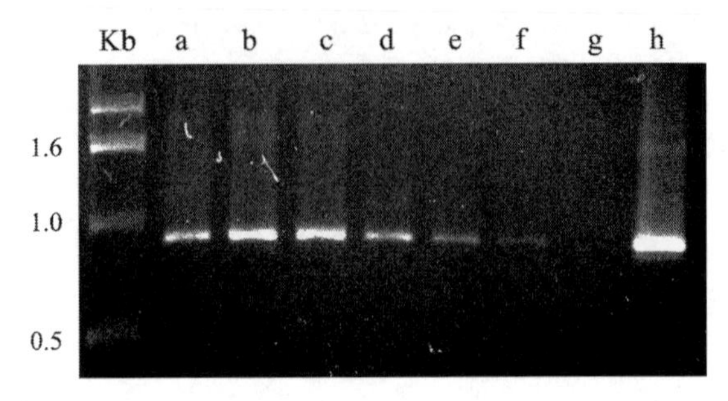

Fig. 15.1 Agarose gel (1.4%) showing PCR products [Amplification conditions were: one cycle of 94°C (30 s) and 25 cycles of 94°C (1 min), 60°C (1 min 30 s) and 72°C (1 min 30 s). A final extension step was run for 5 min at 72°C and the reaction tubes were then cooled to 4°C. All reaction mixtures contained: 1 μl of template DNA, 10 mM Tris-HCl, pH 8.3, 10 mM KCl, 0.2 mM of each deoxynucleoside triphosphate, 1 mM $MgCl_2$, 1 mM of each primer, and 5U/25 μl of *Taq* DNA polymerase. A colony from *P. macerans* LMD 24.10^T grown on trypticase soy agar (TSA) plates was suspended in 100 μl of NaOH 0.05 mM, heated to 95°C for 15 min and, following centrifugation, 1 μl of the supernatant was used as the DNA source] obtained after amplification using a colony and different amounts of DNA from *P. macerans* strain LMD 24.10^T as template and primers MAC 1 and 2 (Vollú et al. 2003). a – 50 μg of DNA, b – 10 μg, c – 5 μg, d – 1 μg, e – 500 pg, f – 100 pg, g – 50 pg, h – lysed LMD 24.10^T colony. Kb, 1 Kb ladder, Invitrogen

15.4 *Paenibacillus peoriae*

P. peoriae was first isolated from soil (Montefusco et al. 1993). However, no studies on the role of this species in the rhizosphere have been reported. Von der Weid et al. (2003) were the first to demonstrate that *P. peoriae* strain NRRL BD-62 produces antimicrobial substances with activity against various phytopathogenic bacteria and fungi. Physico-chemical characterization of the antimicrobial activity showed that it was stable after treatment with organic solvents and hydrolytic enzymes, and that its activity was retained over a wide range of pH. Partial purification carried out by Sephadex G25 gel filtration showed two profiles of inhibition, against Gram-positive bacteria and fungi, and against Gram-negative bacteria, suggesting that at least two different substances with distinct molecular weights were involved. Besides the antimicrobial inhibition capability, the strain NRRL BD-62 was also able to fix molecular nitrogen effectively (von der Weid et al. 2002; Coelho et al. 2003), and produce chitinases and proteases as well (von der Weid et al. 2003), thus demonstrating the potential of this strain as a plant growth promoter and/or as a biocontrol agent in field experiments. Alvarez et al. (2006) also showed that the same strain (NRRL BD-62) produced proteases that were active in the neutral-to-alkaline pH range, but which were totally inhibited by 1,10-phenanthroline, a zinc-metalloprotease inhibitor. The most elevated protease activity was measured at 96 h, when the highest number of spores and a low concentration of viable cells were observed. In the same year, Toljander et al. (2006) reported the attachment of *gfp* (encoding green fluorescent protein, GFP)-tagged

P. peoriae NRRL BD-62 (pnf8) to non-vital, extraradical hyphae of the arbuscular mycorrhizal fungi *Glomus* sp. MUCL 43205 and *Glomus intraradices* MUCL 43194. Lorentz et al. (2006), working with the soil isolate *P. peoriae* strain 57, demonstrated that it was able to inhibit 13 of 14 filamentous fungi (including *Aspergillus niger*, *A. fumigatus*, *A. oryzae* and *Bipolaris sorokiniana*) and all six of the yeasts tested. The strain *P. peoriae* 57 was the only one of 55 *Paenibacillus* strains tested that inhibited *Candida guilliermondii*. Moreover, this strain showed activity against a range of bacteria that included *Xanthomonas sorokiniana*, *Ralstonia solanacearum* and *Citrobacter freundii*, suggesting that it could be used for biocontrol of phytopathogens.

15.5 *Paenibacillus durus*

The nitrogen-fixing species *Bacillus azotofixans* was proposed on the basis of 16 soil and root-associated strains that exhibited acetylene-reducing ability (Seldin et al. 1984). Later, the species was transferred to the genus *Paenibacillus* (Ash et al. 1993) and then renamed *P. durus*, after the reclassification of the species *P. durum* (formerly *Clostridium durum*) as a member of the species *P. azotofixans* (Rosado et al. 1997), but with the name *P. durum* (corrected to *P. durus*) taking priority. Among the members of this species, nitrogen fixation is not dependent on yeast extract or thiamin plus biotin and all strains possessed the ability to fix nitrogen efficiently in the presence of nitrate ($NaNO_3$ at concentration up to 0.5%, Seldin et al. 1984), making this process possible even in soils where there are inputs of other nitrogen sources.

The first studies on the genetics of nitrogen fixation in *Paenibacillus* were performed in *P. durus* by determining whether nitrogenase-coding genes (*nif* genes) from *Klebsiella pneumoniae* share homology with DNA from *P. durus* (Seldin et al. 1989). Southern hybridization experiments with heterologous *nif* probes showed that *P. durus* strains carry homologous sequences only to *K. pneumoniae* structural *nifDH* genes. The results obtained also suggested that the *P. durus nifDH* are continuous and that *nifH* (encoding dinitrogenase reductase) probably underwent a rearrangement during evolution (Seldin et al. 1989). The presence of multiple copies of *nifH* in the genome of *P. durus* was later demonstrated by different authors (Rosado et al. 1998b; Choo et al. 2003). Moreover, Choo et al. (2003) also cloned three *P. durus* DNA regions containing genes involved in nitrogen fixation and reported the linkage of a *nifB* open reading frame upstream of the structural *nif* genes.

To obtain information about the diversity of *P. durus nifH* gene sequences, a set of degenerate primers described by Zehr and McReynolds (1989) was first used by Rosado et al. (1998b) for PCR amplification of *nifH* gene fragments of *P. durus*. These PCR fragments were then cloned and sequenced and two main clusters (designated I and II) of *nifH* sequences could be distinguished within the *P. durus* strains tested. Phylogenetic analysis showed that, while cluster I *nifH* sequences were quite similar to corresponding sequences of the common dinitrogenase iron protein, all *nifH* cluster II sequences exhibited high levels of similarity with the gene sequence of the alternative (*anf*) nitrogenase system (Rosado et al. 1998b). Based on these conserved regions of *P. durus nifH* cluster I sequences obtained from nine strains, specific primers (NHA1 – 5′TCCACTCGTCTGATCCTG3′ and NHA2 – 5′CTCGCGGATTGGCATTGCG3′) were further designed (Rosado et al. 1998b). As expected, a band of 360 bp was obtained with *P. durus* strains (including those belonging to cluster II), whereas no PCR products were obtained with DNA from any of the other bacterial strains used. Then, a DGGE protocol (using a nested PCR – with degenerate and specific primers as outer and inner primers, respectively) was developed to study the genetic diversity of this region of *nifH* in *P. durus*, as well in soil and rhizosphere samples. Cluster I *nifH*-specific DGGE was first used to separate the *P. durus nifH* fragments amplified from pure cultures. For all strains tested, patterns consisting of more than one band were found in DGGE analysis. Moreover, the patterns were characteristic for each strain tested, indicating that there was interstrain sequence divergence in addition to the presence of multiple copies of *nifH* (Rosado et al. 1998b). The PCR-DGGE approach using bulk and rhizosphere soils allowed the observation of similarities, as well as differences, among the different profiles. However, although the DGGE assay provided a rapid way to assess the intraspecific genetic diversity of an important functional gene, such as *nifH*, it was impossible to determine whether all *P. durus nifH* gene copies revealed by DGGE were functional copies or not.

As mentioned above, data about the expression of *nif* genes in *P. durus* were not available in the literature. Therefore, Teixeira et al. (2008) developed a reverse transcriptase-polymerase chain reaction/denaturing gradient gel electrophoresis (RT-PCR/DGGE)-based approach for the detection of the specific group of functional genes *nifH* and *anfH* in *P. durus*. First, the presence of an alternative system for nitrogen fixation in *P. durus* was confirmed with the design and application of a set of specific *anfH* primers: ANF1 – 5′GCAAGAAACATTGATGGATAC 3′ and ANF2 – 5′GAGGTCCTCAGTGTAAGC 3′. Amplification products of the expected size, approximately 200 bp, were obtained with *P. durus* strains, whereas no PCR products were obtained with DNA from any of the other bacterial strains used. The RT-PCR DGGE analysis of *nifH* and *anfH* gene sequences revealed banding patterns consisting of more than one band for both genes, indicating the presence of multiple copies and sequence divergence for each gene, as previously stated by Rosado et al. (1998b) and Choo et al. (2003). In order to study the regulation of *nifH* and *anfH* expression at transcription level, *P. durus* ATCC 35681 was grown under eight different conditions in which the effects of the addition of ammonium, molybdenum and tungsten could be analysed. The results, shown in Fig. 15.2, demonstrated that *P. durus* mRNA of *nifH* and *anfH* could be detected at high ammonium concentrations by RT-PCR/DGGE; however, a decrease of intensity or band absence was observed at 60 mmol/l (or higher) NH_4^+ for both genes. This fact could explain to a certain extent why nitrogen fixation in *P. durus* is not as sensitive to the presence of available nitrogen (nitrate) as it is observed to be for the majority of diazotrophs (Seldin et al. 1984). Finally, it

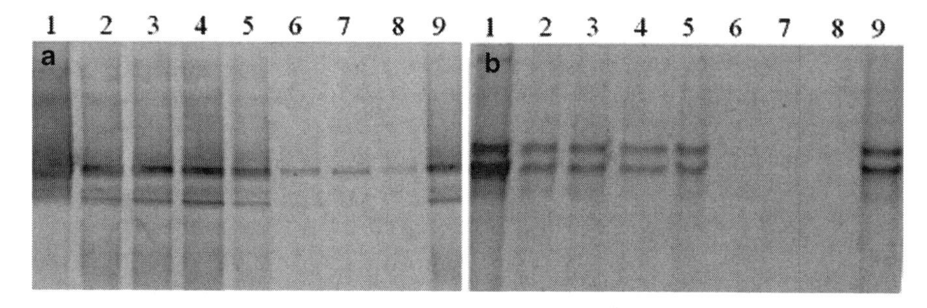

Fig. 15.2 RT-PCR/DGGE analysis of nitrogenase genes in *P. durus* ATCC 35681 under different growth conditions. DGGE of PCR products amplified with (**a**) primers NHA1 and NHA2 (Rosado et al. 1998b) and (**b**) primers ANF1 and ANF2 (Teixeira et al. 2008). Each lane corresponds to a different set of growth conditions: system 1 – TBN [Thiamine, Biotin, Nitrogen medium for *P. durus* growth containing yeast extract, according to Seldin and Penido (1986)], 2 – TBNR [TBN for acetylene-reduction medium used for studying nitrogen fixation by *P. durus* and containing yeast extract and molybdenum (Na_2MoO_4 = 20 μM), according to Seldin et al. (1984)], 3 – TBNR modified [TBN and TBNR modified – without yeast extract and supplemented with ammonium acetate (NH_4Ac)] + 40 mM ammonium acetate, 4 – TBN modified [TBN and TBNR modified – without yeast extract and supplemented with ammonium acetate (NH_4Ac)] + 40 mM ammonium acetate, 5 – TBN modified + 60 mM ammonium acetate, 6 – TBN modified + 100 mM ammonium acetate, 7 – TBN modified + 40 mM ammonium acetate + 1 mM Na_2WO_4 and 8 – TBN modified + 40 mM ammonium acetate + 2 mM Na_2WO_4. Lane 1 is a PCR-DGGE of DNA extracted from system 1 while lanes 2–9 are RT-PCR of mRNA samples from systems 1–8, respectively

was demonstrated that at least three copies of the *anfH* gene are present in *P. durus*, and that at least one copy of *anfH* gene is absent in mRNA samples, suggesting it is a non-functional gene (Teixeira et al. 2008).

Besides the capacity to fix atmospheric nitrogen, strains of *P. durus* produce antimicrobial substances against different Gram-positive and Gram-negative bacteria (Seldin and Penido 1990) and solubilize organic phosphates (Seldin et al. 1998). *P. durus* strains have been already isolated from the rhizospheres of maize, sorghum, sugarcane, wheat, banana and forage grasses (Seldin et al. 1984; Rosado et al. 1998a) and this species was prevalent when PCR products based on *nifH* gene were obtained from sorghum rhizospheres, and cloned and sequenced (Coelho et al. 2009). A similar result was obtained by Silva et al. (2003) when a *Paenibacillus*-specific PCR system was tested and used to amplify specific fragments of the 16S rRNA gene from rhizosphere DNA obtained from different maize cultivars and soil types. Clone libraries were generated from the PCR-generated 16S rDNA fragments, and selected clones were sequenced. The results of the bacterial community analyses showed, at the level of clone libraries, that sequences closely affiliated with *P. durus* were found in all DNA samples.

Rosado et al. (1996) developed a molecular method for the detection of *P. durus* in soil and the wheat rhizosphere which consisted of PCR amplification of part of the variable V1 to V4 regions of the 16S rRNA gene, followed by hybridization with a specific oligonucleotide probe homologous to part of the intervening region. Most Probable Number (MPN)-PCR was also used for assessing the dynamics of *P. durus* target numbers in soil and in rhizosphere. The results demonstrated a gradual decline of the *P. durus* population in bulk soil, whereas persistence of high numbers was found in the wheat rhizosphere, indicating that the organism might be selected for in this habitat.

Other molecular approaches have been used to determine diversity among populations of *P. durus* associated with a variety of different cereals and forage grasses, as well to investigate whether there is a link between taxonomic grouping and habitat. Rosado et al. (1998a) assessed the diversity of *P. durus* strains by using a *nifKDH* probe in hybridization experiments, primers homologous to repetitive DNA sequences (BOX element), and a 20-mer primer that produced randomly amplified polymorphic DNA (RAPD) to produce genomic fingerprints. In addition, all *P. durus* strains were studied with respect to their fermentation patterns using the API 50 CH gallery, as described by Seldin and Penido (1986). A high level of diversity was observed among the strains tested and the diversity was independent of the origins of strains, since a variety of different groups was isolated from each plant studied. On the basis of six carbohydrates (sorbitol, dulcitol, tagatose, starch, glycogen and D-arabitol) from the API 50CH gallery, the strains could be divided into five groups of related strains. In a subsequent study, Albuquerque et al. (2006) investigated whether there was a genetic link between carbohydrate metabolism and the plant or soil where the strains were isolated. For that, PCR-RFLP analysis of parts of the genes encoding 16S rRNA (ARDRA) and DNA gyrase subunit B (*gyrB*-RFLP) were used to produce genetic fingerprints. Besides a high degree of diversity among the *P. durus* strains a linkage to plant type, as revealed by the multivariate ordering of the data presented, was demonstrated in this study. Moreover, to confirm the identification of all strains used in the study, a multiplex PCR was developed based on two sets of primers specific for *P. durus*: BAZO1 (5′GAGTTGTGATGGAGCT3′) and BAZO2 (5′AGGAGCCCATGGTT3′) and NHA1 and NHA2, both systems described by Rosado et al. (1996, 1998b).

When *P. durus* strains isolated from rhizoplanes, rhizospheres and non-root-associated soils from maize plants in two different Brazilian soils were compared by phenotypic and genetic characteristics, differences in the populations of *P. durus* during the four stages of plant growth were demonstrated. Also, it was shown that the extent of diversity among *P. durus* isolates was affected by the type of soil, and that populations from rhizoplane, rhizosphere and non-root-associated soils were significantly different (Seldin et al. 1998).

All these studies provided data that strengthen the taxonomy of *P. durus,* and that also emphasize the importance of the selection of *P. durus* strains for use as inoculants.

15.6 *Paenibacillus brasilensis*

P. brasilensis was proposed in 2002 following the study of 16 maize rhizosphere isolates that showed morphological and biochemical characteristics similar to gas-forming *Paenibacillus* spp. (von der Weid et al. 2000, 2002). All the strains showed

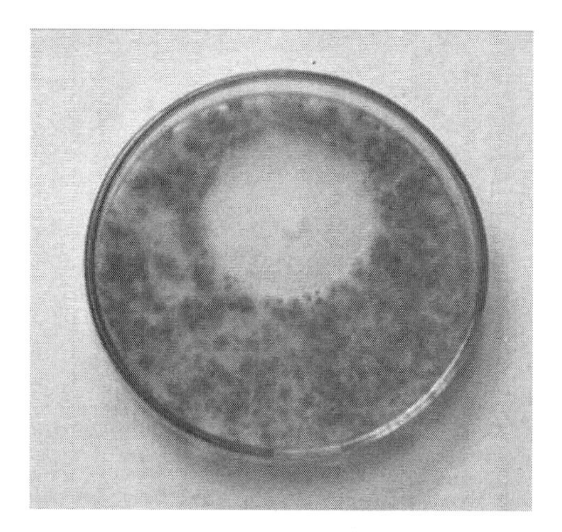

Fig. 15.3 Production by *P. brasilensis* PB177 of antimicrobial substance active against *Fusarium moniliforme*. The overlay method described by Rosado and Seldin (1993) was used to detect antifungal activity. *P. brasilensis* PB177 was inoculated onto a glucose broth agar plate as a 5 μl spot from an overnight culture. After incubation at 28°C for 48 h, the cells were killed by exposure to chloroform vapour for 15 min. The plate was then flooded with a suspension of *Fusarium moniliforme*. Presence of antimicrobial substance active against *Fusarium moniliforme* is indicated by a clear zone of inhibition around the bacterial spot. Similar results were observed with the other maize phytopathogenic fungi *Diplodia macrospora*, *Fusarium oxysporum* and *Verticillium dahliae*

nitrogenase activity when they were tested by the acetylene-reduction assay, and the amounts of ethylene produced were equivalent to those obtained for some *P. durus* strains (von der Weid et al. 2002).

Besides their nitrogen-fixing abilities, some *P. brasilensis* strains produce antimicrobial substances active against bacteria and fungi (Fortes et al. 2008; Lorentz et al. 2006). Von der Weid et al. (2005) demonstrated that *P. brasilensis* strain PB177 was able to inhibit phytopathogenic fungi such as *Fusarium moniliforme* and *Diplodia macrospora* that commonly cause diseases of maize. Figure 15.3 illustrates an antimicrobial substance assay performed as described by Rosado and Seldin (1993).

Von der Weid et al. (2005) also evaluated the potential of *P. brasilensis* strain PB177 to colonize maize plants; the strain was tagged with the *gfp* gene, and the GFP-producing bacteria attached to maize roots were detected by stereo- and confocal microscopy. The GFP-tagged bacteria were also used to treat maize seeds before challenging the seeds with two phytopathogenic fungi. The results demonstrated that the bacterial cells were attracted to the maize roots in the presence of the fungal pathogens. The ability of *P. brasilensis* PB177 to inhibit fungal growth in vitro and its capability of colonizing maize roots in vivo suggest a potential application of this strain as a biological control agent.

The attachment of *gfp*-tagged *P. brasilensis* PB177 (pnf8) to vital and non-vital extraradical hyphae of the arbuscular mycorrhizal fungi *Glomus* sp. MUCL 43205 and *Glomus intraradices* MUCL 43194 was further examined by Toljander et al. (2006). Strain PB177 showed greater attachment to vital hyphae than non-vital hyphae of both *Glomus* species tested. The attachment of PB177 to living arbuscular mycorrhizal fungal extraradical hyphae may be important for nutrient supply and plant health. In addition, the interactions between *P. brasilensis* PB177, two arbuscular mycorrhizal fungi (*Glomus mosseae* and *Glomus intraradices*) and one pathogenic fungus (*Microdochium nivale*) were investigated on winter wheat in a greenhouse trial (Jäderlund et al. 2008). Strain PB177 showed strong inhibitory effects on *M. nivale* in dual culture plate assays. However, in the presence of *P. brasilensis* PB177 and *G. intraradices*, no positive effect on the plant dry weight could be observed when it was infested with *M. nivale*.

In conclusion, *P. brasilensis* may have an important role in the rhizosphere of different grasses, not only by fixing nitrogen but also by preventing plant diseases. However, in order to achieve satisfactory plant growth benefits, different bacterial strains, fungi and plants have to be tested further before they are used as inoculants.

15.7 *Paenibacillus graminis* and *P. odorifer*

The species *P. graminis* and *P. odorifer* were described by Berge et al. (2002); the ten *P. graminis* strains studied were isolated from the rhizospheres of maize and wheat, sown in France, and from an Australian soil, while the five *P. odorifer* strains were isolated in France from wheat roots or from pasteurized purees of leeks and courgettes. All strains showed nitrogenase activity; but only *P. graminis* showed a level of ethylene production comparable to that of *P. durus* (Berge et al. 2002). In Brazil, this species was shown to be prevalent in maize and sorghum rhizospheres and in Cerrado soil (da Mota et al. 2005; Vollú et al. 2006; Coelho et al. 2007), and it was also found in wheat fields (Beneduzi et al. 2008). Furthermore, *nifH* gene sequences related to *P. odorifer* could be detected in DNA samples obtained from seven sweet potato varieties collected in Africa (Reiter et al. 2003).

The effect of *P. graminis* strain MC 22.13 on the symbiosis of the cowpea cultivar IPA-205 with *Bradyrhizobium* sp. was studied in Leonard jars (unpublished results). Co-inoculation with a *P. graminis* strain resulted in an increase of the root system, as demonstrated in Fig. 15.4.

P. graminis isolates from Brazil, Australia and France were shown to produce cyclodextrin in starch-containing medium, in addition to their capacity to fix nitrogen, indicating their agronomic and biotechnological potential (Vollú et al. 2006). However, auxin production, another characteristic that contributes to plant growth, was not observed in any of the 13 *P. graminis* strains tested (da Mota et al. 2008).

To study the diversity of *P. graminis* strains isolated from France, Australia and Brazil, RFLP analysis of parts of genes encoding RNA polymerase (*rpoB*-RFLP) and DNA gyrase subunit B (*gyrB*-RFLP) was performed to investigate whether

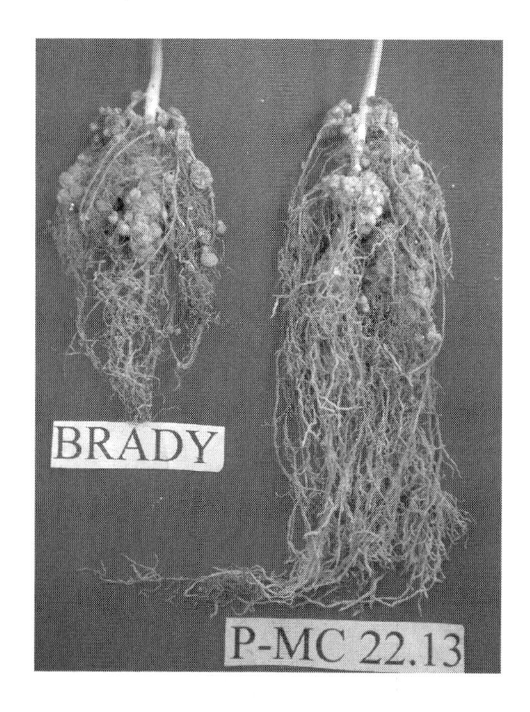

Fig. 15.4 Roots of cowpea (*Vigna unguiculata*) cultivar IPA 205 inoculated with *Bradyrhizobium* sp. BR 3267 (Brady) and co-inoculated with *Bradyrhizobium* sp. plus *P. graminis* strain MC 22.13 (P-MC 22.13). Courtesy of Dr. Márcia do Vale Barreto Figueiredo – Instituto Agronômico de Pernambuco-IPA, Brazil

there is a correlation between strains and the soil type, plant or country of isolation. Furthermore, a comparative analysis of the sequences of the *rpoB* gene was performed for the first time in *P. graminis* strains. The isolates originating from Brazil could be separated from those from Australia and France, when data from the *rpoB*-based phylogenetic tree or *gyrB*-RFLP were considered. These analyses also allowed the separation of all *P. graminis* strains studied into four clusters, and it was suggested that the diversity of these *P. graminis* strains was more affected by the soil type than by the plant from where they had been isolated (Vollú et al. 2006).

15.8 *Paenibacillus borealis*

Seven spore-forming, nitrogen-fixing bacterial strains that could degrade carboxy-methyl cellulose and quitine and could solubilize organic phosphate were isolated from spruce forest humus in Finland; they were proposed as *P. borealis* (Elo et al. 2001).

In Brazil, Beneduzi et al. (2008) isolated a total of 311 putative nitrogen-fixing, aerobic endospore-formers from seven distinct wheat production zones of the Rio Grande do Sul State. The genus *Paenibacillus* was the most prominent group in both the rhizosphere (77.8%) and soil (79%) and *P. borealis* was the most frequently identified species among the isolates.

15.9 *Paenibacillus wynnii, P. massiliensis, P. sabinae, "P. donghaensis", P. zanthoxyli, P. forsythiae, P. riograndensis* and *P. sonchi*

These eight nitrogen-fixing *Paenibacillus* species were described quite recently and only a few data are available concerning their abilities to promote plant growth; in most cases, information about the species is restricted to their taxonomic descriptions. *P. wynnii* was isolated from soil taken from different locations at Mars Oasis on Alexander Island, Antarctica, and the species harbours *nifH* gene-containing strains (Rodríguez-Díaz et al. 2005).

The type strain of *P. massiliensis* was isolated from a blood culture (Roux and Raoult 2004). Later, Zhao et al. (2006) suggested that there might be two copies of *nifH* in *P. massiliensis* strain T7; they sequenced part of the nitrogenase operon and the predicted proteins of *nifBHDKENX* showed high homology with those from other nitrogen-fixing bacteria.

Five nitrogen-fixing bacterial strains isolated from the rhizosphere soils of plants of the shrubs *Sabina squamata*, *Weigela florida* and *Zanthoxylum simulans* (Chinese prickly ash) shared the major phenotypic character of being unable to produce acid and gas from various carbohydrates such as glucose, sucrose, lactose and fructose. They showed high levels of 16S rRNA gene sequence similarity with *P. durus*, but on the basis of phenotypic and genetic features they were proposed as the novel species *P. sabinae* (Ma et al. 2007a). In the same year, Ma et al. (2007b) described another novel species, *P. zanthoxyli*, comprising five nitrogen-fixing strains isolated from rhizosphere soils of *Zanthoxylum simulans* growing in Beijing, China. Highest 16S rRNA gene sequence similarities were found between these novel strains and *P. durus* ATCC 35681 (97.8–98.5% similarity) and *P. stellifer* DSM 14472 (95.4–96.3%). A significant feature of the novel strains, that differentiated them from *P. durus* and most other *Paenibacillus* species, was that none of them could produce acid or gas from various carbohydrates. On the basis of phenotypic properties, 16S rRNA gene sequences, DNA G+C content, DNA–DNA hybridization, chemotaxonomic properties and the *nifH* gene sequence, the five novel strains were considered to represent a novel species (Ma et al. 2007b).

The species *P. forsythiae* was described on the basis of a single nitrogen-fixing isolate (strain T98) isolated from rhizosphere soil of *Forsythia mira* (Ma and Chen 2008). The highest levels of 16S rRNA gene similarity were found between strain T98 and *P. durus* ATCC 35681 (97%), *P. sabinae* DSM 17841 (98.3%) and *P. zanthoxyli* DSM 18202 (96.8%). However, based on the level of DNA–DNA relatedness and other genetic and phenotypic characteristics, the strain was proposed as a new species.

Based on phylogenetic, phenotypic, and chemotaxonomic characteristics, strain JH8T isolated from deep sediment of the East Sea (Korea) was proposed as the novel species, "*P. donghaensis*" (Choi et al. 2008). This strain was able to degrade xylan and its DNA was amplified by *nifH* primers (Poly et al. 2001), suggesting that JH8T is a nitrogen-fixing strain.

Finally, the new nitrogen-fixing species *P. sonchi* and *P. riograndensis* have been recently described. While *P. sonchi* was isolated from the rhizosphere of *Sonchus oleraceus* (Hong et al. 2009), *P. riograndensis* was isolated from the rhizosphere of *Triticum aestivum* (Beneduzi et al. 2010). *P. riograndensis* is formed by only one strain (SBR5^T) and, besides the ability to fix nitrogen, strain SBR5^T displays other PGPR characteristics such as siderophores and IAA production. Based on the 16S rRNA gene sequence *P. riograndensis* is most closely related to *P. graminis* species, showing 98.1% of similarity with *P. graminis* RSA19^T (Beneduzi et al. 2010).

15.10 Concluding Remarks

The nitrogen-fixing *Paenibacillus* species have been studied for their phenotypic, genetic and phylogenic characteristics – some species more thoroughly than others – and properties that fulfill requirements for considering many of their strains as PGPR have also been described. On the other hand, however, in vivo experiments have been few and the real and unquestionable contribution of nitrogen-fixing *Paenibacillus* strains to either the nitrogen input to the plants or to plant health is inconclusive. Therefore, further studies should be encouraged to elucidate these aspects within the long-established and the more recently described species of *Paenibacillus*.

References

Aguilera M, Monteoliva-Sanchez M, Suarez A, Guerra V, Lizama C, Bennasar A, Ramos-Cormenzana A (2001) *Paenibacillus jamilae* sp. nov., an exopolysaccharide-producing bacterium able to grow in olive-mill wastewater. Int J Syst Evol Microbiol 51:1687–1692

Aktuganov GE, Melent'ev AI, Galimzianova NF, Shirokov AV (2008) The study of mycolytic properties of aerobic spore-forming bacteria producing extracellular chitinases. Mikrobiologiia 77:788–797

Albuquerque JP, da Mota FF, von der Weid I, Seldin L (2006) Diversity of *Paenibacillus durus* strains isolated from soil and different plant rhizospheres evaluated by ARDRA and *gyrB*-RFLP analysis. Eur J Soil Biol 42:200–207

Alvarez VM, von der Weid I, Seldin L, Santos AL (2006) Influence of growth conditions on the production of extracellular proteolytic enzymes in *Paenibacillus peoriae* NRRL BD-62 and *Paenibacillus polymyxa* SCE2. Lett Appl Microbiol 43:625–630

Ash C, Farrow JAE, Wallbanks S, Collins MD (1991) Phylogenetic heterogeneity of the genus *Bacillus* revealed by comparative analysis of small subunit-ribosomal RNA sequences. Lett Appl Microbiol 13:202–206

Ash C, Priest FG, Collins MD (1993) Molecular identification of rRNA group 3 bacilli (Ash, Farrow, Wallbanks and Collins) using a PCR probe test. Antonie van Leeuwenhoek 64:253–260

Axelrood PE, Chow ML, Arnold CS, Lu K, McDermott JM, Davies J (2002) Cultivation-dependent characterization of bacterial diversity from British Columbia forest soils subjected to disturbance. Can J Microbiol 48:643–654

Beatty PH, Jensen SE (2002) *Paenibacillus polymyxa* produces fusaricidin-type antifungal antibiotics active against *Leptosphaeria maculans*, the causative agent of blackleg disease of canola. Can J Microbiol 48:159–169

Beneduzi A, Peres D, da Costa PB, Bodanese Zanettini MH, Passaglia LM (2008) Genetic and phenotypic diversity of plant-growth-promoting bacilli isolated from wheat fields in southern Brazil. Res Microbiol 159:244–250

Beneduzi A, da Costa PB, Parma M, Melo IS, Bodanese-Zanettini MH, Passaglia LMP (2010) *Paenibacillus riograndensis*, a nitrogen-fixing species isolated from rhizosphere of *Triticum aestivum* in Brazil. Int J Syst Evol Microbiol 60:128–133

Bent E, Tuzun S, Chanway CP, Enebak S (2001) Alterations in plant growth and in root hormone levels of lodgepole pines inoculated with rhizobacteria. Can J Microbiol 47:793–800

Berge O, Guinebretiere MH, Achouak W, Normand P, Heulin T (2002) *Paenibacillus graminis* sp. nov. and *Paenibacillus odorifer* sp. nov., isolated from plant roots, soil and food. Int J Syst Evol Microbiol 52:607–616

Bezzate S, Aymerich S, Chambert R, Czarnes S, Berge O, Heulin T (2000) Disruption of the *Paenibacillus polymyxa* levansucrase gene impairs its ability to aggregate soil in the wheat rhizosphere. Environ Microbiol 2:333–342

Budi SW, van Tuinen D, Martinotti G, Gianinazzi S (1999) Isolation from *Sorghum bicolor* mycorrhizosphere of a bacterium compatible with arbuscular mycorhiza development and antagonistic towards soilborne fungal pathogens. Appl Environ Microbiol 65:5148–5150

Çakmakçi R, Erat M, Erdoğan U, Dönmez MF (2007) The influence of plant growth-promoting rhizobacteria on growth and enzyme activities in wheat and spinach plants. J Plant Nutr Soil Sci 170:288–295

Chanway CP, Holl FB, Turkington R (1988) Genotypic coadaptation in plant growth promotion of forage species by *Bacillus polymyxa*. Plant Soil 106:281–284

Cho KM, Hong SY, Lee SM, Kim YH, Kahng GG, Lim YP, Kim H, Yun HD (2007) Endophytic bacterial communities in ginseng and their antifungal activity against pathogens. Microb Ecol 54:341–351

Choi J-H, Im W-T, Yoo J-S, Lee S-M, Moon D-S, Kim H-J, Rhee S-K, Roh D-H (2008) *Paenibacillus donghaensis* sp. nov., a xylan-degrading and nitrogen-fixing bacterium isolated from East Sea sediment. J Microbiol Biotechnol 18:189–193

Choo Q-C, Samian M-R, Najimudin N (2003) Phylogeny and characterization of three *nifH*-homologous genes from *Paenibacillus azotofixans*. Appl Environ Microbiol 69:3658–3662

Coelho MRR, von der Weid I, Zahner V, Seldin L (2003) Characterization of nitrogen-fixing *Paenibacillus* species by polymerase chain reaction-restriction fragment length polymorphism analysis of part of genes encoding 16S rRNA and 23S rRNA and by multilocus enzyme electrophoresis. FEMS Microbiol Lett 222:243–250

Coelho MRR, da Mota FF, Carneiro NP, Marriel IE, Paiva E, Rosado AS, Seldin L (2007) Diversity of *Paenibacillus* spp. in the rhizosphere of four sorghum (*Sorghum bicolor*) cultivars sown with two contrasting levels of nitrogen fertilizer accessed by *rpoB*-based PCR-DGGE and sequencing analysis. J Microbiol Biotechnol 17:753–760

Coelho MRR, Carneiro NP, Marriel IE, Seldin L (2009) Molecular detection of *nifH* gene-containing *Paenibacillus* in the rhizosphere of sorghum (*Sorghum bicolor*) sown in Cerrado soil. Lett Appl Microbiol 48:611–617

da Mota FF, Nóbrega A, Marriel IE, Paiva E, Seldin L (2002) Diversity of *Paenibacillus polymyxa* strains isolated from the rhizosphere of four maize genotypes plants in Cerrado soil. Appl Soil Ecol 20:119–132

da Mota FF, Gomes EA, Paiva E, Rosado AS, Seldin L (2004) Use of *rpoB* gene analysis for identification of nitrogen-fixing *Paenibacillus* species as an alternative to the 16S rRNA gene. Lett Appl Microbiol 39:34–40

da Mota FF, Gomes EA, Paiva E, Seldin L (2005) Assessment of the diversity of *Paenibacillus* species in environmental samples by a novel *rpoB*-based PCR-DGGE method. FEMS Microbiol Ecol 53:317–328 da Mota FF, Gomes EA, Seldin L (2008) Auxin production and detection of the gene coding for the Auxin Efflux Carrier (AEC) protein in *Paenibacillus polymyxa*. J Microbiol 46:257–264

Daane LL, Harjono I, Barns SM, Launen LA, Palleroni NJ, Häggblom MM (2002) PAH-degradation by *Paenibacillus* spp. and description of *Paenibacillus naphtalenovorans* sp. nov., a naphthalene-degrading bacterium from the rhizosphere of salt marsh plants. Int J Syst Bacteriol 52:131–139

Dahllöf I, Baillie H, Kjelleberg S (2000) *rpoB*-based microbial community analysis avoids limitations inherent in 16S rDNA gene intraspecies heterogeneity. Appl Environ Microbiol 66:3376–3380

Demba Diallo M, Willems A, Vloemans N, Cousin S, Vandekerckhove TT, de Lajudie P, Neyra M, Vyverman W, Gillis M, Van der Gucht K (2004) Polymerase chain reaction denaturing gradient gel electrophoresis analysis of the N_2-fixing bacterial diversity in soil under *Acacia tortilis* ssp. *raddiana* and *Balanites aegyptiaca* in the dryland part of Senegal. Environ Microbiol 6:400–415

Dijksterhuis J, Sanders M, Gorris LG, Smid EJ (1999) Antibiosis plays a role in the context of direct interaction during antagonism of *Paenibacillus polymyxa* towards *Fusarium oxysporum*. J Appl Microbiol 86:13–21

Elo S, Suominen I, Kampfer P, Juhanoja J, Salkinoja-Salonen M, Haahtela K (2001) *Paenibacillus borealis* sp. nov., a nitrogen-fixing species isolated from spruce forest humus in Finland. Int J Syst Evol Microbiol 51:535–545

Fogarty WM (1983) Microbial enzymes and biotechnology. Applied Science, Essex, England

Fortes TO, Alviano DS, Tupinambá G, Padrón TS, Antoniolli AR, Alviano CS, Seldin L (2008) Production of an antimicrobial substance against *Cryptococcus neoformans* by *Paenibacillus brasilensis* Sa3 isolated from the rhizosphere of *Kalanchoe brasiliensis*. Microbiol Res 163:200–207

Garbeva P, van Overbeek LS, van Vuurde JWL, van Elsas JD (2001) Analysis of endophytic bacterial communities of potato by planting and denaturing gradient gel electrophoresis (DGGE) of 16S rDNA based PCR fragments. Microb Ecol 41:369–383

Garbeva P, van Veen JA, van Elsas JD (2003) Predominant *Bacillus* spp. in agricultural soil under different management regimes detected via PCR-DGGE. Microb Ecol 45:302–316

Gordon RE, Haynes WC, Pang H-N (1973) *The genus* Bacillus. In: Agriculture Handbook no. 427. Agricultural Research Service, US Department of Agriculture, Washington, DC

Grau FH, Wilson PW (1962) Physiology of nitrogen-fixation by *Bacillus polymyxa*. J Bacteriol 83:490–496

Halsall DM, Gibson AH (1985) Cellulose decomposition and associated nitrogen fixation by mixed cultures of *Cellulomonas gelida* and *Azospirillum* species or *Bacillus macerans*. Appl Environ Microbiol 50:1021–1026

He Z, Kisla D, Zhang L, Yuan C, Green-Church KB, Yousef AE (2007) Isolation and identification of a *Paenibacillus polymyxa* strain that coproduces a novel lantibiotic and polymyxin. Appl Environ Microbiol 73:168–178

Holl FB, Chanway CP, Turkington R, Radley RA (1988) Response of crested wheatgrass (*Agropyron cristatum* L.), perennial ryegrass (*Lolium perenne* L.) and white clover (*Trifolium repens* L.) to inoculation with *Bacillus polymyxa*. Soil Biol Biochem 20:19–24

Hong Y-Y, Ma Y-C, Zhou Y-G, Gao F, Liu H-C, Chen S (2009) *Paenibacillus sonchi* sp. nov., a novel nitrogen-fixing species isolated from the rhizosphere of *Sonchus oleraceus*. Int J Syst Evol Microbiol 59:2656–2661

Jäderlund L, Arthurson V, Granhall U, Jansson JK (2008) Specific interactions between arbuscular mycorrhizal fungi and plant growth-promoting bacteria: as revealed by different combinations. FEMS Microbiol Lett 287:174–180

Jeong H, Kim JF, Park Y-K, Kim S-B, Kim C, Park S-H (2006) Genome snapshot of *Paenibacillus polymyxa* ATCC 842^T. J Microbiol Biotechnol 16:1650–1655

Khammas KM, Kaiser P (1992) Pectin decomposition and associated nitrogen fixation by mixed cultures of *Azospirillum* and *Bacillus* species. Can J Microbiol 38:794–797

Kim MK, Kim YA, Park MJ, Yang DC (2008) *Paenibacillus ginsengihumi* sp. nov., a bacterium isolated from soil in a ginseng field. Int J Syst Evol Microbiol 58:1164–1168

Lebuhn M, Heulin T, Hartmann A (1997) Production of auxin and other indolic and phenolic compounds by *Paenibacillus polymyxa* strains isolated from different proximity to plant roots. FEMS Microbiol Ecol 22:325–334

Lee JC, Yoon KH (2008) *Paenibacillus woosongensis* sp. nov., a xylanolytic bacterium isolated from forest soil. Int J Syst Evol Microbiol 58:612–616

Lee M, Ten LN, Baek SH, Im WT, Aslam Z, Lee ST (2007) *Paenibacillus ginsengisoli* sp. nov., a novel bacterium isolated from soil of a ginseng field in Pocheon Province, South Korea. Antonie Van Leeuwenhoek 91:127–135

Li J, Beatty PK, Shah S, Jensen SE (2007) Use of PCR-targeted mutagenesis to disrupt production of fusaricidin-type antifungal antibiotics in *Paenibacillus polymyxa*. Appl Environ Microbiol 73:3480–3489

Lorentz RH, Artico S, da Silveira AB, Einsfeld A, Corção G (2006) Evaluation of antimicrobial activity in *Paenibacillus* spp. strains isolated from natural environment. Lett Appl Microbiol 43:541–547

Lovell CR, Piceno YM, Quattro JM, Bagwell CE (2000) Molecular analysis of diazotroph diversity in the rhizosphere of the smooth cordgrass, *Spartina alterniflora*. Appl Environ Microbiol 66:3814–3822

Ma YC, Chen SF (2008) *Paenibacillus forsythiae* sp. nov., a nitrogen-fixing species isolated from rhizosphere soil of *Forsythia mira*. Int J Syst Evol Microbiol 58:319–323

Ma Y, Xia Z, Liu X, Chen S (2007a) *Paenibacillus sabinae* sp. nov., a nitrogen-fixing species isolated from the rhizosphere soils of shrubs. Int J Syst Evol Microbiol 57:6–11

Ma Y, Zhang J, Chen S (2007b) *Paenibacillus zanthoxyli* sp. nov., a novel nitrogen-fixing species isolated from the rhizosphere of *Zanthoxylum simulans*. Int J Syst Evol Microbiol 57:873–877

Mansfeld-Giese K, Larsen J, Bødker L (2002) Bacterial populations associated with mycelium of arbuscular mycorrhizal fungus *Glomus intraradices*. FEMS Microbiol Ecol 41:133–140

Mavingui P, Heulin T (1994) In vitro chitinase antifungal activity of a soil, rhizosphere and rhizoplane populations of *Bacillus polymyxa*. Soil Biol Biochem 26:801–803

Montefusco A, Nakamura LK, Labeda DP (1993) *Bacillus peoriae* sp. nov. Int J Syst Bacteriol 43:388–390

Nielsen P, Sorensen J (1997) Multi-target and medium independent fungal antagonisms by hydrolytic enzymes in *Paenibacillus polymyxa* and *Bacillus pumilus* strains from barley rhizosphere. FEMS Microbiol Ecol 22:183–192

Oliveira SS, Seldin L, Bastos MCF (1993) Identification of structural nitrogen-fixation (*nif*) genes in *Bacillus polymyxa* and *Bacillus macerans*. World J Microbiol Biotechnol 9:387–389

Peoples MB, Craswell ET (1992) Biological nitrogen fixation: investments, expectations and actual contributions to agriculture. Plant Soil 141:13–39

Petersen DJ, Srinivasan M, Chanway CP (1996) *Bacillus polymyxa* stimulates increased *Rhizobium etli* populations and nodulation when co-resident in the rhizosphere of *Phaseolus vulgaris*. FEMS Microbiol Lett 142:271–276

Pettersson B, Rippere KE, Yousten AA, Priest FG (1999) Transfer of *Bacillus lentimorbus* and *Bacillus popilliae* to the genus *Paenibacillus* with emended descriptions of *Paenibacillus lentimorbus* comb. nov. and *Paenibacillus popilliae* comb. nov. Int J Syst Bacteriol 49: 531–540

Piuri M, Sanchez-Rivas C, Ruzal SM (1998) A novel antimicrobial activity of a *Paenibacillus polymyxa* strain isolated from regional fermented sausages. Lett Appl Microbiol 27:9–13

Poly F, Ranjard L, Nazaret S, Gourbière F, Monrozier LJ (2001) Comparison of *nifH* gene pools in soil and soil microenvironments with contrasting properties. Appl Environ Microbiol 67:2255–2262

Priest FG (1993) Systematics and ecology of *Bacillus*. In: Sonenshein A, Hoch JA, Losick R (eds) Bacillus subtilis and other Gram-positive bacteria. American Society for Microbiology, Washington, DC, pp 3–16

Qi Q, Zimmermann W (2005) Cyclodextrin glucanotransferase: from gene to applications. Appl Microbiol Biotechnol 66:475–485

Reiter B, Bürgmann H, Burg K, Sessitsch A (2003) Endophytic *nifH* gene diversity in African sweet potato. Can J Microbiol 49:549–555

Reynaldi FJ, De Giusti MR, Alippi AM (2004) Inhibition of the growth of *Ascosphaera apis* by *Bacillus* and *Paenibacillus* strains isolated from honey. Rev Argent Microbiol 36:52–55

Rodríguez-Díaz M, Lebbe L, Rodelas B, Heyrman J, De Vos P, Logan NA (2005) *Paenibacillus wynnii* sp. nov., a novel species harbouring the *nifH* gene, isolated from Alexander Island, Antarctica. Int J Syst Evol Microbiol 55:2093–2099

Rosado AS, Seldin L (1993) Production of a potentially novel anti-microbial substance by *Bacillus polymyxa*. World J Microbiol Biotechnol 90:521–528

Rosado AS, Seldin L, Wolters AC, van Elsas JD (1996) Quantitative 16S rDNA-targeted polymerase chain reaction and oligonucleotide hybridization for the detection of *Paenibacillus azotofixans* in soil and the wheat rhizosphere. FEMS Microbiol Ecol 19:153–164

Rosado AS, van Elsas JD, Seldin L (1997) Reclassification of *Paenibacillus durum* (formerly *Clostridium durum*, Smith and Cato 1974) Collins et al. 1994 as a member of the species *P. azotofixans* (formerly *Bacillus azotofixans* Seldin et al. 1984) Ash et al. 1994. Int J Syst Bacteriol 47:569–572

Rosado AS, de Azevedo FS, da Cruz DW, van Elsas JD, Seldin L (1998a) Phenotypic and genetic diversity of *Paenibacillus azotofixans* strains isolated from the rhizoplane or rhizosphere soil of different grasses. J Appl Bacteriol 84:216–226

Rosado AS, Duarte GF, Seldin L, van Elsas JD (1998b) Genetic diversity of *nifH* gene sequences in *Paenibacillus azotofixans* strains and soil samples analyzed by denaturing gradient gel electrophoresis (DGGE) of PCR-amplified gene fragments. Appl Environ Microbiol 64:2770–2779

Roux V, Raoult D (2004) *Paenibacillus massiliensis* sp. nov., *Paenibacillus sanguinis* sp. nov. and *Paenibacillus timonensis* sp. nov., isolated from blood cultures. Int J Syst Evol Microbiol 54:1049–1054

Sakiyama CCH, Paula EM, Pereira PC, Borges AC, Silva DO (2001) Characterization of pectin lyase produced by an endophytic strain isolated from coffee cherries. Lett Appl Microbiol 33:117–121

Santos SCC, Coelho MRR, Seldin L (2002) Evaluation of the diversity of *Paenibacillus polymyxa* strains by using the DNA of bacteriophage IPy1 as a probe in hybridization experiments. Lett Appl Microbiol 35:52–56

Seldin L, Penido EGC (1986) Identification of *Bacillus azotofixans* using API tests. Antonie van Leeuwenhoek 52:403–409

Seldin L, Penido EGC (1990) Production of a bacteriophage, a phage tail-like bacteriocin and an antibiotic by *Bacillus azotofixans*. Ann Acad Bras Ci 62:85–94

Seldin L, van Elsas JD, Penido EGC (1983) *Bacillus* nitrogen fixers from Brazilian soils. Plant Soil 70:243–255

Seldin L, van Elsas JD, Penido EGC (1984) *Bacillus azotofixans* sp. nov., a nitrogen-fixing species from brazilian soils and grass roots. Int J Syst Bacteriol 34:451–456

Seldin L, Bastos MCF, Penido EGC (1989) Identification of *Bacillus azotofixans* nitrogen fixation genes using heterologous *nif* probes. In: Skinner FA et al (eds) Nitrogen fixation with non-legumes. Kluwer Academic, Dordrecht, The Netherlands, pp 179–187

Seldin L, Rosado AS, Cruz DW, Nobrega A, van Elsas JD, Paiva E (1998) Comparison of *Paenibacillus azotofixans* strains isolated from rhizoplane, rhizosphere and non-rhizosphere soil from maize planted in two different Brazilian soils. Appl Environ Microbiol 64:3860–3868

Seldin L, de Azevedo FS, Alviano DS, Alviano CS, Bastos MCF (1999) Inhibitory activity of *Paenibacillus polymyxa* SCE2 against human pathogenic micro-organisms. Lett Appl Microbiol 28:423–427

Selim S, Negrel J, Govaerts C, Gianinazzi S, van Tuinen D (2005) Isolation and partial characterization of antagonistic peptides produced by *Paenibacillus* sp. strain B2 isolated from the sorghum mycorrhizosphere. Appl Environ Microbiol 71:6501–6507

Shishido M, Breuil C, Chanway CP (1999) Endophytic colonization of spruce by plant growth-promoting rhizobacteria. FEMS Microbiol Ecol 29:191–196

Silva KRA, Salles JF, Seldin L, van Elsas JD (2003) Application of a novel *Paenibacillus*-specific PCR-DGGE method and sequence analysis to assess the diversity of *Paenibacillus* spp. in the maize rhizosphere. J Microbiol Methods 54:213–231

Silva V, da Silva LES, Martínez CR, Seldin L, Burity HA, Figueiredo MVB (2007) Strains of *Paenibacillus* promoters of the specific nodulation in the symbiosis *Bradyrhizobium*-caupi. Acta Sci Agron 29:331–338

Takano T, Fukada M, Monma M, Kobayashi S, Kainuma K, Yamane K (1986) Molecular cloning, DNA nucleotide sequencing, and expression in *Bacillus subtilis* cells of the *Bacillus macerans* cyclodextrin glucanotransferase gene. J Bacteriol 166:1118–1122

Teixeira RLF, von der Weid I, Seldin L, Rosado AS (2008) Differential expression of *nifH* and *anfH* genes in *Paenibacillus durus* analyzed by RT-PCR and DGGE. Lett Appl Microbiol 46:344–349

Timmusk S, Wagner EGH (1999) The plant-growth-promoting rhizobacterium *Paenibacillus polymyxa* induces changes in *Arabidopsis thaliana* gene expression: a possible connection between biotic and abiotic stress responses. Mol Plant Microbe Interact 12:951–959

Timmusk S, Nicander B, Granhall U, Tillberg E (1999) Cytokinin production by *Paenibacillus polymyxa*. Soil Biol Biochem 31:1847–1852

Timmusk S, Grantcharova N, Wagner EG (2005) *Paenibacillus polymyxa* invades plant roots and forms biofilms. Appl Environ Microbiol 71:7292–7300

Timmusk S, van West P, Gow NA, Huffstutler RP (2009) *Paenibacillus polymyxa* antagonizes oomycete plant pathogens *Phytophthora palmivora* and *Pythium aphanidermatum*. J Appl Microbiol 106:1473–1481

Toljander JF, Artursson V, Paul LR, Jansson JK, Finlay RD (2006) Attachment of different soil bacteria to arbuscular mycorrhizal fungal extraradical hyphae is determined by hyphal vitality and fungal species. FEMS Microbiol Lett 254:34–40

Tupinambá G, Alviano CS, da Silva AJR, Souto-Padron TCBS, Seldin L, Alviano DS (2008) Antimicrobial activity of *Paenibacillus polymyxa* SCE2 against mycotoxin-producing fungi. J Appl Microbiol 105:1044–1053

Vandeputte O, Oden S, Mol A, Vereeke D, Goethals K, Jaziri M, Prinsen E (2005) Biosynthesis of auxin by the gram-positive phytopathogen *Rhodococcus fascians* is controlled by compounds specific to infected plant tissues. Appl Environ Microbiol 71:1169–1177

Vollú RE, Santos SCC, Seldin L (2003) 16S rDNA targeted PCR for the detection of *Paenibacillus macerans*. Lett Appl Microbiol 37:415–420

Vollú RE, Fogel R, Santos SCC, da Mota FF, Seldin L (2006) Evaluation of the diversity of cyclodextrin-producing *Paenibacillus graminis* strains by different molecular methods. J Microbiol 44:591–599

von der Weid I, Paiva E, Nóbrega A, van Elsas JD, Seldin L (2000) Diversity of *Paenibacillus polymyxa* strains isolated from the rhizosphere of maize planted in Cerrado soil. Res Microbiol 151:369–381

von der Weid I, Duarte GF, van Elsas JD, Seldin L (2002) *Paenibacillus brasilensis* sp. nov., a novel nitrogen-fixing species isolated from the maize rhizosphere in Brazil. Int J Syst Evol Microbiol 52:2147–2153

von der Weid I, Alviano DS, Santos ALS, Soares RMA, Alviano CS, Seldin L (2003) Antimicrobial activity of *Paenibacillus peoriae* against a broad spectrum of phytopathogenic bacteria and fungi. J Appl Microbiol 95:1143–1151

von der Weid I, Artursson V, Seldin L, Jansson JK (2005) Antifungal and root surface colonization properties of GFP-tagged *Paenibacillus brasilensis* PB177. World J Microbiol Biotech 21:1591–1597

Walker R, Powel AA, Seddon B (1998) *Bacillus* isolates from the spermosphere of peas and dwarf French beans with antifungal activity against *Botrytis cinerea* and *Pythium* species. J Appl Microbiol 84:791–801

Wartiainen I, Eriksson T, Zheng W, Rasmussen U (2008) Variation in the active diazotrophic community in rice paddy – *nifH* PCR-DGGE analysis of rhizosphere and bulk soil. Appl Soil Ecol 39:65–75

Widmer F, Shaffer BT, Porteous LA, Seidler RJ (1999) Analysis of *nifH* gene pool complexity in soil and litter at Douglas fir forest site in the Oregon cascade mountain range. Appl Environ Microbiol 65:374–380

Witz DF, Detroy EW, Wilson PW (1967) Nitrogen fixation by growing cells and cell-free extracts of the *Bacillaceae*. Arch Microbiol 55:369–381

Yao WL, Wang YS, Han JG, Li LB, Song W (2004) Purification and cloning of an antifungal protein from the rice diseases controlling bacterial strain *Paenibacillus polymyxa* WY110. Yi Chuan Xue Bao 31:878–887

Zehr J, McReynolds L (1989) Use of degenerate oligonucleotide primers for amplification of the *nifH* gene from the marine cyanobacterium *Trichodesmium thiebautii*. Appl Environ Microbiol 55:2522–2526

Zhao H, Xie B, Chen S (2006) Cloning and sequencing of *nifBHDKENX* genes of *Paenibacillus massiliensis* T7 and its *nif* promoter analysis. Sci China C Life Sci 49:115–122

Zhou WW, Huang JX, Niu TG (2008) Isolation of an antifungal *Paenibacillus* strain HT16 from locusts and purification of its medium-dependent antagonistic component. J Appl Microbiol 105:912–919

Chapter 16
Halophilic and Haloalkaliphilic, Aerobic Endospore-forming Bacteria in Soil

M. Carmen Márquez, Cristina Sánchez-Porro, and Antonio Ventosa

16.1 Introduction

Extremophilic microorganisms capable of growing at extremes of salinity, acidity, alkalinity, temperature or pressure, just to cite some environmental factors, have been studied in detail and many researchers have focused their interest on the features and applications of these microorganisms. Among the extremophiles, halophiles are microorganisms that are adapted to high salt concentrations, and they are found in different habitats over a wide range of salinities (Ventosa 2006). Moderately halophilic bacteria are capable of growing optimally under conditions of 3–15% NaCl (Ventosa et al. 1998), and constitute a very heterogeneous physiological group, including both Gram-positive and Gram-negative bacteria, with great potential uses in biotechnology (Ventosa and Nieto 1995; Margesin and Schinner 2001; Mellado and Ventosa 2003).

Most studies of saline or hypersaline environments have been carried out on aquatic habitats, especially salt lakes and salterns (Ventosa 2006). Fewer studies have focused on saline soils and, in fact, many recently described species of terrestrial origin were obtained from sediments or soil samples collected from the surface sediments or soil layers of aquatic lakes that dried as a consequence of the natural evaporation of the water. Frequently, salinity is associated with alkalinity, and thus many saline or hypersaline environments have alkaline or very alkaline pH values, influencing the microbial populations that must be adapted to these two environmental factors, salinity and alkalinity.

In this chapter we review the aerobic, endospore-forming moderately halophilic and haloalkaliphilic (and some halotolerant) bacteria that have been reported to be isolated from saline soils or sediment samples, as well as their activities and potential applications. Some other related reviews that could be consulted are those of Ventosa et al. (1998, 2008) and Arahal and Ventosa (2002).

M.C. Márquez, C. Sánchez-Porro, and A. Ventosa (✉)
Dept Microbiology and Parasitology, Faculty of Pharmacy, University of Sevilla, Calle Profesor Garcia Gonzalez 2, 41012 Sevilla, Spain
e-mail: ventosa@us.es

N.A. Logan and P. De Vos (eds.), *Endospore-forming Soil Bacteria*, Soil Biology 27,
DOI 10.1007/978-3-642-19577-8_16,

16.2 Taxonomy of the Family *Bacillaceae*

Currently, the family *Bacillaceae*, which belongs to the Phylum *Firmicutes*, Class "*Bacilli*", Order *Bacillales*, comprises 15 genera that include halophilic or haloalkaliphilic, aerobic endospore-forming species isolated from soil or sediment samples. These are *Alkalibacillus*, *Bacillus*, *Gracilibacillus*, *Filobacillus*, *Halalkalibacillus*, *Halobacillus*, *Lentibacillus*, *Ornithinibacillus*, *Paraliobacillus*, *Salirhabdus*, *Salsuginibacillus*, *Tenuibacillus*, *Terribacillus*, *Thalassobacillus* and *Virgibacillus*. Also, within the family *Bacillaceae* there are some genera containing halophilic species for which no endosporulation has yet been observed, such as *Halolactibacillus* and *Sediminibacillus*.

16.2.1 Genus Alkalibacillus

The genus *Alkalibacillus* was proposed by Jeon et al. (2005b) in order to reclassify *Bacillus haloalkaliphilus* and describe the novel species *Alkalibacillus salilacus*. Five species have been included in this genus up to now and four of them have been isolated from soil sediment samples: *A. haloalkaliphilus* (the type species of the genus) (Fritze 1996; Jeon et al. 2005b), *A. salilacus* (Jeon et al. 2005b), *A. halophilus* (Tian et al. 2007) and *A. silvisoli* (Usami et al. 2007). The other species belonging to this genus, *A. filiformis*, was initially isolated from water samples in the south of Italy (Romano et al. 2005). Members of this genus are represented by Gram-variable, motile, strictly aerobic and haloalkaliphilic rods capable of forming spherical endospores in terminal position and swollen sporangia. As the majority of the other genera of the family *Bacillaceae*, all species of this genus have *meso*-diaminopimelic (*meso*-DAP) acid in the cell-wall peptidoglycan and menaquinone with seven isoprenoid units (MK-7) as the predominant isoprenoid quinone.

16.2.2 Genus Bacillus

The genus *Bacillus* is a large taxonomic entity containing species with many different physiological responses. Species of the genus *Bacillus* have been isolated from a wide variety of aquatic and terrestrial environments, demonstrating their ubiquity. Among these, moderately halophilic *Bacillus* species have been isolated from salterns, estuarine waters, salt lakes, salty foods, sea ice and seawater. The genus *Bacillus* was first described by Cohn in 1872 and since then the number of species has fluctuated widely among the different editions of *Bergey's Manual* (Berkeley 2002). The introduction of molecular methods, especially the use of 16S rRNA gene sequencing, has had a major impact on *Bacillus* taxonomy and has resulted in splitting of the genus. Despite the reduction in the number of species in the genus *Bacillus*, the phylogenetic and physiological heterogeneities of what can now be considered as *Bacillus sensu stricto* are still far too large, and the need for further splitting is supported by its widely ranging DNA G+C content, from 31 to 66 mol%.

Based on early molecular analyses of 16S rRNA gene sequences, the genus *Bacillus* comprised six phylogenetically distinct groups (Ash et al. 1991; Stackebrandt and Liesack 1993; Spring et al. 1996; Wainø et al. 1999; Schlesner et al. 2001) and they are attracting interest because these groups of bacteria have biotechnological potentials for the production of compatible solutes or hydrolytic enzymes (Margesin and Schinner 2001). Several early studies with halophilic endospore-forming organisms considered *Bacillus* as a broad group and, on the other hand, some species originally described as members of the genus *Bacillus* have subsequently been reclassified as members of other closely related genera; this is the case of *Bacillus halophilus* (Ventosa et al. 1989) or *Bacillus salexigens* (Garabito et al. 1997), that are currently placed within the genera *Salimicrobium* and *Virgibacillus*, respectively.

The presence of sodium ions (Na^+) in the medium has been considered to be very important for the environmental adaptation of alkaliphilic *Bacillus* species to high pH (Krulwich et al. 2001). Physiological studies on their alkali adaptation revealed two types of Na^+/H^+ antiporter, Mrp (Sha) and NhaC, for lowering cytoplasmic pH. Alkaliphilic *Bacillus* species use Na^+ for the adjustment of intracellular pH, solute transport and flagella rotation. The reason behind the existence of these antiporters in *Bacillus* species might be the avoidance of the H^+ cycle in their solute transport system. However, not every strain of alkaliphilic *Bacillus* shows an obvious NaCl requirement. This might be explained by the variation of affinity for NaCl observed among the alkaliphilic *Bacillus* species.

Six moderately halophilic species belonging currently to the genus *Bacillus* have been isolated from soil samples: *B. patagoniensis* (Olivera et al. 2005), *B. oshimensis* (Yumoto et al. 2005), *B. taeanensis* (Lim et al. 2006), *B. isabeliae* (Albuquerque et al. 2008), *B. aurantiacus* (Borsodi et al. 2008) and *B. aidingensis* (Xue et al. 2008). Some characteristics that allow the differentiation of these species are shown in Table 16.1.

B. patagoniensis was isolated from Patagonia (Argentina) (Olivera et al. 2005). Desert soils, such as the arid soils in north-eastern Patagonia (Argentina), are exposed to wind and water erosion, as well as salinization and alkalinization processes associated with non-irrigated lands. There is very limited knowledge about the microbial diversity of the arid soils of Patagonia, especially from vegetal soil microsites characterized by alkaline and saline conditions. During the characterization of proteolytic microorganisms from such soils, the strain PAT 05 was isolated from the rhizosphere of *Atriplex lampa*, a perennial shrub that is able to colonize alkaline and saline areas. This strain is a producer of alkaline proteases and considering their characteristics such as high optimum pH, high stability and residual activity in the presence of denaturing and chelating agents,

Table 16.1 Characteristics of different moderately halophilic *Bacillus* species isolated from soils or sediment samples

Characteristic	1	2	3	4	5	6
Motility	+	–	+	–	–	+
Nitrate reduction	–	–	+	–	–	+
Oxidase	+	+	+	+	–	–
Hydrolysis of:						
Casein	+	+	–	–	–	–
Gelatin	+	+	–	–	ND	+
Starch	+	+	+	+	–	–
Tween 80	–	+	–	–	+	–
Aesculin	ND	–	+	+	–	+
Acid production from:						
Glycerol	+	+	–	–	ND	+
D-Glucose	+	ND	+	–	ND	+
D-Mannitol	+	+	+	–	ND	+
D-Maltose	+	+	+	–	ND	+
D-Mannose	+	+	+	–	ND	+
Melibiose	–	+	+	–	ND	–
D-Raffinose	+	+	+	–	ND	–
D-Xylose	–	+	+	+	ND	–
Major fatty acids	ND	Iso $C_{15:0}$, anteiso $C_{15:0}$	Iso-$C_{15:0}$, anteiso $C_{15:0}$, iso-$C_{14:0}$	Iso-$C_{15:0}$, anteiso $C_{15:0}$, $C_{16:0}$	Iso $C_{15:0}$, anteiso-$C_{15:0}$, anteiso-$C_{17:0}$	$C_{15:0}$, $C_{16:0}$, anteiso-$C_{17:0}$, iso $C_{15:0}$, iso-$C_{16:0}$
DNA G + C content (mol%)	39.7	40.8	36	37.9	42.9	48.1

Symbols: + positive; – negative; *ND* not determined.
Taxa: (1) *B. patagoniensis* (Olivera et al. 2005); (2) *B. oshimensis* (Yumoto et al. 2005); (3) *B. taeanensis* (Lim et al. 2006); (4) *B. isabeliae* (Albuquerque et al. 2008); (5) *B. aurantiacus* (Borsodi et al. 2008); (6) *B. aidingensis* (Xue et al. 2008).

they could be promising extracellular enzymes for detergent formulation (Olivera et al. 2003). This species is able to produce oval endospores at subterminal position, growth occurs at pH 7–10 with an optimum at about pH 8.0, 5–40°C and with 15% NaCl.

B. oshimensis (Yumoto et al. 2005) constitutes the unique example of non-motile haloalkaliphilic *Bacillus* species. The strain may require NaCl for pH homeostasis for adaptation in an alkaline environment or for energy production through the respiratory chain (Tokuda and Unemoto 1981, 1984) or ATPase (Ueno et al. 2000). *B. oshimensis* was isolated from soil samples obtained from Hokkaido, Japan and produces terminally located ellipsoidal endospores which do not swell the sporangium. It grows in media with 0–20% NaCl, with an optimum concentration of 7% NaCl. The optimum growth temperature is 28–32°C at pH 10.

During the course of screening the surface sediment of a solar saltern in the Tea-An area of Korea, in order to isolate halophilic bacteria, an aerobic, Gram-positive, moderately halophilic bacterium, designated strain BH030017 was isolated and subjected to taxonomic characterization. As a result of this study *B. taeanensis* was described (Lim et al. 2006). This microorganism produces ellipsoidal endospores that are formed terminally in swollen sporangia. Optimum growth occurs at 35°C, pH 7.5 and 2–5% (w/v) NaCl. On the other hand, during a survey of the bacterial diversity of a sea salt evaporation pond on the Island of Sal in the Cape Verde Archipielago, several halophilic, Gram-positive bacteria were isolated and characterized. One of the isolates, designated as strain CVS-8, was found to be phylogenetically related to species of the genus *Bacillus*. This strain shared several physiological and biochemical characteristics with the strains belonging to the species *B. acidicola* (Albert et al. 2005) and *B. shackletonii* (Logan et al. 2004). Nevertheless this novel organism has optimum growth at NaCl concentrations between 4 and 6% (w/v), no growth occurs in media without NaCl and its optimum pH was between 7.5 and 8.0. In contrast, the type strains of *B. acidicola* and *B. shackletonii* grew in media without NaCl, showed a narrow range of salt tolerance and had a lower optimum pH for growth. Furthermore, this new isolate can be clearly distinguished from these two species of *Bacillus* on the basis of its different fatty acid composition and several phenotypic traits and was classified as a new species of the genus, as *B. isabeliae* (Albuquerque et al. 2008).

In 2008, two new halophilic *Bacillus* species isolated from soil samples were described: *B. aurantiacus* (Borsodi et al. 2008) and *B. aidingensis* (Xue et al. 2008). *B. auranticus* was described on the basis of three strains, K1-5 (type strain), K1-10 and B1-1, collected from the upper 3–5 cm sediment layers of two extremely shallow soda lakes located in Hungary. Strains K1-5 and K1-10 were isolated from sediments of Kelemem-szék lake, while B1-1 was isolated from Böddi-skék lake situated in the Kiskunság National Park, Hungary. These shallow soda lakes are the most western representatives of such lakes that lie across Eurasia and are structurally dissected by their extended reed coverage. All three strains grew in nutrient broth medium supplemented with up to 15% NaCl (strain K1-5), 13% NaCl (strain K1-10) or 12% NaCl (strain B1-1).

Finally, *B. aidingensis* (Xue et al. 2008) was isolated from sediment of Lake Ai-Ding in Xin-Jiang Province (China), a typical chloride–sulphate saline lake with neutral pH and a salt concentration of 20–26% (w/v). Previous studies on the microbial diversity of Ai-Ding salt lake have demonstrated the presence of a variety of halophilic microorganisms (Cui et al. 2006a, b; Ren and Zhou 2005a, b). *B. aidingensis* was isolated from a sediment sample of this lake and grows in the range 22–44°C (optimally at 37°C) and at NaCl concentrations in the range 8–33% (w/v) (optimally at 12%). No growth occurs in the absence of NaCl and the pH range for growth is 6.0–9.5 (optimum at pH 7.2).

16.2.3 Genus Filobacillus

The genus *Filobacillus* is phylogenetically located on the periphery of rRNA group 1 of *Bacillus* and is clearly differentiated from other related genera on the basis of its cell-wall peptidoglycan type, based on L-Orn-D-Glu (Schlesner et al. 2001). At present, this genus includes a single species, *Filobacillus milosensis*, which was isolated from beach sediment from Palaeochori Bay, Milos, Greece.

Cells of the type strain of this species stain Gram-negative, but the cell wall is of Gram-positive type. It is moderately halophilic (the NaCl range is approximately 2–23% with an optimum of 8–14%), alkalitolerant (the pH range of growth is 6.5–8.9, with an optimum of 7.3–7.8) and mesophilic (the temperature optimum is 33–38°C, with maximum growth temperature of 40–42°C). It is motile, with spherical endospores located in a terminal position and its DNA G + C is 35 mol%.

16.2.4 Genus *Gracilibacillus*

This genus was created by Wainø et al. (1999) to accommodate Gram-positive, motile, spore-forming rods or filaments with MK-7 as the predominant respiratory quinone. It currently comprises six recognized species. With the exception of *Gracilibacillus dipsosauri* (formerly *Bacillus dipsosauri*), which was isolated from the nasal cavity of a desert iguana (*Dipsosaurus dorsalis*) (Lawson et al. 1996), *Gracilibacillus* species are halophilic or halotolerant bacteria isolated from saline soils. *Gracilibacillus halotolerans*, the type species of the genus, was proposed on the basis of a single strain, designated NN, isolated from surface mud obtained from the Great Salt Lake, UT (USA). It was capable of growing on Tris-medium containing 0–20% (w/v) NaCl at 30°C but did not require NaCl for growth, being the first extremely halotolerant species described that grows optimally without NaCl in this habitat (Wainø et al. 1999).

During the course of a broad study of moderately halophilic bacteria from several salt lakes in China, three moderately halophilic Gram-positive rods were isolated from water and sediment samples of two lakes located near Xilin Hot and Ejinor, in Inner Mongolia. They produced spherical endospores located at terminal position in swollen sporangia similar to those produced by *G. dipsosauri*, the species most closely related phylogenetically (Fig. 16.1). On the basis of a polyphasic study, they were assigned to the genus *Gracilibacillus*, as *G. orientalis* (Carrasco et al. 2006).

In 2007, Ahmed and co-workers isolated the novel species *G. boraciitolerans* from a soil with naturally high boron minerals in the Hisarcik area of Turkey (Ahmed et al. 2007b). Organisms that grow on soils naturally high in a particular element such as boron, are of great interest biologically as a source of tolerance gene(s) for other microorganisms and also for their ability to function under such extreme conditions (Ahmed et al. 2007a). *G. boraciitolerans* was described as moderately halotolerant (with a NaCl range of 0–11% NaCl), alkalitolerant (with a pH range for growth of 6.0–10.0 and an optimum of pH 7.5–8.5) and highly borotolerant (it can tolerate 0–450 mM boron, but grows optimally in the absence of boron).

Fig. 16.1 Phase-contrast micrograph of *Gracilibacillus orientalis* XH-63^{T}. Bar 10 μm

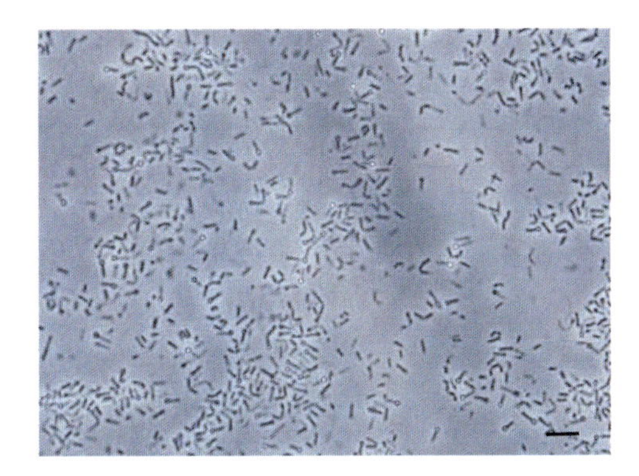

Recently, two Gram-positive, moderately halophilic, endospore-forming species belonging to this genus, *G. lacisalsi* and *G. halophilus*, have been isolated from saline sediment and soil samples in China (Jeon et al. 2008; Chen et al. 2008a). Both species are motile, oxidase and catalase positive and able to reduce nitrate to nitrite. The proposal of *G. lacisalsi* was based on two isolates capable of growing in media with 1–18% NaCl (optimum at 5–7% NaCl), at temperatures of 15–50°C (optimum of 40°C) and pH values of 5.5–10.0 (optimum at pH between 7.5 and 8.0). The DNA G + C content was 38.8–39.0 mol%. This species was most closely related to *G. orientalis*. On the other hand, *G. halophilus* is markedly different from other *Gracilibacillus* species because it is a thermophilic organism with a temperature range for growth of 28–60°C and an optimum in the range 45–50°C. In addition, it has a comparatively high NaCl concentration for optimum growth (15%) and a different fatty acid profile with significant amounts of unbranched saturated components and smaller amounts of anteiso-$C_{15:0}$ than other species of this genus.

16.2.5 *Genus Halalkalibacillus*

It has tacitly been believed that habitats of halophiles capable of growing in media containing more than 20% NaCl are restricted to saline environments, and very few reports have been published on the isolation of halophiles from ordinary garden soil samples. In 2005, Echigo et al. reported that halophilic bacteria phylogenetically related with the members of the family *Bacillaceae* inhabit different non-saline habitats in an area surrounding Tokyo, Japan. Subsequently, these authors proposed the genus *Haloalkalibacillus*, with the species *H. halophilus*, on the basis of a moderately halophilic and alkaliphilic bacterium isolated from ordinary non-saline garden soil, in Japan. This species grows in 5.0–25% NaCl (optimum at 10–15% NaCl and pH 8.5–9.0), is motile, forms spherical endospores located terminally in swollen sporangia and has A1γ *meso*-diaminopimelic-type murein. Phylogenetically, this species is most closely related to three species of the genus *Alkalibacillus*

(*A. haloalkaliphilus*, *A. filiformis* and *A. salilacus*), *Thalassobacillus devorans*, *Filobacillus milosensis* and *Tenuibacillus multivorans* (Echigo et al. 2007).

16.2.6 Genus Halobacillus

The increasing number of publications on bio-applications and other aspects of the genus *Halobacillus* and the large number of 16S rRNA gene sequences deposited in databases for unidentified strains reflect the wide distribution of these bacteria and their considerable scientific interest (Burja et al. 1999; Pinar et al. 2001; Yang et al. 2002; Rivadeneyra et al. 2004).

The genus *Halobacillus* was created by Spring and co-workers in 1996 to accommodate two novel species, *Halobacillus litoralis* and *Halobacillus trueperi*, and transfer *Sporosarcina halophila* (Claus et al. 1983) to this genus as *Halobacillus halophilus*. Currently, this genus comprises 17 species with validly published names, 7 of them being extremely halotolerant or moderately halophilic bacteria isolated from saline soil or sediment samples. For a long time, this genus could be differentiated clearly from other related genera of the family *Bacillaceae* by having cell-wall peptidoglycan based on L-Orn-D-Asp (Spring et al. 1996; Nunes et al. 2006; An et al. 2007a). However, two recently described species, *H. campisalis* and *H. seohaensis*, were found to contain *meso*-diaminopimelic acid instead of L-ornithine as the amino acid in their cell-wall peptidoglycan (Yoon et al. 2007, 2008). These and other characteristics that differentiate between the validly published halophilic *Halobacillus* species isolated from soil or sediment samples are shown in Table 16.2.

H. halophilus, the type species of the genus, was originally described on the basis of 22 endospore-forming cocci isolated from salt marsh soils at different parts of the North Sea coast in Germany (Claus et al. 1983). It has also been isolated from saline soils near Alicante (Spain) (Ventosa et al. 1983). This species can tolerate up to 18% NaCl (its optimal growth rate is between 3 and 12% NaCl). In 1998, Roeβler and Müller demonstrated, for the first time in a bacterium, that *H. halophilus* requires chloride for growth. It was later shown that chloride not only supports growth in *H. halophilus*, but is also essential for germination of endospores, flagellar synthesis and motility, and glycine betaine transport, as well as for the regulation of a large number of proteins (Dohrmann and Muller 1999; Roeβler et al. 2000; Roeβler and Müller 2001, 2002).

H. halophilus compensates the osmotic burden by the accumulation of compatible solutes (Roeβler and Müller 1998, 2001; Müller and Saum 2005; Saum et al. 2006; Saum and Müller 2007, 2008). It has been demonstrated that this bacterium produces different compatible solutes depending on the salinity in the external environment; thus, in the presence of high salt concentrations, ectoine is produced predominantly along with proline, whereas at moderate salinities glutamine and glutamate are the major compatible solutes accumulated (Saum and Müller 2007, 2008).

Table 16.2 Differential properties of halophilic *Halobacillus* species isolated from saline soil or sediment samples

Characteristics	1	2	3	4	5	6	7
Cell morphology	Cocci or oval-shaped	Cocci or oval-shaped	Rods	Rods	Rods	Rods	Rods
Flagellation	Single or peritrichous	Peritrichous	Absent	Absent	Peritrichous	Single	Peritrichous
Gram staining	+	+	+	+	+	+(V)	+
Endospore shape	Spherical	Spherical	Ellipsoidal or spherical	Ellipsoidal or spherical	Ellipsoidal or spherical	Ellipsoidal	Ellipsoidal or spherical
Endospore position	Central or lateral	Central	Central	Central or subterminal	Central or subterminal	Central or subterminal	Central or subterminal
Colony pigmentation	Orange	Light yellow	Orange	Cream or white	Orange	Yellowish-white	Orange
Maximum temperature for growth (°C)	40	41	45	49	43	38	44
Growth at:							
4°C	−	+	−	−	−	+	−
pH 5.5	−	+	+	−	−	−	−
0.5% NaCl	−	+	+	−	+	−	+
25% NaCl	−	−	−	−	+	−	+
Hydrolysis of:							
Aesculin	−	+	ND	+	−	−	−
Casein	+	+	ND	+	−	+	−
Gelatin	+	−	+	+	+	−	+
Starch	+	+	ND	+	−	+	−
Acid production from:							
D-Fructose	−	+	+	+	+	−	+
D-Galactose	−	+	+	−	−	−	+
Maltose	−	−	+	+	+	−	+
Sucrose	−	+	+	−	+	+	+
D-Xylose	−	−	−	−	+	+	−
D-Glucose	−	+	+	+	+	−	+

(*continued*)

Table 16.2 (continued)

Characteristics	1	2	3	4	5	6	7
D-Mannitol	−	−	+	+	+	+	−
Trehalose	−	+	+	ND	+	+	+
Cell-wall type	L-Orn-D-Asp	*meso*-DAP	L-Orn-D-Asp	L-Orn-D-Asp	L-Orn-D-Asp	*meso*-DAP	L-Orn-D-Asp
DNA G + C content (mol%)	40.1–40.9	42.1	46.5	41.3	42	39.3	43

Symbols: + positive reaction; − negative reaction; *V* variable; *ND* not determined.
Taxa: (1) *H. halophilus* (Claus et al. 1983; Spring et al. 1996); (2) *H. campisalis* (Yoon et al. 2007); (3) *H. faecis* (An et al. 2007c); (4) *H. karajensis* (Amoozegar et al. 2003); (5) *H. litoralis* (Spring et al. 1996); (6) *H. seohaensis* (Yoon et al. 2008); (7) *H. trueperi* (Spring et al. 1996). All were positive for catalase and oxidase and negative for nitrite reduction (not determined for *H. karajensis*), urease, anaerobic growth and hydrolysis of tyrosine (not determined for *H. karajensis*).

Very recently, a study of the carotenoids in *H. halophilus* was reported (Köcher et al. 2009). The analysis included the study of the structure, function and the organization of the genes involved in the biosynthesis of these pigments. Carotenoids are widespread fat-soluble pigments with important roles in several physiological functions such as lipophilic antioxidations as well as providing photoprotection during photosynthesis. In contrast to photoautotrophic organisms, for which the presence of carotenoids as photoprotectans is essential, formation of these pigments is found only in some heterotrophic microorganisms (Goodwin 1980). Köcher et al. (2009) showed that the carotenoids that accumulate in *H. halophilus* were C_{30} compounds structurally related to staphyloxanthins. As lipophilic antioxidants, they promote survival of the cells under oxidative stress. These authors identified a carotenogenic gene cluster in which all genes necessary for the synthesis of staphyloxanthins were organized in two operons. The knowledge on the biosynthetic pathway in *H. halophilus* and the identification of the initial genes for C_{30} carotenoid synthesis will allow future studies on carotenoid protective function, analysis of salt-dependent carotenoid synthesis and carotenoid pathway regulation in this halophilic bacterium.

Halobacillus litoralis and *H. trueperi* are heterotrophic bacteria with high tolerance of a wide range of salinities (from 0.5 to 25% and 0.5 to 30% (w/v) of salt, respectively), which may be due to the adaptation of these bacteria to environments characterized by fluctuations in the salt concentrations. This is the case with the Great Salt Lake, in Utah, a hypersaline lake from whose sediments these two species were originally isolated (Spring et al. 1996).

In 2003, Amoozegar et al. proposed *Halobacillus karajensis*. It is non-motile and grows at salinities of 1–24% NaCl, at pH values between 6.0 and 9.6 and at 10–49°C. The G + C content of its DNA is 41.4 mol%. This species, isolated from surface saline soils of Karaj (Iran), produces two extracellular enzymes, an amylase and a protease, that may possess commercial value due to their thermophilic and haloalkaline properties, respectively. The maximum amylase activity was achieved at 50°C, pH 7.5–8.5 and 5% (w/v) NaCl while the maximum protease activity was observed at 50°C, pH 9 and 12% NaCl (Amoozegar et al. 2003; Karbalaei-Heidari et al. 2009). In another investigation, the presence of a DNA-binding protein HU, a kind of histone-like protein (HLP), was assayed in *H. karajensis*, this protein being the first HLP studied in any *Halobacillus* species. This protein has the same molecular weight of that described for *Bacillus subtilis* but the genes encoding the HU protein showed some differences from those of *B. subtilis* (Ghadam et al. 2007).

In the course of an environmental study of a mangrove area on Ishagaki Island (Japan), a Gram-positive, endospore-forming, non-motile, rod-shaped extremely halotolerant bacterial strain, designated IGA-7, was isolated from a sediment sample. This strain was characterized taxonomically using a polyphasic approach. On the basis of phenotypic, chemotaxonomic and phylogenetic data, the isolate was proposed as a novel species of the genus *Halobacillus*, with the name *H. faecis* (An et al. 2007c).

Finally, two other *Halobacillus* species isolated from saline sediments, collected from marine solar salterns in Korea, are *H. campisalis* and *H. seohaensis* (Yoon

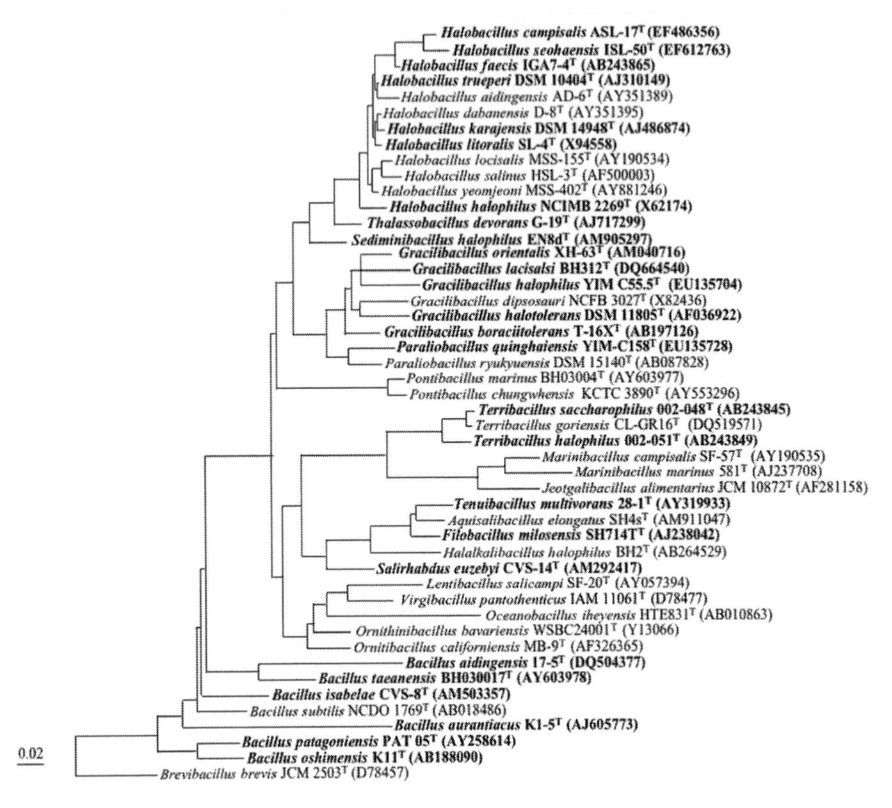

Fig. 16.2 Maximum-parsimony tree, based on 16S rRNA gene sequence comparisons, showing the relationship of species of the genus *Halobacillus* with related species. In *bold* are shown the halophilic species isolated from sediment or soil samples. The accession numbers of the sequences are shown in *parentheses* after the strain designations. *Brevibacillus brevis* JCM 2503^T was used as outgroup. The *scale bar* represents 0.02 substitutions per nucleotide position

et al. 2007, 2008). As previously commented, unlike other *Halobacillus*, the cell wall peptidoglycan of these two species is based on *meso*-diaminopimelic acid. A comparative 16S rRNA gene sequence analysis showed that the type strains of *H. campisalis* and *H. seohaensis* form a fairly stable clade, together with *H. halophilus*, within the *Halobacillus* cluster, with a 16S rRNA sequence similarity value of 98.5%. However, the percentage of DNA–DNA hybridization between them was only 19% (Yoon et al. 2008). Figure 16.2 shows the phylogenetic relationship of species of the genus *Halobacillus* with other related species.

16.2.7 Genus *Lentibacillus*

The genus *Lentibacillus* was proposed by Yoon et al. (2002) in order to classify a Gram-variable, aerobic, endospore-forming moderately halophilic rod isolated from a salt field in Korea. This microorganism was designated as *Lentibacillus salicampi* (Yoon et al. 2002). This genus currently comprises nine species, three of them isolated from fermented fish sauce: *Lentibacillus juripiscarius* (Namwong et al. 2005), *Lentibacillus halophilus* (Tanasupawat et al. 2006) and *Lentibacillus kapialis* (Pakdeeto et al. 2007). The following species of the genus *Lentibacillus* were isolated from soil sediments of different hypersaline habitats: *Lentibacillus salarius* (Jeon et al. 2005a), *Lentibacillus lacisalsi* (Lim et al. 2005), *Lentibacillus halodurans* (Yuan et al. 2007), *Lentibacillus salinarum* (Lee et al. 2008a) and *Lentibacillus salis* (Lee et al. 2008b).

The description of the genus *Lentibacillus* was emended by Jeon et al. (2005a), and it includes Gram-variable rods, capable of producing spherical or oval endospores at terminal positions in swollen sporangia. They are catalase positive and urease negative. Their cell-wall peptidoglycan contains *meso*-diaminopimelic acid; the predominant menaquinone is MK-7 and the major polar lipids are phosphatidylglycerol and diphosphatidylglycerol. The major fatty acids are anteiso-$C_{15:0}$ and iso-$C_{16:0}$ (Jeon et al. 2005a). Four species, *L. salarius*, *L. lacisalsi*, *L. halodurans* and *L. salis* were isolated from soil sediments of salt lakes located in Xinjiang Province, China (Jeon et al. 2005a; Lim et al. 2005; Yuan et al. 2007; Lee et al. 2008b). Finally, *Lentibacillus salinarum* was isolated from a sediment sample collected from a marine saltern of the Yellow Sea in Korea (Lee et al. 2008a). The studies carried out with the species of the genus *Lentibacillus* were focused on the taxonomic characterization and descriptions of the species but nothing is known about their biodiversity, ecological distribution or roles that they may have in the habitats from which they were isolated. Some features that differentiate the soil and sediment species of *Lentibacillus* are shown in Table 16.3.

16.2.8 Genus *Ornithinibacillus*

This genus contains currently the species *O. bavariensis*, isolated by Francis and Tebo (2002) from pasteurized milk form Bavaria, Germany, and *O. californiensis*, isolated from coastal surface sediments in California, USA (Mayr et al. 2006). The latest species was described as a moderately halophilic rod with a range for growth between 0.5 and 12% NaCl (optimum growth at 0.5–8% NaCl). It has a peptidoglycan type A4β (L-Orn←D-Asp) in the cell wall, MK-7 as the predominant respiratory quinone, and the presence of iso-$C_{15:0}$ and anteiso-$C_{15:0}$ as the major cellular fatty acids (Mayr et al. 2006).

16.2.9 Genus *Paraliobacillus*

The genus *Paraliobacillus* was originally described by Ishikawa et al. (2002) and constitutes an independent lineage within the halophilic/halotolerant/alkaliphilic

Table 16.3 Characteristics that distinguish the halophilic species of the genus *Lentibacillus* isolated from soils or sediment samples

Characteristic	1	2	3	4	5
Cell size (μm)	0.7–1.2 × 2.0–4.0	0.5 × 1.5–2.5	0.2–0.3 × 1.5–3.0	0.4–0.6 × 0.8–2.5	0.4–0.6 × 1.2–3.0
Endospore shape	Oval	Spherical/oval	Spherical/oval	Spherical	Spherical
Pigmentation	Cream-yellow	White	Cream	Light yellow	Cream
Motility	+	–	+	+	+
NaCl range (%, w/v)	3–24	5–30	1–20	5–15	5–25
NaCl optimum (%, w/v)	10–12	8–12	12–14	10	12–15
pH range	6–9.5	6–9	6–8.5	7–9.2	7–9.5
pH optimum	6.5–7	7–7.5	7–7.5	8	8
Temperature range (°C)	15–45	22–45	15–50	20–45	15–40
Temperature optimum (°C)	37–40	30	30–35	37	30–32
Anaerobic growth	+	–	–	–	–
Nitrate reduction	+	–	+	+	+
Hydrolysis of aesculin	+	–	+	–	–
Acid production from:					
D-Fructose	–	+	+	+	+
D-Glucose	+	+	+	+	–
Lactose	–	–	+	–	–
D-Maltose	–	–	+	–	–
D-Mannitol	–	–	+	+	–
D-Mannose	–	+	+	–	–
D-Ribose	+	–	+	–	+
D-Trehalose	–	–	+	+	–
D-Xylose	–	–	+	+	+
DNA G + C content (mol%)	49	43.4	43	46.2	44

Symbols: + positive; – negative.

Taxa: (1) *Lentibacillus salinarum* (Lee et al. 2008a); (2) *Lentibacillus halodurans* (Yuan et al. 2007); (3) *Lentibacillus salarius* (Jeon et al. 2005a); (4) *Lentibacillus salis* (Lee et al. 2008b); (5) *Lentibacillus lacisalsi* (Lim et al. 2005).

and/or alkalitolerant group in rRNA group 1 of the phyletic clade classically defined as the genus *Bacillus*. Currently, this genus comprises two recognized species; one of them, *Paraliobacillus quingaiensis*, was isolated in a recent study of the microbial diversity of the Qaidam Basin (north-west China), from a sediment sample. The type strain of this species was moderately halophilic, as its optimum NaCl concentration for growth was 5%, with a NaCl concentration range for growth of 1–20%. It contains *meso*-diamonopimelic acid in the cell wall murein, phosphatidylmethylethanolamine and phosphatidylcholine as the polar lipids and MK-7 as the sole respiratory quinone (Chen et al. 2009).

16.2.10 Genus Salirhabdus

At present, this genus contains the single species *Salirhabdus euzebyi*, which is phylogenetically most closely related to species of the genus *Salinibacillus* but has a distinctly lower NaCl requirement for optimal growth and a characteristic fatty acid composition. The description of this species is based on a single strain, designated CVS-14, isolated from a soil sample of a sea salt evaporation pond collected on the island of Sal in the Cape Verde Archipielago. Strain CSV-14 produces oval endospores at a terminal position within swollen sporangium, giving the cells the appearance of spermatozoids. It grows in media without added salt and in media containing 16% NaCl and the optimum NaCl concentration for growth is between 4 and 6% (Albuquerque et al. 2007).

16.2.11 Genus Salsuginibacillus

As previously commented, halophilic microorganisms are also often alkaliphilic or alkali-tolerant. The genus *Salsuginibacillus* was created by Carrasco et al. (2007) and, at the time of writing, only comprises the species *Salsuginibacillus kocurii*, proposed on the basis of a single isolate. It is an alkali-tolerant, moderately halophilic, Gram-positive, endospore-forming rod that was isolated from the sediment of an alkaline, saline lake in Inner Mongolia, China, and is closely related to *Marinococcus* and *Bacillus agaradhaerens*. Its cell wall peptidoglycan contained *meso*-diaminopimelic acid, the major respiratory isoprenoid quinone was MK-7, the predominant cellular fatty acids were anteiso-$C_{15:0}$, anteiso-$C_{17:0}$, iso-$C_{17:0}$ and iso-$C_{15:0}$ and its polar lipid pattern consisted of diphosphatidylglycerol, phosphatidylglycerol, phosphatidylethanolamine and two phospholipids of unknown structure. The G + C content of its DNA was 44.7 mol%.

16.2.12 Genus *Tenuibacillus*

Tenuibacillus multivorans is currently the only described species within this genus (Ren and Zhou 2005a). The proposal of this species is based on the isolation of two strains from soil of a salt lake in Xin-Jiang, China. They are moderately halophilic rods phylogenetically related to *Filobacillus milosensis* and *Alkalibacillus haloalkaliphilus*. In contrast to *F. milosensis*, *T. multivorans* contains *meso*-diaminopimelic acid instead of L-ornithine as the amino acid in their cell-wall peptidoglycan, which is common in members of *Bacillus* and related genera. On the other hand, this species can be distinguished from *A. haloalkalophilus* in its Gram reaction (it is positive), optimal pH for growth (it is neutrophilic with no growth above pH 9.0), and fatty acids profile (with the presence of iso $C_{16:0}$ as one of the predominant components).

16.2.13 Genus *Terribacillus*

The genus *Terribacillus* includes Gram-positive, aerobic rods capable of producing ellipsoidal or oval endospores that are formed subterminally within swollen sporangia. Colonies are circular and convex. Strains are catalase positive and urease negative, the predominant menaquinone is MK-7, the major fatty acids are anteiso-$C_{15:0}$ and anteiso-$C_{17:0}$ and the G + C content of the DNA is in the range 44–46 mol%. This genus contains two extremely halotolerant species, both of them isolated from soils, *T. saccharophilus* (the type species) and *T. halophilus*. These two species have an optimum NaCl concentration for growth in the range of 1–5% and are capable of growing in media without NaCl; however, *T. saccharophilus* can grow up to 16% NaCl whereas *T. halophilus* tolerates a NaCl concentration of 19% (An et al. 2007a).

16.2.14 Genus *Thalassobacillus*

This genus comprises a single species, *Thalassobacillus devorans*, that was isolated from a phenol enrichment of samples collected in hypersaline environments in South Spain. This species includes Gram-positive motile rods, capable of producing ellipsoidal endospores in central positions. They are aerobic, non-pigmented and moderately halophilic, growing over a wide range of NaCl concentrations (0.5–20%), showing optimal growth at 7.5–10% NaCl. It is not capable of growing in the absence of NaCl. Its cell-wall peptidoglycan contains *meso*-diaminopimelic acid, the predominant menaquinone is MK-7, and the major fatty acids are anteiso-$C_{15:0}$ and iso-$C_{16:0}$. It is phylogenetically related to species of the genus *Halobacillus* (García et al. 2005a). The most interesting feature of this species is its ability to degrade several aromatic compounds, especially phenol, under saline conditions and thus it could be useful for biotechnological purposes (García et al. 2005b).

16.2.15 Genus *Virgibacillus*

Heyndrickx et al. (1998) proposed the genus *Virgibacillus* on the basis of polyphasic data from phenotypic characterization, amplified rDNA restriction analysis (ARDRA) results, SDS-PAGE patterns of whole-cell proteins and fatty acids profiles. The description of the genus was later emended by Heyrman et al. (2003). Members of the genus *Virgibacillus* are motile, Gram-positive or Gram-variable rods that produce oval to ellipsoidal endospores and have DNA G + C contents ranging from 30.7 to 42.8 mol%. They have cell-walls containing peptidoglycan of the *meso*-diaminopimelic acid type and anteiso-$C_{15:0}$ as the major cellular fatty acid (Chen et al. 2008b; Wang et al. 2008). The phylogenetic relationships of the species of the genus *Virgibacillus* with other related species is shown in Fig. 16.3.

At present, *Virgibacillus* comprises 15 recognized species names with *V. pantothenticus* as the type species (Heyndrickx et al. 1998) but only five of

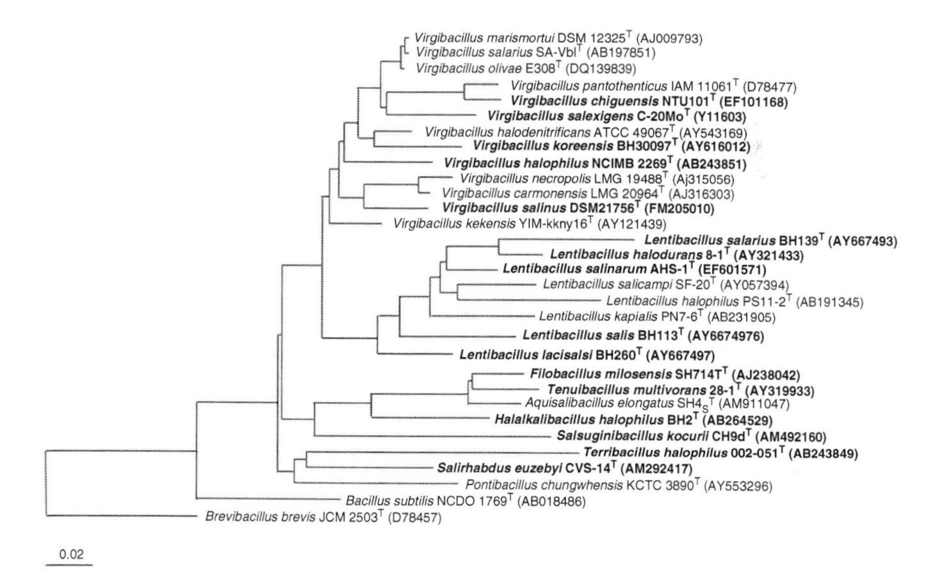

Fig. 16.3 Maximum-parsimony tree, based on 16S rRNA gene sequence comparisons, showing the relationship of species of the genus *Virgibacillus* with related species. In *bold* are shown the moderately halophilic or extremely halotolerant *Virgibacillus* species isolated from sediment or soil samples. The accession numbers of the sequences are shown in *parentheses* after the strain designations. *Brevibacillus brevis* JCM 2503^T was used as outgroup. The *scale bar* represents 0.02 substitutions per nucleotide position

Table 16.4 Characteristics useful to distinguish the extremely halotolerant or moderately halophilic *Virgibacillus* species isolated from sediment or soil samples

Characteristic	1	2	3	4	5
Oxidase	–	+	+	+	ND
Colony pigmentation	Cream	–	–	Yellow	–
Anaerobic growth	–	–	+	–	+
Temperature range (°C)	10–40	15–45	10–45	5–45	ND
Optimum temperature (°C)	37	37	25	ND	ND
Nitrate reduced to nitrite	+	–	–	+	+
Hydrolysis of:					
Aesculin	+	+	+	ND	+
Casein	–	+	ND	ND	+
Gelatin	–	+	–	ND	+
Acid production from:					
D-Galactose	+	–	–	w	+
D-Glucose	+	+	w	+	+
D-Mannitol	–	+	–	+	w
D-Rhamnose	+	–	ND	–	–
D-Trehalose	+	–	w	+	–
DNA G + C content (mol%)	38.8	36.3–39.5	41.0	42.6	37.3

Symbols: + positive; – negative; *w* weakly positive; *ND* no data available.
Taxa: (1) *V. salinus* (Carrasco et al. 2009); (2) *V. salexigens* (Garabito et al. 1997); (3) *V. koreensis* (Lee et al. 2006); (4) *V. halophilus* (An et al. 2007c); (5) *V. chiguensis* (Wang et al. 2008).

them have been described as extremely halotolerant or moderately halophilic bacteria isolated from saline soil or sediment samples (Table 16.4).

Virgibacillus salexigens was originally described as *Bacillus salexigens* and was based on six moderately halophilic bacteria, some of them isolated from hypersaline soils in Spain (Garabito et al. 1997). This species was firstly allocated to the genus *Salibacillus* by Wainø et al. (1999) and then transferred to *Virgibacillus* by Heyrman et al. (2003) who, on the basis of genotypic and phenotypic data, proposed the merger of *Virgibacillus* and *Salibacillus* in the single genus *Virgibacillus*. The type strain of *V. salexigens* was most closely related to *V. pantothenticus* (formerly *Bacillus pantothenticus*), within the phylogenetic group I of the genus *Bacillus* as defined by Ash et al. (1991).

Two other halophilic *Virgibacillus* species, isolated from soils in Japan, are *V. koreensis* (Lee et al. 2006) and *V. halophilus* (An et al. 2007b). *V. koreensis* is a moderately halophilic rod that grows anaerobically and at a NaCl concentration of 0.5–20% with an optimum at 5–10% NaCl, whereas *V. halophilus* is an extremely halotolerant microorganism capable of growing both in the absence of NaCl and in the presence of 18% NaCl. On the other hand, the recently described species *V. chiguensis* has been reported as growing at higher NaCl concentrations; the type strain of this species is capable of growing between 0 and 30% NaCl, with optimal growth at 5–10% (Wang et al. 2008).

Very recently, during the course of a broad study of moderately halophilic bacteria from a saline lake in Inner Mongolia (China), our group isolated a strain, designated XH-22, from a sediment sample. This strain had *meso*-diaminopimelic acid in the cell wall peptidoglycan, MK-7 as the predominant menaquinone and anteiso-$C_{15:0}$, $C_{16:0}$, and iso-$C_{14:0}$ as the major fatty acids. The polar lipids consisted of diphosphatidylglycerol, phosphatidylglycerol, a glycolipid and two different unidentified phospholipids. The DNA G + C content was 38.8 mol%. Analysis of 16S rRNA gene sequence revealed that the sequence similarities between strain XH-22 and the type strains of recognized *Virgibacillus* species ranged from 97.6% (with *V. carmonensis*) to 94.9 (with *V. koreensis*). The DNA–DNA hybridization between strain XH-22 and *V. carmonensis* DSM 14868^T and *V. necropolis* DSM 14866^T were 32% and 28%, respectively. This strain was also different in several phenotypic features from the species of *Virgibacillus* previously described, and we thus proposed the creation of a novel species, with the name *Virgibacillus salinus* (Carrasco et al. 2009).

16.2.16 Other, Non-endospore forming, Members of Bacillaceae

The ability to form endospores has long been used as a mandatory characteristic for the inclusion of novel isolates into different genera of the family *Bacillaceae* (Hippe et al. 1992; Slepecky and Hemphill 1992; Sneath 1984); however, 16S rRNA gene sequence analyses revealed that *Bacillus* was not a coherent genus and was interspersed with genera partly or exclusively consisting of species for which endospore formation has not been observed (Ash et al. 1991).

Two examples of moderately halophilic bacteria described very recently, for which no endospore formation has been observed, are *Halolactibacillus alkaliphilus* (Cao et al. 2008) and *Sediminibacillus halophilus* (Carrasco et al. 2008). These two species are facultative aerobes and were isolated from sediment samples in Inner Mongolia, China. *H. alkaliphilus* is a Gram-positive and non-motile rod with MK-9H_4 and MK-9H_2 as the predominant quinones, and $C_{16:0}$ and anteiso-$C_{13:0}$ as the main cellular fatty acids. *Sediminibacillus halophilus* is currently the only species described in the genus *Sediminibacillus*. This genus includes motile rod-shaped, oxidase positive and able to reduce nitrate and nitrite. The cell-wall peptidoglycan type is A1γ with *meso*-diaminopimelic acid and the major cellular fatty acids are anteiso-$C_{15:0}$ and anteiso-$C_{17:0}$. Phylogenetically it is related to the genera *Thalassobacillus* and *Halobacillus*, although represents a clearly separate line of descent within the radiation of *Firmicutes*.

Following the recommendations of the Subcommittee on the Taxonomy of the genus *Bacillus* and related organisms of the International Committee on Systematics of Prokaryotes, in order to describe new genera and species of aerobic, endospore-forming bacteria, it is important to study cultures grown for 24 h and up to 7 days on a medium that encourages sporulation. Sometimes, the utilization of too rich media inhibits the endospore formation and thus, the reduction of the nutrient content can facilitate the observation of endospores. The use of media supplemented with 5 mg/l $MnSO_4$ also stimulates the endosporulation (Logan et al. 2009). On the other hand, the potential to form endospores may be also detected using a PCR method based upon certain genes for sporulation (Brill and Wiegel 1997; Onyenwoke et al. 2004).

16.3 Ecology

Very few studies have been carried out in order to determine the diversity and ecological behaviour of endospore-forming bacteria in saline soils. A pioneering study was carried out by Quesada et al. (1982) on a hypersaline soil (with NaCl concentrations from 5.0 to 10.7%) located in Alicante, Spain. This study showed that the bacteria isolated from this habitat were mainly halophilic and that they had a much more euryhaline (capable of living in waters of a wide range of salinity) response than those isolated from hypersaline aquatic habitats. In fact, most isolates were capable of growing over a wide range of salt concentrations, from 0.9 to 20–25% NaCl, a fact that might reflect the heterogeneous structure and differences in salinity of saline soils in contrast with saline aquatic systems that are in general more homogeneous and not submitted to wide salinity changes. Another observation of this study was that the bacterial distribution was more similar to that of non-saline soils rather than to those of saline aquatic habitats such as salterns or lakes. It has been claimed that salinity might not be as important an environmental factor as the structure of the natural habitat, and that the latter might select for the bacterial distributions in different saline habitats (Rodriguez-Valera 1988). Although Gram-negative bacteria were abundant, this study showed that Gram-positive organisms represented a large proportion of the total population, with representatives of the *Bacillus* group as the most frequently isolated (19% of total). Also, Gram-positive cocci, related to the species *H. halophilus* and other species, have been isolated from these soils (Ventosa et al. 1983). However, their roles and contributions to the biogeochemistry of the soils are unknown. Besides halophilic bacteria, aerobic endospore-forming halotolerant bacteria have been isolated from hypersaline soils. In a study carried out by Garabito et al. (1998) they identified the isolates as members of several species of the genus *Bacillus*. These results were obtained on the basis of traditional isolation methods and only a few studies have been carried out using more recent molecular culture-independent methods.

Members of *Bacillaceae* and also *Rhizobiaceae* and actinomycetes were isolated from cultivated and non-cultivated salt-affected soils of Egypt by Zahran et al. (1992). The high populations of bacteria and actinomycetes closely corresponded with the relatively high levels of organic matter, whatever the degree of soil salinity. *B.* (now *Geobacillus*) *stearothermophilus* and *B. subtilis* were more frequently isolated than other *Bacillus* species.

Bacterial diversity associated with the Baer Soda Lake, located in the Autonomous Region of Inner Mongolia in China, has been investigated using a culture-independent method. Bacterial 16S rRNA gene libraries were generated using bacterial oligonucleotide primers, and 16S rRNA gene sequences of 58 clones were analysed phylogenetically. The library was dominated by 16S rRNAs of Gram-negative bacteria with a lower percentage of clones corresponding to Gram-positive bacteria (Ma et al. 2004). Another study using culture-independent methods was the work carried out by Rees et al. (2004) in Kenya. The Kenyan-Tanzanian Rift Valley contains a number of lakes which range from 5% to 35% (saturation) salts and have pH values of 8.5 to >11.5. In this study, the samples were collected from five soda lakes of this region in January 1999 (Lake Magadi, Crater Lake, Lake Elmenteita, Lake Nakuru and Lake Bogoria). DNA was extracted from water and sediment samples and also from microbial enrichment cultures of sediment samples. 16S rRNA genes were amplified by PCR and microbial diversity was studied using DGGE of 16S rDNA amplicons. Phylogenetic analysis of the sequenced amplicons revealed that these sequences were related to different genera, including several *Bacillus*-like species (Rees et al. 2004).

Echigo et al. (2005) carried out an study of non-saline environments such as ordinary garden soils, yards, fields and roadways the area surrounding Tokyo, Japan. Analyses of partial 16S rRNA gene sequences of 176 isolates suggested that they were halophiles belonging to genera of the family *Bacillaceae*, *Bacillus* (11 isolates), *Filobacillus* (19 isolates), *Gracilibacillus* (6 isolates), *Halobacillus* (102 isolates), *Lentibacillus* (1 isolate), *Paraliobacillus* (5 isolates) and *Virgibacillus* (17 isolates). Sequences of 15 isolates showed similarities lower than 92% with respect to those of previously described species, suggesting that they may represent novel taxa within the family *Bacillaceae*. It has been suggested that a possible source of these halophilic endospore-forming strains could be their transportation by Kosa event (Asian dust storms).

Caton et al. (2004) studied the cultivable aerobic heterotrophic bacteria isolated from the Great Salt Plains, a hypersaline unvegetated, barren salt flat that is part of the Salt Plains National Wildlife Refuge in Oklahoma (USA). Besides a variety of Gram-negative representatives, the Gram-positive isolates were identified as members of the genera *Bacillus*, *Salibacillus*, *Oceanobacillus*, *Virgibacillus* and *Halobacillus*. Most isolates showed a wide range of halotolerance and were thermotolerant; in fact a 64% of the isolates were capable of growing at or above 50°C.

The microbial diversity of cultivable bacteria isolated from the ancient salt deposits from the Yipinglang Salt Mine in the Yunnan Province, China, was investigated by using a conventional culture-dependent method and phylogenetic analyses based on 16S rRNA gene sequence comparisons. A total of 38 bacterial strains were isolated from the brine, halite and saline soil samples. The results showed that the isolates were members of 24 genera including *Bacillus* (Chen et al. 2007).

Another study in China focused on the isolation of bacteria from sediment samples of 22 sites at the Nansha area on the South China Sea. Bacterial isolation was conducted, followed by 16S rRNA sequencing and phylogenetic analysis. In total 349 bacteria were obtained, belonging to 87 species. Analyses of 16S rRNA sequences showed that *Bacillus* and other endospore-forming bacteria comprised the majority of isolates from 10 sites. Representatives of *Bacillus* were the most abundant bacteria and showed high diversity, with 34 species and 8 possible novel species. *Halobacillus* spp. also occurred frequently while other endospore-forming bacteria including *Brevibacillus*, *Paenibacillus*, *Pontibacillus* and *Thalassobacillus* were also found, but less frequently (Wang et al. 2008).

To study the ecology and diversity of Lonar Lake (India), water and sediment were screened in the winter season of January 2002. To study the bacterial diversity and to select the bacterial strains for further characterization, the screening was done on the basis of pH and salt tolerance of the isolates. The 64 isolates were subjected to phenotypic characterization and 16S rRNA gene sequencing. Phylogenetic analysis indicated that most of the Lonar Lake isolates were related to the phylum *Firmicutes*, containing low G + C, Gram-positive bacteria, with close relationships to different genera including *Bacillus*, *Paenibacillus* and *Alkalibacillus* (Joshi et al. 2008).

A total of 89 isolates were obtained from the sediments of four deep-sea, hypersaline anoxic brine lakes in the Eastern Mediterranean Sea: l'Atalante, Bannock, Discovery and Urania basins. Screening by Amplified Ribosomal DNA Restriction Analysis (ARDRA) and partial sequencing of the 16S rRNA genes revealed that these isolates were mostly representatives of the genus *Bacillus* and close relatives (90% of all isolates). Most of these *Bacillus*-like isolates are closely related to previously cultured organisms, many of which are moderately halophilic or alkaliphilic. Six strains (from l'Atalante, Urania and Bannock, but not from Discovery basin) belong to a cluster accommodating genera with many halotolerant representatives like *Halobacillus*, *Virgibacillus* and *Pontibacillus*. All of these strains were isolated on a medium of high salt concentration (12% NaCl) (Sass et al. 2008).

Finally, we refer to two recent studies. One was the study performed by Ettoumi et al. (2009), based on the phylogenetic diversity of a collection of 96 bacilli, isolated from 17 distinct stations of five oceanographic campaigns. This diversity was analysed by phenotypic and molecular approaches based on ARDRA, amplification of the internal transcribed spacers (ITS-PCR) and on 16S rRNA sequencing. Intra-specific polymorphism was efficiently detected by biochemical analysis and ARDRA while results of ITS-PCR were in agreement with 16S rRNA sequencing. The identification results assigned 68% of the isolates to the species *B. subtilis*, *B. licheniformis*, *B. pumilus* and *B. cereus*. Other isolates showed close affiliations to the genera *Virgibacillus*, *Gracilibacillus* and *Paenibacillus*. The other study was carried out by Valenzuela-Encinas et al. (2009). The flooding of an extreme alkaline-saline soil decreased alkalinity and salinity, which implied changes on the bacterial populations. Bacterial 16S rDNA libraries were generated from three soils with different electrolytic conductivity using universal bacterial oligonucleotide primers, and 463 clone 16S rDNA sequences were analysed phylogenetically. Clones belonging to *Firmicutes* were only found in one soil (Valenzuela-Encinas et al. 2009).

16.4 Biotechnological Applications

Moderately halophilic bacteria are capable of producing many compounds and have the capability to degrade molecules under a wide range of salt concentrations, thus they are claimed to be of great biotechnological interest. Some publications that have reviewed these aspects in detail are Ventosa et al. (1998), Margesin and Schinner (2001), and Mellado and Ventosa (2003). In spite of the intensive studies that have been carried out with respect to the biodiversity and characterization of new halophilic microorganisms, the number of studies concerning the biotechnological applications of halophiles are limited and in the near future further efforts will be necessary in order to find suitable ways to use these extremophiles in industrial processes. We will now review some publications in which halophilic endospore-formers are involved.

One of the most interesting applications of halophilic bacteria is the production of compatible solutes, which are organic osmolytes of low molecular weight that can be used to protect biological macromolecules and whole cells from damage by external stresses. These natural compounds have been designated as "extremolytes" (Lentzen and Schwarz 2006) and some of them, especially the ectoines, have been produced at a large scale and are currently used as cell protectants in skin care and as protein-free stabilizers of proteins and cells. Also, a wide range of new applications have been reported and are under development (Lentzen and Schwarz 2006). Most ectoine producers are Gram-negative bacteria, but species of *Bacillus*, *Halobacillus*, *Virgibacillus* and probably other endospore-formers, are able to accumulate ectoine (Ventosa et al. 1998) and could be used in the future for the commercial production of this osmolyte.

Another interesting application of the moderately halophilic bacteria is their use in biodegradation processes that are carried out under saline conditions. In a recent study focused on the aromatic compound-degrading halophilic bacteria isolated from water and sediment of salterns as well as from hypersaline soils in different areas of South Spain, García et al. (2005b) showed that *Halobacillus* sp. strain G19.1 was able to degrade phenol and other aromatic compounds. This strain was later described as a new species, *Thalassobacillus devorans* (García et al. 2005a). Some halotolerant *Bacillus* strains that were isolated from soils and bottom sediments contaminated by waste industrial products, and that are able to degrade polycyclic aromatic hydrocarbons such as naphthalene, phenanthrene and biphenyl, have also been reported (Plotnikova et al. 2001). Heavy metal-tolerant and halotolerant bacteria identified as members of the genus *Bacillus* have been isolated from hypersaline soils in different areas of Spain (Ríos et al. 1998). Ahmed et al. (2007a, b) described two new halotolerant species, *Bacillus boroniphilus* and *Gracilibacillus boraciitolerans*, isolated from a naturally high boron-containing soil in Turkey that require boron for growth and can tolerate up to 450 mM B.

Some interesting further studies are related to the production of extracellular enzymes by moderately halophilic bacteria (reviewed by Ventosa et al. 2005). Screening of bacteria from different hypersaline environments in Spain resulted in the isolation of 29 *Bacillus* and other moderately halophilic, endospore-forming bacteria, as well as other Gram-negative representatives, capable of producing amylases, DNases, lipases, proteases and pullulanases (Sánchez-Porro et al. 2003). In fact, most culture collection endospore-forming species assayed presented higher percentages of hydrolytic activities than the Gram-negative species studied. As we stated earlier, *Halobacillus karajensis* is able to produce two extracellular enzymes, an amylase and a protease, with interesting biotechnological features (Amoozegar et al. 2003; Karbalaei-Heidari et al. 2009). Also, Kiran and Chandra (2008) described a new moderately halophilic and alkalitolerant *Bacillus* sp. able to produce an extracellular surfactant and detergent-stable amylase isolated from a soil in India. The maximum amylase production was achieved in a medium with 10% NaCl, at pH 8.0 and 30°C. Glucose, maltose and maltotriose were the main end-products of starch hydrolysis, suggesting that the extracellular enzyme is an alpha-amylase.

16.5 Concluding Remarks and Future Prospects

Most studies on moderately halophilic and haloalkaliphilic bacteria have been focused on hypersaline aquatic habitats, while few studies have been carried out on saline or hypersaline soils. In fact, there is fragmentary information about halophilic bacteria from sediments in lakes and other aquatic habitats. Over the coming years it is necessary to increase our knowledge of the microbial diversity of saline terrestrial environments, as well as of the activities and roles that the microbial communities may play in such habitats. In this way, moderately halophilic endospore-forming bacteria could constitute excellent models for the study of the bacterial adaptation to extreme conditions. The knowledge of the compounds that are produced, and the activities of these bacteria, will be essential in order to design future biotechnological applications and their industrial production. Besides, studies at the molecular level, including sequencing of the genomes, genomics and proteomics studies, will be important in order to elucidate their adaptative mechanisms to the changing conditions of extremely heterogeneous habitats such as the saline and hypersaline soils. The detailed knowledge of these mechanisms would be very helpful for their use on bioremediation of polluted saline soils and the recovery of arid and saline soils for agriculture.

Acknowledgements We thank I.J. Carrasco for supplying some unpublished material. The work of the authors was supported by grants from the Spanish Ministerio de Ciencia y Tecnología (CGL2010-19303), National Science Foundation (Grant DEB-0919290) and Junta de Andalucía (P06-CVI-01829).

References

Ahmed I, Yokota A, Fujiwara T (2007a) A novel highly boron tolerant bacterium, *Bacillus boroniphilus* sp. nov., isolated from soil, that requires boron for its growth. Extremophiles 11:217–224

Ahmed I, Yokota A, Fujiwara T (2007b) *Gracilibacillus boraciitolerans* sp. nov., a highly boron-tolerant and moderately halotolerant bacterium isolated from soil. Int J Syst Evol Microbiol 57:796–802

Albert RA, Archambault J, Rosselló-Mora R, Tindall BJ, Matheny M (2005) *Bacillus acidicola* sp. nov., a novel mesophilic, acidophilic species isolated from acidic Sphagnum peat bogs in Wisconsin. Int J Syst Evol Microbiol 55:2125–2130

Albuquerque L, Tiago I, Rainey FA, Taborda M, Nobre MF, Veríssimo A, da Costa MS (2007) *Salirhabdus euzeby* gen. nov., sp. nov., a Gram-positive, halotolerant bacterium isolated from a sea salt evaporation pond. Int J Syst Evol Microbiol 57:1566–1571

Albuquerque L, Tiago I, Taborda M, Nobre MF, Veríssimo A, da Costa MS (2008) *Bacillus isabeliae* sp. nov., a halophilic bacterium isolated from a sea salt evaporation pond. Int J Syst Evol Microbiol 58:226–230

Amoozegar MA, Malekzadeh F, Malik KA, Schumann P, Spröer C (2003) *Halobacillus karajensis* sp. nov., a novel moderate halophile. Int J Syst Evol Microbiol 53:1059–1063

An S-Y, Asahara M, Goto K, Kasai H, Yokota A (2007a) *Terribacillus saccharophilus* gen nov., sp. nov., and *Terribacillus halophilus* sp. nov., spore-forming bacteria isolated from field soil in Japan. Int J Syst Evol Microbiol 57:51–55

An S-Y, Asahara M, Goto K, Kasai H, Yokota A (2007b) *Virgibacillus halophilus* sp. nov., spore-forming bacteria isolated from soil in Japan. Int J Syst Evol Microbiol 57:1607–1611

An S-Y, Kanoh K, Kasai H, Goto K, Yokota A (2007c) *Halobacillus faecis* sp. nov., a spore-forming bacterium isolated from a mangrove area on Ishigaki Island, Japan. Int J Syst Evol Microbiol 57:2476–2479

Arahal DR, Ventosa A (2002) Moderately halophilic and halotolerant species of *Bacillus* and related genera. In: Berkeley RCW, Heyndrickx M, Logan NA, de Vos P (eds) Applications and systematics of *Bacillus* and relatives. Blackwell, Oxford, pp 83–99

Ash C, Farrow JAE, Wallbanks S, Collins MD (1991) Phylogenetic heterogeneity of the genus *Bacillus* as revealed by comparative analysis of small-subunit ribosomal-RNA sequences. Lett Appl Microbiol 13:202–206

Berkeley RCW (2002) Whither *Bacillus*? In: Berkeley RCW, Heyndrickx M, Logan NA, de Vos P (eds) Applications and systematics of *Bacillus* and relatives. Blackwell, Oxford, pp 1–7

Borsodi AK, Márialigeti K, Szabó G, Palatinszky M, Pollák B, Kéki Z, Kovács AL, Schumann P, Tóth EM (2008) *Bacillus aurantiacus* sp. nov., an alkaliphilic and moderately halophilic bacterium isolated from Hungarian soda lakes. Int J Syst Evol Microbiol 58:845–851

Brill JA, Wiegel J (1997) Differentiation between spore-forming and asporogenic bacteria using a PCR and Southern hybridization based method. J Microbiol Methods 31:29–36

Burja AM, Webster NS, Murphy PT, Hill RT (1999) Microbial symbionts of Great Barrier Reef sponges. Mem Queensl Mus 44:63–75

Cao S-J, Qu J-H, Yang J-S, Sun Q, Yuan H-L (2008) *Halolactibacillus alkaliphilus* sp. nov., a moderately alkaliphilic and halophilic bacterium isolated from a soda lake in Inner Mongolia, China, and emended description of the genus *Halolactibacillus*. Int J Syst Evol Microbiol 58:2169–2173

Carrasco IJ, Márquez MC, Xue Y, Ma Y, Cowan DA, Jones BJ, Grant WD, Ventosa A (2006) *Gracilibacillus orientalis* sp. nov., a novel moderately halophilic bacterium isolated from a salt lake in Inner Mongolia, China. Int J Syst Evol Microbiol 56:599–604

Carrasco IJ, Márquez MC, Xue Y, Ma Y, Cowan DA, Jones BJ, Grant WD, Ventosa A (2007) *Salsuginibacillus kocurii* gen. nov., sp. nov., a moderately halophilic bacterium from soda-lake sediment. Int J Syst Evol Microbiol 57:2381–2386

Carrasco IJ, Márquez MC, Xue Y, Ma Y, Cowan DA, Jones BJ, Grant WD, Ventosa A (2008) *Sediminibacillus halophilus* gen. nov., sp. nov., a moderately halophilic, Gram-positive bacterium from a hypersaline lake. Int J Syst Evol Microbiol 58:1961–1967

Carrasco IJ, Márquez MC, Ventosa A (2009) *Virgibacillus salinus* sp. nov., a novel moderately halophilic bacterium from sediment of a saline lake. Int J Syst Evol Microbiol 59:3068–3073

Caton TM, Witte LR, Ngyuen HD, Buchheim JA, Buchheim MA, Schneegurt MA (2004) Halotolerant aerobic heterotrophic bacteria from the Great Salt Plains of Oklahoma. Microb Ecol 48:449–462

Chen YG, Li HM, Li QY, Chen W, Cui XL (2007) Phylogenetic diversity of culturable bacteria in the ancient salt deposits of the Yipinglang Salt Mine, P. R. China. Wei Sheng Wu Xue Bao 47:571–577

Chen Y-G, Cui X-L, Zhang Y-Q, Li W-J, Wang Y-X, Xu L-H, Peng Q, Wen M-L, Jiang C-L (2008a) *Gracilibacillus halophilus* sp. nov., a moderately halophilic bacterium isolated from saline soil. Int J Syst Evol Microbiol 58:2403–2408

Chen Y-G, Cui X-L, Fritze D, Chai L-H, Schumann P, Wen M-L, Wang Y-X, Xu L-H, Jiang C-L (2008b) *Virgibacillus kekensis* sp. nov., a moderately halophilic bacterium isolated from a salt lake in China. Int J Syst Evol Microbiol 58:647–653

Chen Y-G, Cui X-L, Zhang Y-Q, Li W-J, Wang Y-X, Xu L-H, Wen M-L, Peng Q, Jiang C-L (2009) *Paraliobacillus quinghaiensis* sp. nov., isolated from salt-lake sediment in China. Int J Syst Evol Microbiol 59:28–33

Claus D, Fahmy F, Rolf HJ, Tosunoglu N (1983) *Sporosarcina halophila* sp. nov., an obligate, slightly halophilic bacterium from salt marsh soils. Syst Appl Microbiol 4:496–506

Cohn F (1872) Untersuchungen über Bakterien. Beitr Biol Pflanz 1:127–244

Cui HL, Tohty D, Zhou PJ, Liu SJ (2006a) *Halorubrum lipolyticum* sp. nov. and *Halorubrum aidingense* sp. nov., isolated from two salt lakes in Xin-Jiang, China. Int J Syst Evol Microbiol 56:1631–1634

Cui HL, Yang Y, Dilbr T, Zhou PJ, Liu SJ (2006b) Biodiversity of halophilic archaea isolated from two salt lakes in Xin-Jiang region of China. Wei Sheng Wu Xue Bao 46:171–176

Dohrmann AB, Muller V (1999) Chloride dependence of endospore germination in *Halobacillus halophilus*. Arch Microbiol 172:264–267

Echigo A, Hino M, Fukushima T, Mizuki T, Kamekura M, Usami R (2005) Endospores of halophilic bacteria of the family *Bacillaceae* isolated from non-saline Japanese soil may be transported by Kosa event (Asian dust storm). Saline Systems 1:8. doi:10.1186/1746144818, http://salinesystems.org/content/1/1/8

Echigo A, Hino M, Fukushima T, Mizuki T, Kamekura M, Usami R (2007) *Halalkalibacillus halophilus* gen. nov., sp. nov., a novel moderately halophilic and alkaliphilic bacterium isolated from a non-saline soil sample in Japan. Int J Syst Evol Microbiol 57:1081–1085

Ettoumi B, Raddadi N, Borin S, Daffonchio D, Boudabous A, Cherif A (2009) Diversity and phylogeny of culturable spore-forming Bacilli isolated from marine sediments. J Basic Microbiol 49:1–11

Francis CA, Tebo BM (2002) Enzymatic manganese (II) oxidation by metabolically dormant spores of diverse *Bacillus* species. Appl Environ Microbiol 68:874–880

Fritze D (1996) *Bacillus haloalkalophilus* sp. nov. Int J Syst Bacteriol 46:98–101

Garabito MJ, Arahal DR, Mellado E, Márquez MC, Ventosa A (1997) *Bacillus salexigens* sp. nov., a new moderately halophilic *Bacillus* species. Int J Syst Bacteriol 47:735–741

Garabito MJ, Márquez MC, Ventosa A (1998) Halotolerant *Bacillus* diversity in hypersaline environments. Can J Microbiol 44:95–102

García MT, Gallego V, Ventosa A, Mellado E (2005a) *Thalassobacillus devorans* gen. nov., sp. nov., a moderately halophilic, phenol-degrading, Gram-positive bacterium. Int J Syst Evol Microbiol 55:1789–1795

García MT, Ventosa A, Mellado E (2005b) Catabolic versatility of aromatic compound-degrading halophilic bacteria. FEMS Microbiol Ecol 54:97–109

Ghadam P, Shariatian N, Amoozagar MA, Rabbani A, Shahriari SH (2007) Assaying the presence of histone-like protein HU in *Halobacillus karajensis*. Pak J Biol Sci 10:3380–3384

Goodwin TW (1980) The biochemistry of the carotenoids, vol 1. Chapman & Hall, New York

Heyndrickx M, Lebbe L, Kersters K, De Vos P, Forsyth G, Logan NA (1998) *Virgibacillus*: a new genus to accommodate *Bacillus pantothenticus* (Proom and Knight 1950). Emended description of *Virgibacillus pantothenticus*. Int J Syst Bacteriol 48:99–106

Heyrman J, Logan NA, Busse H-J, Balcaen A, Lebbe L, Rodriguez-Díaz M, Swings J, De Vos P (2003) *Virgibacillus carmonensis* sp. nov., *Virgibacillus necropolis* sp. nov. and *Virgibacillus picturae* sp. nov., three novel species isolated from deteriorated mural paintings, transfer of the species of the genus *Salibacillus* to *Virgibacillus*, as *Virgibacillus maismortui* comb. nov. and *Virgibacillus salexigens* comb. nov., and emended description of the genus *Virgibacillus*. Int J Syst Evol Microbiol 53:501–511

Hippe H, Andreesen JR, Gottschalk G (1992) The genus *Clostridium*-nonmedical. In: Balows A (ed) The prokaryotes, vol 2. Springer, Berlin, pp 1800–1866

Ishikawa M, Ishizaki S, Yamamoto Y, Yamasato K (2002) *Paraliobacillus ryukyuensis* gen. nov., sp. nov., a new Gram-positive, slightly halophilic, extremely halotolerant, facultative anaerobe isolated from a decomposing marine alga. J Gen Appl Microbiol 48:269–279

Jeon CO, Lim JM, Lee JC, Lee GS, Lee JM, Xu LH, Jiang CL, Kim CJ (2005a) *Lentibacillus salarius* sp. nov., isolated from saline sediment in China, and emended description of the genus *Lentibacillus*. Int J Syst Evol Microbiol 55:1339–1343

Jeon CO, Lim JM, Lee JM, Xu LH, Jiang CL, Kim CJ (2005b) Reclassification of *Bacillus haloalkaliphilus* Fritze 1996 as *Alkalibacillus haloalkaliphilus* gen. nov., comb. nov. and the description of *Alkalibacillus salilacus* sp. nov., a novel halophilic bacterium isolated from salt lake in China. Int J Syst Evol Microbiol 55:1891–1896

Jeon CO, Lim J-M, Jang HH, Park D-J, Xu L-H, Jiang C-L, Kim C-J (2008) *Gracilibacillus lacisalsi* sp. nov., a halophilic Gram positive bacterium from a salt lake in China. Int J Syst Evol Microbiol 58:2282–2286

Joshi AA, Kanekar PP, Kelkar AS, Shouche YS, Vani AA, Borgave SB, Sarnaik SS (2008) Cultivable bacterial diversity of alkaline Lonar lake, India. Microb Ecol 55:163–172

Karbalaei-Heidari HR, Amoozegar MA, Hajighasemi M, Ziaee A-A, Ventosa A (2009) Production, optimization and purification of a novel extracellular protease from the moderately halophilic bacterium *Halobacillus karajensis*. J Ind Microbiol Biotechnol 36:21–27

Kiran KK, Chandra TS (2008) Production of surfactant and detergent-stable, halophilic, and alkalitolerant alpha-amylase by a moderately halophilic *Bacillus* sp. strain TSCVKK. Appl Microbiol Biotechnol 77:1023–1031

Köcher S, Breitenbach J, Müller V, Sandmann G (2009) Structure, function and byosynthesis of carotenoids in the moderately halophilic bacterium *Halobacillus halophilus*. Arch Microbiol 191:95–104

Krulwich TA, Ito M, Guffanti AA (2001) The Na^{+}-dependence of alkaliphily in *Bacillus*. Biochim Biophys Acta 1505:158–168

Lawson PA, Deutch CE, Collins MD (1996) Phylogenetic characterization of a novel salt-tolerant *Bacillus* species: description of *Bacillus dipsosauri* sp. nov. J Appl Bacteriol 81:109–112

Lee J-S, Lim J-M, Lee KC, Lee J-C, Park Y-H, Kim C-J (2006) *Virgibacillus koreensis* sp. nov., a novel bacterium from a salt field, and transfer of *Virgibacillus picturae* to the genus *Oceanobacillus* as *Oceanobacillus picturae* comb. nov. with emended descriptions. Int J Syst Evol Microbiol 56:251–257

Lee SY, Choi WY, Oh TK, Yoon JH (2008a) *Lentibacillus salinarum* sp. nov., isolated from a marine solar saltern in Korea. Int J Syst Evol Microbiol 58:45–49

Lee JC, Li WJ, Xu LH, Jiang CL, Kim CJ (2008b) *Lentibacillus salis* sp. nov., a moderately halophilic bacterium isolated from a salt lake. Int J Syst Evol Microbiol 58:1838–1843

Lentzen G, Schwarz T (2006) Extremolytes: natural compounds from extremophiles for versatile applications. Appl Microbiol Biotechnol 72:623–634

Lim JM, Jeon CO, Song SM, Lee JC, Ju YJ, Xu LH, Jiang CL, Kim CJ (2005) *Lentibacillus lacisalsi* sp. nov., a moderately halophilic bacterium isolated from a saline lake in China. Int J Syst Evol Microbiol 55:1805–1809

Lim JM, Jeon CO, Kim CJ (2006) *Bacillus taeanensis* sp. nov., a halophilic Gram-positive bacterium from a solar saltern in Korea. Int J Syst Evol Microbiol 56:2903–2908

Logan NA, Lebbe L, Verhelst A, Goris J, Forsyth G, Rodríguez-Díaz M, Heyndrickx M, De Vos P (2004) *Bacillus shackletonii* sp. nov., from volcanic soil on Candlemas Island, South Sandwich archipelago. Int J Syst Evol Microbiol 54:373–376

Logan NA, Berge O, Bishop AH, Busse H-J, De Vos P, Fritze D, Heyndrickx M, Kämpfer P, Rabinovitch L, Salkinoja-Salonen MS, Seldin L, Ventosa A (2009) Proposed minimal standards for describing new taxa of aerobic, endospore-forming bacteria. Int I Syst Evol Microbiol 59:2114–2121

Ma Y, Zhang W, Xue Y, Zhou P, Ventosa A, Grant WD (2004) Bacterial diversity of the Inner Mongolian Baer Soda Lake as revealed by 16S rRNA gene sequence analyses. Extremophiles 8:45–51

Margesin R, Schinner F (2001) Potential of halotolerant and halophilic microorganisms for biotechnology. Extremophiles 5:73–83

Mayr R, Busse H-J, Worliczek HL, Ehling-Schulz M, Scherer S (2006) *Ornithinibacillus* gen. nov., with the species *Ornithinibacillus bavariensis* sp. nov. and *Ornithinibacillus californiensis* sp. nov. Int J Syst Evol Microbiol 56:1383–1389

Mellado E, Ventosa A (2003) Biotechnological potential of moderately and extremely halophilic microorganisms. In: Barredo JL (ed) Microorganisms for health care, food and enzyme production. Research Signpost, Kerala, pp 233–256

Müller V, Saum S (2005) The chloride regulon of *Halobacillus halophilus*: a novel regulatory network for salt perception and signal transduction in bacteria. In: Gunde-Cimerman N, Oren A, Plemenitas A (eds) Adaptation of life at high salt concentrations in Archaea, Bacteria, and Eucarya. Springer, Dordrecht, pp 303–310

Namwong S, Tanasupawat S, Smitinont T, Visessanguan W, Kudo T (2005) Isolation of *Lentibacillus salicampi* strains and *Lentibacillus juripiscarius* sp. nov. from fish sauce in Thailand. Int J Syst Evol Microbiol 55:315–320

Nunes I, Tiago I, Pires AL, da Costa MS, Veríssimo A (2006) *Paucisalibacillus globulus* gen. nov., sp. nov., a Gram-positive bacterium isolated from potting soil. Int J Syst Evol Microbiol 56:1841–1845

Olivera N, Sequeiros C, Marguet ER, Breccia JD (2003) Extracellular proteolytic activity characterization of the alkaliphilic *Bacillus* sp. PAT 5 isolated from Patagonia arid soils, Argentina. In: Schmidell Netto W (ed) SINAFERM, article 148. Universidades Federal de Santa Catarina Press, Florianópolis, Brazil, pp 1–7

Olivera N, Siñeriz F, Breccia JD (2005) *Bacillus patagoniensis* sp. nov., a novel alkalitolerant bacterium from the rhizosphere of *Atriplex lampa* in Patagonia, Argentina. Int J Syst Evol Microbiol 55:443–447

Onyenwoke RU, Brill JA, Farahi JW (2004) Sporulation genes in members of the low G+C Gram-type-positive phylogenetic branch (*Firmicutes*). Arch Microbiol 182:182–192

Pakdeeto A, Tanasupawat S, Thawai C, Moonmangmee S, Kudo T, Itoh T (2007) *Lentibacillus kapialis* sp. nov., from fermented shrimp paste in Thailand. Int J Syst Evol Microbiol 57:364–369

Pinar G, Ramos C, Rolleke S, Schabereiter-Gurtner C, Vybiral D, Lubitz W, Denner EB (2001) Detection of indigenous *Halobacillus* populations in damaged ancient wall painting and building materials: molecular monitoring and cultivation. Appl Environ Microbiol 67:4891–4895

Plotnikova EG, Altyntseva OV, Kosheleva IA, Puntus IF, Filonov AE, Gavrish EIu, Demakov VA, Boronin AM (2001) Bacteria–degraders of polycyclic aromatic hydrocarbons, isolated from soil and bottom sediments in salt-mining areas. Mikrobiologiia 70:61–69

Quesada E, Ventosa A, Rodriguez-Valera F, Ramos-Cormenzana R (1982) Types and properties of some bacteria isolated from hypersaline soils. J Appl Bacteriol 34:287–292

Rees HC, Grant WD, Jones BE, Heaphy S (2004) Diversity of Kenyan soda lake alkaliphiles assessed by molecular methods. Extremophiles 8:63–71

Ren PG, Zhou PJ (2005a) *Tenuibacillus multivorans* gen. nov., sp. nov., a moderately halophilic bacterium isolated from saline soil in Xin-Jiang, China. Int J Syst Evol Microbiol 55:95–99

Ren PG, Zhou PJ (2005b) *Salinibacillus aidingensis* gen. nov., sp. nov. and *Salinibacillus kushneri* sp. nov., moderately halophilic bacteria isolated from a neutral saline lake in Xin-Jiang, China. Int J Syst Evol Microbiol 55:949–953

Ríos M, Nieto JJ, Ventosa A (1998) Numerical taxonomy of heavy metal-tolerant nonhalophilic bacteria isolated from hypersaline environments. Int Microbiol 1:45–51

Rivadeneyra MA, Párraga J, Delgado R, Ramos-Cormenzana A, Delgado G (2004) Biomineralization of carbonates by *Halobacillus trueperi* in solid and liquid media with different salinities. FEMS Microbiol Ecol 48:39–46

Rodriguez-Valera F (1988) Characteristic and microbial ecology of hypersaline environments. In: Rodriguez-Valera F (ed) Halophilic bacteria. CRC, Boca-Raton, pp 3–30

Roeβler M, Müller V (1998) Quantitative and physiological analysis of chloride dependence of growth of *Halobacillus halophilus*. Appl Environ Microbiol 64:3813–3817

Roeβler M, Müller V (2001) Chloride dependence of glycine betaine transport in *Halobacillus halophilus*. FEBS Lett 489:125–128

Roeβler M, Müller V (2002) Chloride, a new environmental signal molecule involved in gene regulation in a moderately halophilic bacterium, *Halobacillus halophilus*. J Bacteriol 184:6207–6215

Roeβler M, Wanner G, Muller V (2000) Motility and flagellum synthesis in *Halobacillus halophilus* are chloride dependent. J Bacteriol 182:532–535

Romano I, Lama L, Nicolaus B, Gambacorta A, Giordano A (2005) *Alkalibacillus filiformis* sp. nov., isolated from a mineral pool in Campania, Italy. Int J Syst Evol Microbiol 55:2395–2399

Sánchez-Porro C, Martín S, Mellado E, Ventosa A (2003) Diversity of moderately halophilic bacteria producing extracellular hydrolytic enzymes. J Appl Microbiol 94:295–300

Sass AM, McKew BA, Sass H, Fichtel J, Timmis KN, McGenity TJ (2008) Diversity of *Bacillus*-like organisms isolated from deep-sea hypersaline anoxic sediments. Saline Syst 9:4–8

Saum SH, Müller V (2007) Salinity-dependent switching of osmolyte strategies in a moderately halophilic bacterium: glutamate induces praline biosynthesis in *Halobacillus halophilus*. J Bacteriol 189:6968–6975

Saum SH, Müller V (2008) Growth phase-dependent switch in osmolyte strategy in a moderate halophile: ectoine is a minor osmolyte but major stationary phase solute in *Halobacillus halophilus*. Environ Microbiol 10:716–726

Saum SH, Sydow JF, Palm P, Pfeiffer F, Oesterhelt D, Müller V (2006) Biochemical and molecular characterization of the biosynthesis of glutamine and glutamate, two major compatible solutes in the moderately halophilic bacterium *Halobacillus halophilus*. J Bacteriol 188:6808–6815

Schlesner H, Lawson PA, Collins MD, Weiss N, Wehmeyer U, Völker H, Thomm M (2001) *Filobacillus milensis* gen. nov., sp. nov., a new halophilic spore-forming bacterium with Orn-D-Glu-type peptidoglycan. Int J Syst Evol Microbiol 51:425–431

Slepecky RA, Hemphill HE (1992) The genus *Bacillus* – nonmedical. In: Balows A (ed) The prokaryotes, vol 2. Springer, Berlin, pp 1663–1698

Sneath PHA (1984) Endospore-forming Gram-positive rods and cocci. In: Sneath PHA (ed) Bergey's manual of systematic bacteriology, vol 2. Williams & Wilkins, Baltimore, pp 1104–1207

Spring S, Ludwig W, Marquez MC, Ventosa A, Schleifer KH (1996) *Halobacillus* gen. nov., with descriptions of *Halobacillus litoralis* sp. nov., and *Halobacillus trueperi* sp. nov., and transfer of *Sporosarcina halophila* to *Halobacillus halophilus* comb. nov. Int J Syst Bacteriol 46:492–496

Stackebrandt E, Liesack W (1993) Nucleic acids and classification. In: Goofellow M, O'Donnell AG (eds) Handbook of new bacterial systematics. Academic, London, pp 152–189

Tanasupawat S, Pakdeeto A, Namwong S, Thawai C, Kudo T, Itoh T (2006) *Lentibacillus halophilus* sp. nov., from fish sauce in Thailand. Int J Syst Evol Microbiol 56:1859–1863

Tian X-P, Dastager SG, Lee J-C, Tang S-K, Zhang Y-Q (2007) *Alkalibacillus halophilus* sp. nov., a new halophilic species isolated from hypersaline soil in Xin-Jiang province, China. Syst Appl Microbiol 30:268–272

Tokuda H, Unemoto T (1981) Respiration-dependent primary sodium extrusion system functioning at alkaline pH in the marine bacterium *Vibrio alginolyticus*. Biochem Biophys Res Commun 102:265–271

Tokuda H, Unemoto T (1984) Na^+ is translocated at NADH: quinone oxidoreductase segment in the respiratory chain of *Vibrio alginolyticus*. J Biol Chem 259:7785–7790

Ueno S, Kaieda N, Koyama N (2000) Characterization of a P-type Na^+-ATPase of a facultatively anaerobic alkaliphile, Exiguobacterium aurantiacum. J Biol Chem 275:14537–14540

Usami R, Echigo A, Fukushima T, Mizuki T, Yoshida Y, Kamekura M (2007) *Alkalibacillus silvisoli* sp. nov., an alkaliphilic moderate halophile isolated from non-saline forest solid in Japan. Int J Syst Evol Microbiol 57:770–774

Valenzuela-Encinas C, Neria-González I, Alcántara-Hernández RJ, Estrada-Alvarado I, Zavala-Díaz de la Serna FJ, Dendooven L, Marsch R (2009) Changes in the bacterial populations of the highly alkaline saline soil of the former lake Texcoco (Mexico) following flooding. Extremophiles 13:609–621

Ventosa A (2006) Unusual micro-organisms from unusual habitats: hypersaline environments. In: Logan NA, Lappin-Scott HM, Oyston PCF (eds) Prokaryotic diversity: mechanisms and significance. Cambridge University Press, Cambridge, pp 223–253

Ventosa A, Nieto JJ (1995) Biotechnological applications and potentialities of halophilic microorganisms. World J Microbiol Biotechnol 11:85–94

Ventosa A, Ramos-Cormenzana A, Kocur M (1983) Moderately halophilic Gram-positive cocci from hypersaline environments. Syst Appl Microbiol 4:564–570

Ventosa A, Garcia MT, Kamekura M, Onishi H, Ruiz-Berraquero F (1989) *Bacillus halophilus* sp. nov., a moderately halophilic *Bacillus* species. Syst Appl Microbiol 12:162–165

Ventosa A, Nieto JJ, Oren A (1998) Biology of moderately halophilic aerobic bacteria. Microbiol Mol Biol Rev 62:504–544

Ventosa A, Sánchez-Porro C, Martín S, Mellado E (2005) Halophilic archaea and bacteria as a source of extracellular hydrolytic enzymes. In: Gunde-Cimerman N, Oren A, Plemenitas A (eds) Cellular origin, life in extreme habitats and astrobiology, adaptation to life at high salt concentrations in Archaea, Bacteria and Eukarya. Springer, Ljubljana, Heildelberg, pp 337–354

Ventosa A, Mellado E, Sánchez-Porro C, Márquez MC (2008) Halophilic and halotolerant microorganisms from soils. In: Dion P, Nautiyal CS (eds) Microbiology of extreme soils. Springer, Heidelberg, pp 85–115

Wainø M, Tindall BJ, Schumann P, Ingvorsen K (1999) *Gracilibacillus* gen. nov., with description of *Gracilibacillus halotolerans* gen. nov., sp. nov.; transfer of *Bacillus dipsosauri* to *Gracilibacillus dipsosauri* comb. nov., and *Bacillus salexigens* to the genus *Salibacillus* gen. nov., as *Salibacillus salexigens* comb. nov. Int J Syst Bacteriol 49:821–831

Wang C-Y, Chang C-C, Ng CC, Chen T-W, Shyu Y-T (2008) *Virgibacillus chiguensis* sp. nov., a novel halophilic bacterium isolated from Chigu, a previously commercial saltern located in southern Taiwan. Int J Syst Evol Microbiol 58:341–345

Xue Y, Ventosa A, Wang X, Ren P, Zhou P, Ma Y (2008) *Bacillus aidingensis* sp. nov., a moderately halophilic bacterium isolated from Ai-Ding salt lake in China. Int J Syst Evol Microbiol 58:2828–2832

Yang L, Tan R-X, Wang Q, Huang W-Y, Yin Y-X (2002) Antifungical cyclopeptides from *Halobacillus litoralis* YS3106 of marine origin. Tetrahedron Lett 43:6545–6548

Yoon J-H, Kang KH, Park Y-H (2002) *Lentibacillus salicampi* gen. nov., sp. nov., a moderately halophilic bacterium isolated from a salt field in Korea. Int J Syst Evol Microbiol 52:2043–2048

Yoon J-K, Kang S-J, Jung Y-T, Oh T-K (2007) *Halobacillus campisalis* sp. nov., containing *meso*-diaminopimelic acid in the cell-wall peptidoglycan, and emended description of the genus *Halobacillus*. Int J Syst Evol Microbiol 57:2021–2025

Yoon J-K, Kang S-J, Oh T-K (2008) *Halobacillus seohaensis* sp. nov., isolated from a marine solar saltern in Korea. Int J Syst Evol Microbiol 58:622–627

Yuan S, Ren P, Liu J, Xue Y, Ma Y, Zhou P (2007) *Lentibacillus halodurans* sp. nov., a moderately halophilic bacterium isolated from a salt lake in Xin-Jiang, China. Int J Syst Evol Microbiol 57:485–488

Yumoto I, Hirota K, Goto T, Nodasaka Y, Nakajima K (2005) *Bacillus oshimensis* sp. nov., a moderately halophilic, non-motile alkaliphile. Int J Syst Evol Microbiol 55:907–911

Zahran HH, Moharram AM, Mohammad HA (1992) Some ecological and physiological studies on bacteria isolated from salt-affected soils of Egypt. J Basic Microbiol 32:405–413

Index

N.A. Logan and P. de Vos (eds.), *Endospore-forming Soil Bacteria*, Soil Biology 27,
DOI 10.1007/978-3-642-19577-8, © Springer-Verlag Berlin Heidelberg 2011